Alfred John North

Descriptive catalogue of the nests & eggs of birds found breeding in

Australia and Tasmania

Alfred John North

Descriptive catalogue of the nests & eggs of birds found breeding in Australia and Tasmania

ISBN/EAN: 9783337271602

Printed in Europe, USA, Canada, Australia, Japan

Cover: Foto ©berggeist007 / pixelio.de

More available books at **www.hansebooks.com**

AUSTRALIAN MUSEUM, SYDNEY.

(CATALOGUE No. 12.)

DESCRIPTIVE CATALOGUE

OF THE

NESTS & EGGS OF BIRDS

FOUND BREEDING

IN

AUSTRALIA AND TASMANIA,

BY

A. J. NORTH, F.L.S.

Printed by order of the Trustees of the Australian Museum.

Dr. E. P. Ramsay, Curator.

SYDNEY :
F. W. WHITE, GENERAL PRINTER,
1889.

❧ DR. E. REY ❧
LEIPZIG
❧ Flossplatz 11 ❧

PREFACE.

THE present Descriptive Catalogue of the Nests and Eggs of Australian Birds, has been prepared by Mr. A. J. North, who has been engaged by the Trustees as an Assistant in the Zoological Department of the Museum. It contains a careful description of such authentic eggs of each species as have been accessible to the author, wherever there was any doubt as to the authenticity of the eggs of any species they have been discarded, typical specimens have always been selected, supplemented with descriptions of such other distinct varieties as occasionally occur. No egg is here described of which the history is unknown, with the exception of a few taken from Gould's "Handbook to the Birds of Australia."

ED. P. RAMSAY,

CURATOR.

INTRODUCTION.

THE object of this Catalogue is to give authentic descriptions of the Nests and Eggs, as far as are known, of the Birds found breeding in Australia, Tasmania and on the adjacent islands, and especially of such specimens as are to be found in the Collection of the Australian Museum.

In the preparation of this work I am deeply indebted to Dr. Ramsay the Curator, and Messrs. Ramsay Bros. of Dobroyde, who kindly placed their MS. notes, which have been kept since 1858 up to the present time, and the whole of the Dobroyde Collection at my disposal; and it is a matter for congratulation that the work so ably done by the late Mr. Gould, should have been followed up in the same accurate and systematic manner by these gentlemen. Of later years, Mr. K. H. Bennett of "Yandembah," whose name frequently occurs through these pages, has by a series of close observations, contributed largely to our knowledge of the habits and nidification of many species which had previously been unobserved, and it is to the perusal of his MS. notes taken on the spot, together with the use of his collection, which he kindly placed at my disposal, that I am enabled to give additional information regarding the breeding of many birds in the interior of New South Wales, and more particularly as to the members of the *Accipitres* and *Psittacidæ*.

Through the courtesy of the Hon. William Macleay, who has always been the foremost to assist the advancement of Natural History in Australia, I have been enabled to examine and describe a number of nests and eggs in the collection of the Macleayan Museum, and I am also indebted to Mr. George Masters, the Curator, for supplying me with information relative to the same, and for a knowledge of the nidification of many of the Western

Australian birds, collected by him during his visits to that colony in 1863, and in 1868.

Mr. George Barnard of Coomooboolaroo, Queensland, and his sons, have contributed largely towards a knowledge of the nidification of a number of the birds of Central Queensland, and I have to acknowledge the assistance and the loan of specimens for description from Dr. James C. Cox, and Dr. George Hurst of Sydney, and Mr. E. D. Atkinson of Tasmania.

It must be borne in mind that this is a Descriptive Catalogue of the Nests and Eggs of Australian Birds, remarks therefore on each species are necessarily very brief, but in order to somewhat relieve the monotony of one technical description following another, and where the eggs have been previously described, I have given in full in some instances, papers contributed at various times by Dr. Ramsay to the "Ibis," entitled "Notes on Birds breeding in the Neighbourhood of Sydney," and in addition given extracts, which I thought might prove of interest, from other papers furnished by the same author to kindred societies and publications.

Of the 469 species here described, with the exception of those of 34 taken from Mr. Gould's "Handbook to the Birds of Australia," I have personally examined those of every species, and need hardly state that they have been taken only from thoroughly authentic specimens.

Appended is a list of those birds of which, as far as is known, no authentic information has been recorded of their having been found breeding in Australia or Tasmania, but which have been found in other parts of the world, and fully described by different authors; these being only visitors or stragglers to Australia, are not included in this work.

For the nomenclature I have followed the authors of the Catalogues of Birds in the British Museum, and Dr. Ramsay's List of the Birds of Australia, the habitats also being taken from the latter work. No systematic arrangement has been adopted except that used by Mr. Gould in his Handbook to the Birds of Australia, and to which reference is made.

In conclusion I would ask those favourably situated, to add to our knowledge of the nidification of those birds which are still a desiderata, and by the contribution of any new or rare specimens to the Museum, to assist in completing as far as possible, the National Collection.

Order ACCIPITRES.

Sub-Family ACCIPITRINÆ.

Genus CIRCUS, *Lacepéde.*

2 - 3.   CIRCUS ASSIMILIS, *Jardine and Selby.*

(*C. Jardinii,* Gould.)

Allied Harrier.

Gould, Handbk. Birds Aust., Vol. i., sp. 27, p. 60.   *ll . 4.*

The nest of this species is a flat structure, outwardly composed
of small sticks and twigs, lined inside with a few green
Eucalyptus leaves, and usually placed among the thick branches
of a low tree. Eggs two to three in number for a sitting, white,
with a bluish-green tinge on the inner surface of the shell.
Two specimens taken on the 30th September, 1884, measure in
length (A) 2 x 1·5 inches ; (B) 2·05 x 1·55 inches.
50,8   38,4   52,1   39,4
This species breeds during September and the two following
months.

Hab. Derby, N. W. Australia, Port Denison, Wide Bay,
Richmond and Clarence River District, New South Wales,
Interior, Victoria, South Australia, Western Australia, Cobar
and Bourke Districts, Dawson River. (*Ramsay.*)

A

3. ## CIRCUS GOULDI, *Bonaparte.*

(*C. assimilis,* Gould, *non* Jardine and Selby.)

Gould's Harrier.

Gould, Handbk. Bds. Aust., Vol. i., sp. 26, p. 58.   II. 3

This species is found breeding in swampy localities, or in the inlets and bays of the coast, constructing a nest of sticks and twigs on the top of a low thick bush, or clump of rushes. A set of eggs, three in number, taken from a nest built in some low bushes at Western Port Bay, Victoria, on the 15th November, 1884, are white with a bluish-green tinge inside. Length (A) 1·98 x 1·5 inches, (B) 2·03 x 1·53 inches, (C) 2·1 x 1·56 inches.
33,1.   51,6   39,9.   53,3. 39,6.
The breeding season of this species extends during September and the three following months.

Hab. Derby, N.W. Australia, Rockingham Bay, Port Denison, Wide Bay District, Dawson River, Richmond and Clarence River District, New South Wales, Interior, Victoria and South Australia, Tasmania. (*Ramsay.*)

Genus ASTUR, *Lacepède.*

2. ## ASTUR CINEREUS, *Vieillot.*

Grey-backed or New Holland Goshawk.

Gould, Handbk. Bds. Aust., Vol. i., sp. 14, p. 37.

Some naturalists consider that the Tasmanian bird, which is purely white in both sexes, is a distinct species from the continental form, in which the male alone is white, the females having an ashy-grey back, and in the young stage both sexes have ashy-grey bars on the under surface of the body. As the Tasmanian form is purely white it will perhaps be better to distinguish these varieties (or species?) under the names of the

Tasmanian Goshawk, and the continental form the New Holland Goshawk, (the name of the Australian Goshawk being retained for *Astur approximans*) which is distributed over the whole continent. A smaller variety of White Goshawk, the female of which has a ashy-grey back and a few bars of the same colour on the under surface of the body has been separated by Mr. R. B. Sharpe under the name of *Astur leucosomus*. It appears to be confined to Cape York and the southern portions of New Guinea, the adult male in this case is also purely white.

The nest of *Astur cinereus* (the large continental form) is an open structure composed of thin sticks, and lined with twigs and leaves. One found near the Cape Otway Forest, Victoria, in October, 1865, was placed in the topmost boughs of a lofty Eucalyptus, and contained two eggs in form nearly oval, slightly swollen at one end, of a dull bluish-white, smeared and blotched with faded markings of yellowish and reddish-brown, particularly towards the larger end, and which, were it not for their size, might be easily mistaken for those of *Astur approximans* which they closely resemble. Length (A) 1·97 x 1·48 inches; (B) 2·05 x 1·51 inches.

Hab. Derby, Rockingham Bay, Port Denison, Wide Bay District, Dawson River, Richmond and Clarence River District, New South Wales, Victoria and South Australia. (*Ramsay.*)

ASTUR APPROXIMANS, *Vigors and Horsfield.*

Australian Goshawk.

Gould, Handbk. Bds. Aust., Vol. i., sp. 17, p. 41.

The nest of the Australian Goshawk is comparatively a large structure, composed of sticks and lined with Eucalyptus leaves, and placed in a lofty tree, usually a Eucalyptus or Casuarina. The eggs are three in number for a sitting, nearly oval in form, being but slightly swollen at the larger end. The eggs (set A) are of a long narrow oval; colour dull white, smeared with yellowish-buff; averuge length 1·74 x 1·3 inches. A second set (B) shows . smears to a less extent, and there are a few scattered spots of a deep reddish-brown ; form a round-oval ; average length 1·75 x 1·4 inches in breadth. (*Ramsay, Note-book*, 1880-1, p. 5 ; *P.L.S.*, *N.S.W., 2nd Series*, Vol. i., p. 1141.)

This species commences to breed in August and continues through the three following months.

Hab. Derby, N.W. Australia, Cape York, Rockingham Bay, Port Denison, Wide Bay District, Dawson River, Richmond and Clarence River District, New South Wales, Interior, Victoria and South Australia, Tasmania, W. and S. W. Australia. (*Ramsay.*)

ASTUR RADIATUS, *Latham.*

Radiated Goshawk.

Gould, Handbk. Bds. Aust., Vol. i., sp. 16, p. 40.   *II   2.*

The following description is from Dr. Ramsay's note-book, under date 11th October, 1884, p. 25 :—

"The egg of *Astur radiatus*, just received from Mr. Barnard, of Coomooboolaroo, in the Dawson River District, Queensland, is much like a large egg of *Astur approximans* or that of *Aquila morphnoides.* It is of a dull white, roundish, with a few blackish-brown smears and blotches, and irregular markings and dots of a slightly darker shade; the shell is slightly rough. Length 2·2 inches, diameter 1·8 inches." (*P.L.S., N.S.W.,* 2nd *Series,* Vol. i., p. 1141. *Ramsay.*)

Hab. Wide Bay District, Dawson River, Richmond and Clarence River Districts, New South Wales, Interior. (*Ramsay.*)

GENUS ACCIPITER, *Brisson.*

(2) - 3 ACCIPITER CIRRHOCEPHALUS, *Vieillot.*

(*A. torquatus*, Vig. and Horsf.)

Collared Sparrow-Hawk.

Gould, Handbk. Bds. Aust., Vol. i., sp. 19, p. 45.   *II. 6.*

The nest is a scanty structure of a few sticks generally placed crosswise over a horizontal bough, where twigs spring to support it, and is lined with leaves. Being often at a considerable distance from the ground and near the extremity of the boughs it is difficult to get at. The eggs are usually three for a sitting but sometimes only two, and are the smallest of any of our Australian Hawk's eggs. The ground colour is greenish-white, with smears and specks of yellowish-buff, with here and there an irregular shaped spot of the same tint. Length (A) 1·74 x 1·43 inches, (B) 1·8 x 1·42 inches. (*Dobr. Mus. P.L.S., N.S.W.*, Vol. vii., p. 53. *Ramsay.*)

Hab. Derby, N.W. Australia, Cape York, Rockingham Bay' Port Denison, Wide Bay District, Dawson River, Richmond and Clarence River Districts, New South Wales, Interior, Victoria and South Australia, Tasmania, W. and S.W. Australia, South Coast New Guinea. (*Ramsay.*)

Sub-Family AQUILINÆ.

GENUS AQUILA, *Brisson.*

2.   AQUILA AUDAX, *Lath.*

Wedge-tailed Eagle.

Gould, Handb. Bds. Aust., Vol. i., sp. 1, p. 8.   *I. 1.*

"The nests of this species are easily found, for, indeed, they are large and conspicuous. They are often three feet high, and consist of a mass of sticks piled up between the forks of the topmost branches of the larger Eucalypti, or placed at the end of a leaning bough. The lower part of the nest is made of thick sticks, smaller ones being used for the top, and the whole lined with twigs and grasses. The first eggs I obtained were taken in August, 1860, and were given to me by Mr. James Ramsay, at Cardington, a station on the Bell River, near Molong. They were taken from the nest by a black boy who had "*stepped* " the tree. The nest was placed upon a fork near the end of one of the main branches of a large Eucalyptus. It was fully 70 feet from the ground, and no easy task to get to it. The structure was about 3½ feet high by 4 or 5 broad, and about 18 inches deep, lined with tufts of grass and with down and feathers plucked from the breasts of the birds, upon which the eggs were placed. The eggs were two in number, nearly round, and very thick and rough in the shell. One egg is 3 inches long by 2⅜ broad; the ground colour white, thickly blotched and minutely freckled with rust-red, light yellowish brown, and obselete spots of a lilac tint. The other egg is nearly all white, having only a few blotches of light yellowish brown, and some fine dots of light rust-red; it is 2⅞ inches in length by 2½ in breadth." (*Ibis*, 1863, Vol. v., p. 446, *Ramsay*.)

Two eggs of this species in the Australian Museum Collection measure as follows:—length (A) 3·01 x 2·18 inches; (B) 3·02 x 2·22 inches.

Hab. Derby, N.W. Australia, Gulf of Carpentaria, Cape York, Rockingham Bay, Port Denison, Wide Bay District, Dawson River, Richmond and Clarence River District, New South Wales, Interior, Victoria and South Australia, Tasmania, W. and S.W. Australia. (*Ramsay.*)

2.   AQUILA MORPHNOIDES, *Gould.*
Little Eagle.
Gould, Handbk. Bds. Aust., Vol. i., sp. 2, p. 11.   II. 1.

The nest of this Eagle is about the size of that of *Corvus coronoides,* and composed of similar materials, sticks and twigs,

and lined with Eucalyptus leaves ; sometimes the birds take
possession of an old crow's nest of the previous year. The eggs
are two in number for a sitting, but not unfrequently only one
is found ; the ground colour is dull white, with a few smears of
buff ; length (A) 2·2 x 1·8 inches, (B) 2·2 x 1·83 inches ; each
taken from different nests of one each.   (*Mr. Bennett's Coll.*)
(*P.L.S.*, *N.S.W.*, Vol. vii., p. 412.   *Ramsay.*)

Specimens of these eggs in my collection, taken by Mr. Geo.
Barnard, of Coomooboolaroo, Queensland, in 1883, are similar in
colour but not quite as rounded in form ; they measure as follows :
length (A) 2·22 x 1·7 inches ; (B) 2·17 x 1·7 inches.

Hab. Port Denison, Wide Bay District, Dawson River,
Richmond and Clarence Rivers Districts, New South Wales,
Interior, W. & S.W. Australia, South Coast N. Guinea. (*Ramsay.*)

Genus HALIAETUS.

2   HALIAETUS LEUCOGASTER, *Gmel.*

White-bellied Sea Eagle.

Gould, Handbk. Bds. Aust., Vol. i., sp. 3, p. 13.   *I. 2.*

The White-bellied Sea Eagle is found at intervals all along the
coast-line of Australia frequenting the bays, inlets, and estuaries
of rivers.   Its nest is a large flat structure composed of sticks,
and lined with finer twigs, the site chosen for its situation is very
variable, at times being placed on a lofty Eucalyptus, the top of
a thick mangrove, and on the summit of a rock.   Although
finding several nests of this species, I was never fortunate enough
to have the pleasure of taking the eggs myself, the nests I saw
at all times being placed out of the way of any one desirous of
taking them.   Two eggs taken by Mr. Ralph Hargrave, at
Wattamolla, New South Wales, from different nests, are white
smeared with light yellowish buff, or stained with dull light
brownish yellow, one specimen (B) has only light yellowish smears

in the centre, the texture of the shell is rough and slightly granular. Length (A) 2·98 x 2·16 inches, taken 9th August, 1875 ; (B) 2·72 x 2·06 inches, taken in August, 1870.

The breeding season commences in July in New South Wales, and I have seen young birds in the nest in Victoria, during the month of November.

Hab. Port Darwin and Port Essington, Gulf of Carpentaria, Cape York, Rockingham Bay, Port Denison, Wide Bay District, Dawson River, Richmond and Clarence Rivers Districts, New South Wales, Interior, Victoria and South Australia, Tasmania, W. and S.W. Australia, South Coast New Guinea. (*Ramsay.*)

Genus HALIASTUR, *Selby*.

HALIASTUR INDUS, *Bodd.*

Sub-Species H. GIRRENERA, *Vieillot.*

(*H. leucosternus*, Gould.)

White-breasted or Red-backed Fish Eagle.

Gould, Handbk. Bds. Aust., Vol. i., sp. 4, p. 17.

Upon the authority of Mr. Rainbird, Dr. Ramsay gives the following account of the nidification of this interesting species :—
"The nest of the Red-backed Fish Eagle is by no means so bulky a structure as that of many of its allies, nor is it so large as one would expect from a member of the family to which it belongs. In almost every instance the examples found by Mr. Rainbird were placed near the tops of the larger trees in belts of mangroves skirting the edges of salt-water swamps and marshes in the neighbourhood of Port Denison. They were composed of twigs and dead branches of mangrove, lined with a finer material. One, from which that gentleman shot the bird, and brought me the egg upon which she was sitting, was lined with tufts of lichen ; and in this instance the egg was placed on

various fish-bones, shells and claws of crabs, &c. ; the edges and sides were beautifully ornamented with long streamers of bleached seaweed, which gave the nest a novel and pleasing appearance. The egg has a rough ground of a bluish-white color, with a few minute spots of brownish-red near the larger end ; it is of an oval form 2 inches by 1 inch 6 lines in breadth. Mr. Rainbird states that this species of Hawk is far from rare about Port Denison. Throughout the whole year many may be seen hovering over the water near the mouth of the creeks, and over the salt-marshes which are invariably edged with dense belts of high mangroves." (*Ramsay, Ibis*, 1865, Vol. i., New Series, p. 83.)

Hab. Derby, N.W.A., Port Darwin and Port Essington, Gulf of Carpentaria, Cape York, Rockingham Bay, Port Denison, Wide Bay District, Richmond and Clarence River Districts, New South Wales, and South Coast New Guinea. (*Ramsay.*)

2   HALIASTUR SPHENURUS, *Vieillot.*

Whistling Eagle.

Gould, Handbk. Bds. Aust., Vol. i., sp. 5, p. 20. II ./.2.

With the exception of the extreme southern portions of the continent, this bird is distributed over the whole of Australia, and is found both on the open plains and in the timbered country. The nest is a large open structure composed of sticks and twigs, lined with a few Eucalyptus leaves, and is generally placed on the horizontal branch of a tree at a great height from the ground. Eggs two in number for a sitting, varying in form from true- to rounded-ovals, of a faint bluish-white, some specimens being heavily blotched with irregular shaped markings of reddish-chestnut, similar to those of *Lophoictinia isura*, others being but slightly marked with dull yellowish-brown, and in some instances almost devoid of markings of any kind, with the exception of a few fine scratches and smears hardly discernible on the smaller end of the egg.

Dimensions of two eggs taken by Mr. K. H. Bennett at Ivanhoe, in October, 1884, length (A) 2·17 inches x 1·68 inch; (B) 2·15 x 1·71 inch.

Three eggs in my collection, taken by Mr. Geo. Barnard of Coomooboolaroo, Duaringa, Queensland, during 1881, measure as follows : length (A) 2·22 inches x 1·68 inch; (B) 2·07 inches x 1·65 inch; (C) 2·3 inches x 1·67 inch; B and C are from the same nest.

This bird commences to breed about the middle of September, and continues the two following months.

Hab. Derby, N.W.A., Port Darwin and Port Essington, Gulf of Carpentaria, Cape York, Rockingham Bay, Port Denison, Wide Bay District, Dawson River, Richmond and Clarence River Districts, New South Wales, Interior, Victoria and South Australia, W. and S. W. Australia, South Coast New Guinea. (*Ramsay.*)

GENUS MILVUS, *Cuvier.*

3   MILVUS AFFINIS, *Gould.*

Allied Kite.

Gould, Handbk. Bds. Aust., Vol. i., sp. 21, p. 49.   *IX. 5. 6.*

Eggs three for a sitting, of a dull white ground colour, with reddish irregular spots and dots. No. 1 has rather large spots, rather evenly dispersed over the surface. No. 2 has only a few spots and smears.   Length (1) 1·84 x 1·48 inch; (2) 1·75 x 1·5 inches. (*Dobr. Mus. P.L.S., N.S.W.*, Vol. vii., p. 413. *Ramsay.*)

Hab. Derby, N.W.A., Cape York, Rockingham Bay, Port Denison, Wide Bay District, Dawson River, Richmond and Clarence River Districts, New South Wales, Interior, Victoria and South Australia, W. and S.W. Australia, South Coast New Guinea. (*Ramsay.*)

Genus LOPHOICTINIA, *Kaup.*

2 - 3   ## LOPHOICTINIA ISURA, *Gould.*

Square-tailed Kite.

Gould, Handbk. Bds. Aust, Vol. i., sp. 22, p. 51.   *IV.*   3. *4.*

Nest, of sticks and twigs rather loosely constructed, and lined with a few Eucalyptus leaves, placed in a fork of some of the higher branches of the trees, or in the interior where the trees are stunted and low, in any suitable branch that will bear its weight. Eggs two to three in number, the ground colour white, on the thicker end are blotches, smudges and scattered irregular spots of reddish-brown or rusty-red, with minute dots here and there sprinkled over the surface, frequently one egg in a set is blotched at the thin end, some are more heavily and deeply marked than others, one specimen is covered (more numerously at the thin end) with irregular freckles only, in many places superimposed. Length (A) 2·03 x 1·57 ; (B) 1·97 x 1·76 ; (C) 2·06 x 1·67 inches. (*Mus. Dobr. P.L.S., N.S.W.,* Vol, vii., p. 53, *Ramsay.)*

Hab. Wide Bay District, Dawson River, Richmond and Clarence River Districts, New South Wales, Interior, Victoria and South Australia, W. and S.W. Australia. (*Ramsay.*)

Genus GYPOICTINIA, *Kaup.*

2   ## GYPOICTINIA MELANOSTERNON, *Gould.*

Black-breasted Buzzard.

Gould, Handbk. Bds. Aust., Vol. i., sp. 20, p. 47.   *V.*   3. *4.*

Respecting the nidification and habits of this species Mr. K. H. Bennett writes as follows :—

"The range of this bird—so far as my experience goes—is confined to the plains which border the banks of the Murrumbidgee and Lachlan Rivers, and the wide expanse of open country on the north bank of the latter stream appears to be its especial habitat, for it is most frequently seen in that locality, and here also on several occasions I have discovered its nests. Its prey to a great extent, consists of various reptiles—such as snakes, frill-necked and sleepy lizards—it also has the singular habit of robbing the nests of Emus and Bustards of their eggs. My first information on this point I obtained from the blacks, and for some time I was inclined to disbelieve their assertion though the same story was told by blacks from all parts of the district, as it was so contrary to my experience of the Accipiter family. At length, however, I was compelled to alter my opinion, for I subsequently found portions of Emu egg shells in the nest of one of these Buzzards. The manner in which they effect the abstraction of the Emu eggs—as told me by the blacks—shows an amount of cunning and sagacity that one would scarcely give the bird credit for, and is as follows:—'On discovering a nest, the Buzzard searches about for a stone, or what is much more frequently found here, a hard lump of calcined earth. Armed with this the Buzzard returns (and should the Emu be on the nest) alights on the ground some distance off, and approaches with outstretched flapping wings, the Emu alarmed at this, to it, strange looking object, hastily abandons the nest and runs away, the Buzzard then takes quiet possession, and with the stone breaks a hole in the side of each egg into which it inserts its claw and carries them off at its leisure; for when the eggs are broken the Emu abandons the nest.' So much for the blacks' story!"

"This however, is in a great measure corroborated by a friend of mine, who lives on the adjoining station, and who told me that in August last, (1881) he found the nest of an Emu containing five eggs, and that all of them had a broken hole in the side, and that the fracture had been done quite recently, and in the nest also was one of these lumps of calcined earth about the size of a man's fist."

"In a nest to which I recently ascended, I found amongst the remains of various reptiles, the shells of a couple of Bustards' eggs. In this nest were a couple of young Buzzards lately hatched."

" I think after all this testimony there can be little doubt of its nest-robbing proclivities, a habit which I think is peculiar to this bird, and is not shared by any other member of the Accipiter family so far as I know. I have often asked the blacks, if the Wedge-tailed Eagle robs nests, but they always say no."

" The nest of this bird is a rough structure, generally placed on a forked horizontal branch, and is often quite as large as that of the Wedge-tailed Eagle. It lays two eggs, which in colour and shape resemble those of the above mentioned bird, but are much smaller. Length 2·16 x 1·85, being strongly blotched with bright rust-red, with spots and dots of the same colour."

" It usually lays about the middle of August, and the young birds leave the nest about the beginning of December. If undisturbed, the old birds resort year after year to the same nest, but should it be robbed, they abandon it for ever, and it is never occupied by birds of the same species again, although other species of hawks, notably the Brown Hawk—*(Hieracidea orientalis)* sometimes takes possession. I have never known the Buzzard to touch carrion, or to feed upon anything it did not capture, and except at the nest I have never seen them perch on a tree, but I have often seen them alight on the ground. The note which is something between a whistle and a scream is only uttered when visiting the nest." *(P.L.S., N.S.W.,* Vol. vi., p. 146, *Bennett.)*

I have a series of these handsome eggs now before me, taken by Mr. K. H. Bennett at different times, they vary considerably, some being heavily and richly blotched with reddish-brown and lilac, others with bright rust-red, and a few being but sparingly marked with freckles and hair lines of purplish-brown. The measurements are as follows :—One specimen taken in September 1884, at Mossgiel, length 2·6 inches x 1·96 inch. Two eggs taken in October 1884, length (A) 2·42 inches x 1·89 inch ; (B) 2·42 inches x 1·91 inch *(dark var.)* An egg taken from a nest which contained a young bird also, measures length 2·36 inches x 1·9 inch. Three light varieties taken from different nests, during November 1885, measure length (A) 2·35 inches x 1·67 inch ; (B) 2·27 inches x 1·68 inch ; (C) 2·5 inches x 1·8 inch.

14 FALCONIDÆ.

Hab. Derby, N.W.A., New South Wales, Interior, Victoria and South Australia, W. and S.W. Australia. (*Ramsay.*)

Genus ELANUS, *Savigny*.

3 -*⁴* ELANUS AXILLARIS, *Latham*.

Black-shouldered Kite.

Gould, Handbk. Bds. Aust., Vol. i., sp. 23, p. 53. **III**. *6.*

Dr. Ramsay writes in the Proceedings of the Linnean Society of New South Wales as follows :—

" During the last six years several pairs of these hawks have been known to breed on the Iindah Estate, on the Mary River in Queensland, but it was only in November last (1877) that a pair gave my brother (Mr. John Ramsay) an opportunity of taking their nest and eggs, which was not lost."

" The nest in question was placed among the topmost forked branches of a *Flindersia*, and as usual, composed of sticks and twigs ; it was not, however a bulky structure, as is often the case with the Australian hawk's nests. The eggs were three in number but my brother assures me that four is the correct number for a sitting. The ground colour, where visible, is of a dull white, but it is mostly obscured by blotches and smears of a dark reddish-chocolate. Length (A) 1·6 x 1·25 inch ; (B) 1·72 x 1·25 inch ; (C) 1·58 x 1·27 inch. One specimen (A) is reddish rusty chocolate smeared and clouded with a darker tinge." (*P.L.S., N.S.W.*, Vol. ii., p. 109, *Ramsay.*)

Dr. Cox has a very handsome set of the eggs of this bird in his collection ; taken on the Hawkesbury River, New South Wales, they are heavily blotched all over with rich, reddish-chocolate markings, and measure as follows. Length (A) 1·65 x 1·23 inch; (B) 1·67 x 1·24 inch ; (C) 1·66 x 1·23 inch.

Hab. Gulf of Carpentaria, Cape York, Rockingham Bay, Port Denison, Wide Bay District, Dawson River, Richmond and Clarence River Districts, New South Wales, Interior, Victoria and South Australia, W. and S.W. Australia. (*Ramsay.*)

3 ## ELANUS SCRIPTUS, *Gould.*

Letter-winged Kite.

Gould, Handbk., Bds. Aust., Vol. i., sp. 24, p. 55.

The Letter-winged Kite was at one time common on the Keilor Plains near Melbourne ; the last specimen I procured was a fine male, it was hovering at dusk in a paddock at Moonee Ponds, catching the field mice which constitute its food.

The mode of nidification, and the colour and disposition of the markings of the eggs are similar to the preceding species. Eggs three in number for a sitting. An average specimen in my own collection, taken at Keilor in September 1881, is of a dull white ground colour, heavily blotched with chocolate-brown markings, the only part of the ground colour clearly visible being at the smaller end. Length 1·7 inch x 1·27 inch in breadth.

This species breeds during the months of September and October.

Hab. Wide Bay District, Richmond and Clarence· Rivers District, New South Wales, Interior, Victoria and South Australia. (*Ramsay.*)

Genus BAZA, *Hodgson.*

3. ### BAZA SUBCRISTATA, *Gould.*

Crested Hawk.

Gould, Handbk. Bds. Aust., Vol. i., sp. 25, p. 56. *II. 5.*

The nidification of this species, the single representative of the genus found in Australia is described by Dr. Ramsay, as follows :

" During my recent trip to the North Richmond River, viâ Grafton, I met with this rare species upon several occasions. I found it giving preference to the edges of the scrubs on the Richmond and Clarence Rivers, whence it would sally out to the more open parts in the mornings and evenings in search of food, and not unfrequently visit the slaughtering establishments in the vicinity of the townships."

"I was fortunate enough to procure three eggs of this species, taken by Mr. Macgillivray's blackfellow "Daddy." Mr. Macgillivray informs me that, when Daddy was taking the eggs, the female dashed so close to him that he killed it with his tomahawk. The male bird belonging to the nest had been shot the day before. The nest was a comparatively small structure of sticks placed upon a horizontal bough, at a considerable distance from the ground. It contained three eggs, much smaller than I expected to find them. They have the peculiarity of being very much rounded at the larger end, are short upon the whole, and have the thin end pointed abruptly. Their average length is $1\frac{7}{10}$ by $1\frac{1}{10}$ inch in breadth; the ground colour greenish-white (the inside of the shell being of a pale sea-green), very sparingly smeared and blotched or spotted with light brown; one specimen has but two or three spots of a light brownish tint." (*P.Z.S.*, 1867, p. 392, *Ramsay*).

Mr. Geo. Barnard of Coomooboolaroo, has been successful in taking sexeral nests of this species in the Dawson River District. Two eggs received from him in 1884, measure as follows :—length (A) 1·73 x 1·41 inch ; (B) 1·75 x 1·4 inch.

Hab. Port Darwin and Port Essington, Cape York, Rockingham Bay, Port Denison, Wide Bay District, Dawson River, Richmond and Clarence Rivers District, New South Wales.

Genus FALCO,

2 - 3.   FALCO MELANOGENYS, *Gould.*

Black-cheeked Falcon.

Gould, Handbk. Bds. Aust., Vol. i., sp. 8, p. 26.   III.   4.

This species builds its nest upon the side of a cliff or precipitous rock. Mr. K. H. Bennett found a nest of this species at Mount Manara, in the Wilcannia District on the 9th September, 1885, which contained three eggs, the nest was about 70 feet from the ground, and very difficult to obtain, being placed upon the face of an almost perpendicular rock; upon visiting the same place the following year in the month of October, he found that the same(?) pair of birds had repaired the old nest, and that it contained a single fresh egg, but when disturbed again by his climbing to it, they abandoned it, and built a new nest a few yards higher up, out of reach, the rock on which it was placed completely overhanging the site of the old nest.

The eggs are two to three in number for a sitting, of a warm flesh or ruddy ground colour, almost obscured by freckles, spots, and confluent patches of a rich reddish-brown, in a few places the spots are nearly black. Length (A) 2·1 x 1·64 inch; (B) 1·97 x 0·65 inch; (C) 2·1 x 1·57 inch.

An egg taken by Mr. Bennett from the same nest the following year, and which he kindly gave me, is of a very light variety, the ground colour being pinky-white, with reddish-brown or light chestnut markings, confined to the smaller end of the egg. Length 2·1 x ·6 inch.

The eggs of this species are usually laid during the months of August and September.

Hab. Derby, N. W. A., Port Denison, Wide Bay District, Richmond and Clarence Rivers Districts, New South Wales, Interior, Victoria and South Australia, Tasmania, W. and S.W. Australia. (*Ramsay.*)

3   FALCO HYPOLEUCUS, *Gould.*

Grey Falcon.

Gould, Handbk. Bds. Aust., Vol. i., sp. 7, p. 24.   *III.* 3.

" This is a rare species not plentiful in any part of Australia, but occasionally obtained in the northern portion of the interior of Queensland, and Mr. Gould records it from Western Australia. I am indebted to Mr. J. B. White for specimens of the eggs taken on the upper Thomson River in Queensland."

B

"The nest, from his description is like that of an *Hieracidea*, composed of sticks and twigs, and placed on a horizontal bough ; the eggs were three in number, length (A) 2·07 inches x 1·51 inch ; (B) 2 inches x 1·52 inch ; they are oblong ovals, the whole of the ground colour obscured by minute dots and freckles of rusty-red ; there is in one an indistinct band on the larger end, the shell is smooth and slightly glossy ; the bird was seen on the nest. (*Mus. Dobr., from Mr. J. B. White's Collection.*") *P.L.S.*, *N.S. W.*, Vol. vii., p. 414, *Ramsay*.

Hab. Gulf of Carpentaria, Wide Bay District, Dawson River, Interior, W. and S.W. Australia. (*Ramsay.*)

4.   FALCO SUBNIGER, *G. R. Gray*.

Black Falcon.

Gould, Handbk. Bds. Aust., Vol. i., sp. 9, p. 28.   Ⅲ. /.

"It is through Mr. K. H. Bennett's exertions and liberality that I am enabled to describe the eggs of this rare species, taken by himself on the 27th of September, 1884. There were four laid for a sitting, which closely resemble large specimens of the Merlin's, and are not unlike finely freckled eggs of *Hieracidea orientalis*, but of a richer or brighter red, the ground colour being obscured with rich reddish dots and freckles all over the surface ; in some these dots form confluent markings on one end of the egg, or patches on the side ; they are almost identical in colour and shape with those of *F. hypoleucus*, but larger ; the shell is of finer grain than is shown in those of the *Hieracidea*. In form they are almost true ovals, being but slightly swollen at the thicker end ; one is rather elongate in form. Length (A) 2·1 x 1·6 inches ; (B) 2·13 x 1·58 inches ; (C) 2·18 x 1·55 inches." (*P.L.S., N.S. W., 2nd Series*, Vol. i., p. 1146, *Ramsay.*)

From a most interesting and exhaustive account of the habits of this bird, given by Mr. K. H. Bennett in the P.L.S., N.S.W., Vol. x., p. 167, I also extract the following :—

"The Black Falcon sometimes breeds here in the interior of New South Wales, but not often, for during my long residence (over 20 years) in this locality, I have only met with their nests on four occasions, (one only a few days since). It breeds in September, and lays four eggs which much resemble those of *Hieracidea orientalis.*"

" The nest I recently found was a clear case of appropriation, as last year it was occupied by a pair of *Gypoictinia melanosternon*, the Falcons had possession however this year, and the nest contained four eggs. Immediately beneath the edge of the nest a pair of *Xerophila leucopsis* had constructed theirs, which contained young, and the old birds were flying in and out, apparently quite unconcerned at the proximity of their dangerous neighbours, well knowing that the interstices between the large sticks of which the upper nest was composed, afforded them a secure asylum."

Hab. Wide Bay District, New South Wales, Interior, Victoria and South Australia. (*Ramsay.*)

3   FALCO LUNULATUS, *Latham.*
White-fronted Falcon.
Gould, Handbk. Bds. Aust., Vol. i., sp. 10, p. 29.

This species is almost entirely dispersed over the continent of Australia, specimens having been recently received in the collections formed by the late Mr. T. H. Boyer-Bower, from the vicinity of Derby, North Western Australia ; the only districts it has not yet been reported from being Port Darwin and Port Essington.

The nest is rather a large structure, outwardly composed of sticks securely woven together, and lined with fine strips of fibrous bark, and usually placed in the topmost branches of a lofty Eucalyptus.

Eggs three in number for a sitting, oval in form, buffy-white, thickly freckled and blotched with very light reddish-brown markings, which in some places are confluent, not unfrequently being on the smaller end of the egg. Specimens received from Mr. Geo. Barnard of Coomooboolaroo, Queensland, in 1883, measure as follows, length (A) 1·83 x 1·33 inch ; (B) 1·85 x 1·34 inch.

Hab. Derby, N.W. Australia, Gulf of Carpentaria, Cape York, Rockingham Bay, Port Denison, Wide Bay District, Dawson River, Richmond and Clarence Rivers Districts, New South Wales, Interior, Victoria and South Australia, Tasmania, West and South-West Australia. (*Ramsay.*)

Genus HIERACIDEA, *Gould.*

2 - 3   HIERACIDEA ORIENTALIS, *Schlegel.* •

(H. berigora, Gray.)

Brown Hawk.

Gould, Handbk. Bds. Aust., Vol. i., sp. 11, p. 31.   *III.   2.*

The Brown Hawk is without exception, the most common species of the Falconidæ inhabiting New South Wales and Victoria. It builds its nest sometimes in the topmost branches of a Eucalyptus or Casuarina, or at other times within a few feet of the ground ; it is an open structure composed of sticks, lined with fine twigs, leaves, &c. The eggs are two to three in number for a sitting, and are very variable, both in their size and markings ; from a number of sets now before me, I give the description of two, which are of the most usual form.

Var. (A). Eggs two in number, nearly oval in form, being slightly swollen at the larger end, of a warm reddish-white ground colour, minutely freckled, spotted, and blotched with rich reddish-brown, in some places these markings are confluent, on one specimen

(A) a coalesced patch of markings extends nearly over half the entire surface of the egg on the larger end ; in the other specimen (B), the markings are more uniform and evenly dispersed over the surface of the shell. Length (A) 2 inches x 1·45 inch ; (B) 1·94 x 1·47 inch. Taken at Mossgiel, by Mr. K. H. Bennett, September 9th, 1885.

Var. (B). Eggs two in number, oval in form, of a buffy-white ground colour, with minute freckles of reddish-brown, all over the surface, in one specimen (A) intermingled with dark reddish-black blotches, closely resembling the markings of *Pandion leucocephalus*, in the other (B) the markings are neither so large, dark, nor numerous, and are confined entirely to the smaller end of the egg. Length (A) 2 inches x 1·47 inch ; (B) 1·97 x 1·46 inch. Taken at Mossgiel, by Mr. K. H. Bennett, September 16th, 1885.

This species breeds during the months of September, October, and November.

Hab. Derby, N.W. Australia, Port Darwin and Port Essington, Gulf of Carpentaria, Cape York, Rockingham Bay, Port Denison, Wide Bay District, Dawson River, Richmond and Clarence Rivers Districts, New South Wales, Interior, Victoria and South Australia, Tasmania, West and South-West Australia. (*Ramsay.*)

3 HIERACIDEA BERIGORA, *Vigors and Horsfield.*

(*H. occidentalis*, Gould.)

Western Brown Hawk.

Gould, Handbk. Bds. Aust., Vol. i., sp. 12, p. 33.

The nidification of the Western Brown Hawk is similar to that of the preceding species. Eggs two to three in number for a sitting, usually more rounded in form than those of *H. orientalis*, the ground colour being almost obscured with small rich reddish-brown markings, uniformly dispersed over the whole surface of the shell. A set taken by Mr. Joseph Hill, at Kewell, in the Wimmera District, Victoria, on the 9th September, 1887, measures length (A) 1·98 x 1·55 inch ; (B) 2 inches x 1·6 inch.

The breeding season commences during the latter end of August and continues the two following months.

Hab. Derby, N.W. Australia, Interior, Victoria and South Australia, West and South-West Australia. (*Ramsay.*)

Genus TINNUNCULUS, *Vieillot.*

TINNUNCULUS CENCHROIDES, *Vigors and Horsfield.*

Nankeen Kestrel.

Gould, Handbk. Bds. Aust., Vol. i., sp. 13, p. 35.   *III. 5*

This species usually deposits its eggs, four in number for a sitting, on the decaying wood in a hollow limb of a tree. The eggs vary in form, from short to rounded ovals, and are of a reddish-white ground colour, minutely freckled, and heavily blotched all over with rich reddish-brown markings ; in some instances the blotches are confined entirely to the smaller end. On two occasions I have seen eggs of this species nearly white. A set taken at Cardington, on the Bell River, New South Wales, in September 1867, measure as follows :—length (A) 1·55 x 1·15 inch ; (B) 1·48 x 1·18 inch ; (C) 1·5 x 1·2 inch ; (D) 1·48 x 1·19 inch.

The breeding season comprises the months of September, October, and November.

Hab. Derby, N.W. Australia, Port Denison, Wide Bay District, Dawson River, New South Wales, Interior, Victoria and South Australia. (*Ramsay.*)

Sub-Order PANDIONES.

Genus PANDION, *Savigny*.

3   PANDION LEUCOCEPHALUS, *Gould*.

White-headed Osprey.

Gould, Handbk. Bds. Aust., Vol. i., sp. 6, p. 22.   *Ⅴ. /. 2.*

This species constructs a large nest of sticks lined with twigs and seaweed, and is sometimes placed on the summit of a rock, but more often in the top of some high tree.   Mr. John S. Ramsay took a nest of this species, containing two eggs, (Var. A.) in the Wide Bay district, on the 15th of August, 1880, and another with three eggs (Var. B.) at Iindah, on the Mary River, Queensland, in 1882.

Var. A. Eggs two in number, elongated in form, of a deep yellowish-white ground colour, heavily blotched and spotted with rich reddish-brown markings, but particularly on the larger end, where they form a coalesced patch.   Length (A) 2·56 x 1·73 inches; (B) 2·57 x 1·72 inches.

Var. B. Eggs three in number, oval in form, of a rich flesh-white ground colour, spotted and blotched with deep reddish-brown, and minutely freckled with small dots of the same colour, intermingled with obsolete splashes of purple appearing as if beneath the surface of the shell.   Length (A) 2·27 x 1·7 inches; (B) 2·27 x 1·68 inches; (C) 2·31 x 1·68 inches.

The breeding season which commences early in July, continues during the three following months.

Hab. Derby, N.W. Australia, Port Darwin and Port Essington, Gulf of Carpentaria, Cape York, Rockingham Bay, Port Denison, Wide Bay District, Richmond and Clarence Rivers Districts, New South Wales, Victoria and South Australia, Tasmania, West and South-West Australia, South Coast of New Guinea. (*Ramsay.*)

Sub-Order STRIGES.

Family STRIGIDÆ.

GENUS STRIX, *Linnæus.*

(2)   STRIX CANDIDA, *Tickell.*

Grass Owl.

Gould, Suppl. Bds. Aust., Pl. i.   II. 5.

Respecting this rare Australian Owl, Dr. Ramsay remarks:—
"I am indebted to my friend Mr. J. A. Boyd, for a set of the
eggs of this species, taken in the Herbert District, Queensland;
like those of all others of the genus, they are white, but have a
slight bluish tinge; in form oval, rather swollen about the centres;
length (A) 1·68 x 1·25 inches in breadth; (B) 1·7 x 1·25 inches."
(*P.L.S., N.S.W., 2nd Series,* Vol. i., p. 1060, *Ramsay.*)

Hab. Rockingham Bay, Wide Bay District, New South Wales.
(*Ramsay.*)

6.   STRIX FLAMMEA, *Linnæus.*

Sub-Species STRIX DELICATULA, *Gould.*

Delicate Owl.

Gould, Handbk. Bds. Aust., Vol. i., sp. 31, p. 66.

On two occasions, Mr. K. H. Bennett found nests of this
species at Ivanhoe, in the interior of New South Wales. The
eggs were in both instances six in number, of a dull uniform
white, and were deposited on the decayed wood in the hollow
limb of a Box tree. An average specimen measures 1·55 inch in
length x 1·28 inch in breadth,

Hab. Derby, N.W. Australia, Gulf of Carpentaria, Cape York,
Rockingham Bay, Port Denison, Wide Bay District, Richmond
and Clarence River Districts, New South Wales, Interior, Victoria

and South Australia, West and South-West Australia, South Coast New Guinea. (*Ramsay.*)

Sub-Family BUBONINÆ.

Genus NINOX, *Hodgson.*

3. NINOX BOOBOOK, *Latham.*

Boobook Owl.

Gould, Handbk. Bds. Aust., Vol. i., sp. 36, p. 74. *XI. 6.*

This species makes no nest, but lays its eggs, three in number, on the decayed wood, in some hole or spout of a Gum tree. Egg white, the shell minutely pitted as if with the point of a pin; one of a set of three taken in South Gippsland, measures :—length 1·42 inch x 1·2 inch in width.

The breeding season commences in October and lasts during the two following months.

Hab. Port Darwin and Port Essington, Cape York, Rockingham Bay, Port Denison, Wide Bay District, Dawson River, Richmond and Clarence Rivers Districts, New South Wales, Interior, Victoria and South Australia, West and South-West Australia. (*Ramsay.*)

NINOX CONNIVENS, *Latham.*

Winking Owl.

Gould, Handb. Bds. Aust., Vol. i., sp. 34, p. 71.

"Gilbert procured an egg of this species in Western Australia, it was pure white, somewhat round in form, and large for the size of the bird, measuring two inches in length by one and five-eights in breadth." (*Gould.*)

Hab. Gulf of Carpentaria, Cape York, Rockingham Bay, Port Denison, Wide Bay District, Richmond and Clarence Rivers Districts, New South Wales, Interior, Victoria and South Australia. (*Ramsay.*)

Order INSESSORES.
Family CAPRIMULGIDÆ.

GENUS ÆGOTHELES, *Vigors and Horsfield.*

ÆGOTHELES NOVÆ-HOLLANDIÆ, *Vigors and Horsfield.*

The Owlet Nightjar.

Gould, Handbk. Birds Aust., Vol. i., sp. 38, p. 79. XIV *11.*

This species deposits its eggs, which are white and from three to four in number on the debris at the bottom of a hollow branch of a tree, usually a Eucalyptus. A set of three taken at Cardington, in December, 1867, by Mr. John S. Ramsay, measure as follows :—length (A) 1·15 x 0·87 inch ; (B) 1·1 x 0·85 inch ; (C) 1·15 x 0·87 inch. The shell is thick and hard in texture.

The breeding season commences in October and lasts during the three following months.

Hab. Rockingham Bay, Wide Bay District, Dawson River, Richmond and Clarence Rivers Districts, New South Wales, Interior, Victoria and South Australia, Tasmania, West and South-West Australia. (*Ramsay.*)

GENUS PODARGUS, *Cuvier.*

PODARGUS STRIGOIDES, *Latham.*

Tawny-shouldered Podargus.

Gould, Handbk. Bds. Aust., Vol. i., sp. 40, p. 84. XX. 3.

This bird is universally dispersed throughout New South Wales and Victoria ; it builds a flat nest of sticks loosely placed together on the horizontal branch of any suitable tree. Eggs three in number for a sitting, perfectly white, elongated in form, shell finely granulate. Dimensions of a set taken at Dobroyde, October 16th, 1862. Length (A) 1·88 x 1·32 inch ; (B) 1·85 x 1·35 inch ; (C) 1·94 x 1·32 inch,

The breeding season commences in September, and is at its height in October, and continues the two following months.

Hab. Port Denison, Wide Bay District, Dawson River, Richmond and Clarence Rivers Districts, New South Wales, Interior, Victoria and South Australia. *(Ramsay.)*

PODARGUS CUVIERI, *Vigors and Horsfield.*

Cuvier's Podargus.

Gould, Handbk. Bds. Aust., Vol. i , sp. 41, p. 87. ⅩⅠ 4.

The nidification of this species is similar to that of *P. strigoides,* but the nest is more compactly and securely built. Eggs white, of a uniform size at both ends ; specimens taken in Tasmania, in October, 1885, measure as follows :—length (A) 1·71 x 1·23 inch ; (B) 1·73 x 1·23 inch.

Hab. Victoria and Tasmania. *(Ramsay.)*

GENUS EUROSTOPODUS,

EUROSTOPODUS GUTTATUS, *Vigors and Horsfield.*

Spotted Nightjar.

Gould, Handbk. Bds. Aust., Vol. i., sp. 49, p. 98. ⅩⅠ 2.

" Its single egg is laid on the ground without any preparation for it, usually near some stone or stump on the side of a stony ridge ; the ground colour is of a light greenish creamy-white, sparingly marked all over with dots and oval spots of blackish and slaty-brown, a few markings appearing as if beneath the surface of the shell. Length 1·38 x 1 inch in breadth, both ends nearly equal." (*P.Z.S.*, 1875, p. 581, *Ramsay.*)

An egg of this species in my possession taken by Mr. George Barnard in 1886, measures 1·31 inch in length by 0·97 inch in breadth.

Hab. Derby, N.W. Australia, Port Darwin and Port Essington, Gulf of Carpentaria, Cape York, Rockingham Bay, Port Denison, Wide Bay District, Dawson River, Richmond and Clarence Rivers Districts, New South Wales, Interior, Victoria and South Australia, West and South-West Australia. (*Ramsay.*)

EUROSTOPODUS ALBIGULARIS, *Vigors and Horsfield.*

White-throated Nightjar.

Gould, Handbk. Bds. Aust., Vol. i., sp. 48, p. 96. XI 3.

" The egg resembles that of *E. guttatus*, without the greenish tinge on the ground-colour, which is of a rich, light cream-colour, spotted sparingly all over with round dots and oval-shaped marks of black, blackish-brown, and slaty-black, which latter appear as if beneath the surface of the shell; length from 1·41 to 1·5 inch, breadth from 1·03 to 1·13, equal at both ends."

"I found this species breeding both at Brisbane and on the Mary River. My brother, Mr. James Ramsay, of Nanama, forwarded to me authentic eggs from the Merule in the Riverina district, of New South Wales."

" The Australian Museum having recently received from Mr. A. Clarke, a very fine specimen of the egg of this species, which differs in size and colour from all I have before examined, I cannot let the occasion slip by without describing so fine a specimen. The eggs taken at Gootchy by Mr. D. Helsham, were evidently quite authentic, as the birds were flushed off them, and several specimens shot by myself at the time, The Gootchy eggs are smaller than the present specimen, but slightly larger than those of *E. guttatus*. The egg at present under consideration, in size is almost as large as that of *Podargus humeralis*. An egg from Gootchy is 1·53 x 1·05 inch ; colour light cream with black round dots. That received from Mr. A. Clarke, is of a rich deep cream-yellow, having on one side a cluster of round black spots, which touch one another here and there, and a few similar dots sprinkled over the remaining surface ; length 1·55 inches x 1·15 inches in breadth. In consequence of the great width of the egg it appears to be larger than it really is." (*Ramsay*, *P.Z.S.*, 1875, p. 581 ; and *P.L.S.*, *N.S.W.*, Vol. i., *2nd Series*, p. 1142.)

Hab. Wide Bay District, Dawson River, Richmond and Clarence Rivers Districts, New South Wales, Interior, Victoria and South Australia. (*Ramsay*.)

Genus CAPRIMULGUS, *Linnæus*.

2   CAPRIMULGUS MACRURUS, *Horsfield*.
Large-tailed Nightjar.
Gould, *Handbk. Bds. Aust.*, Vol. i., sp. 50, p. 100.

" I am indebted to Inspector Robert Johnstone for a fine pair of these eggs ; they were found on the ground on the side of a ridge near the Herbert River, Queensland ; and are of a light rich cream colour, fading to whitish after being emptied, clouded all over with fleecy markings of pale slaty lilac, which appear as if beneath the surface of the shell ; length 1·1 x ·81 inch in breadth. (*Ramsay*, *P.Z.S.*, 1875, p. 581.)

Hab. Port Darwin and Port Essington, Gulf of Carpentaria, Cape York, Rockingham Bay, Port Denison, Wide Bay District, New South Wales, Interior, Victoria and South Australia, West and South-West Australia, South Coast New Guinea. (*Ramsay*.)

Family HIRUNDINIDÆ.

Genus HIRUNDO, *Linnæus*.

— 5.   HIRUNDO NEOXENA, *Gould*.

(H. frontalis, Quoy et Gaimard.)

Welcome Swallow.

Gould, Handbk., B.Is. Aust., Vol. i., sp. 53, p. 107.

"Although this species is strictly migratory, yet it is no easy task to determine the exact date of its arrival in the neighbourhood of Sydney, or its departure therefrom, owing to the number of stragglers which remain with us during the whole of the year. I believe however, that the visitants arrive early in July, or perhaps late in June, and leave us again in the end of January and February. After their arrival, and again just before their departure, they may be seen in great numbers flying to and fro over the fields, and often skimming over the water-holes and lagoons; but keeping very high sometimes almost out of sight, during the middle of the day."

"I have frequently observed them, in company with the Fairy Martin *(Lagenoplastes ariel)* flying over the lawn of the inner Domain in Sydney. Tree Swallows *(Petrochelidon nigricans)* also accompany this species in search of food. We met with all three species mixed up in one immense flock during December 1864, at Lake Bathurst; here they were following in our wake as we walked through the rushes on a small island, obtaining a rich feast on the small *Libellulæ,* which flew up in countless numbers at every step we took. The pupa-cases of these insects were lying piled up between the rushes to the height of two, and even three feet, while the edges of the island at dusk were alive with the pupæ crawling out of the water. The proper breeding season of *Hirundo neoxena* is during the months of August and throughout to the end of December; stragglers, however, may be found breeding at almost any time. I have found them building in the Dobroyde stables, both in the months of February and June; and on April 17th, 1864, I took a nest with fresh eggs from the same buildings."

"In choosing a site for the nest they seem to be even less particular than in their time of breeding. Almost any building will serve them where they can obtain a horizontal beam or ledge. On this they place their round bowl-shaped nest, the wall of which is composed of pellets of mud, mingled with grass, and securely fastened on the beam. As soon as the mud-work is dry, it is warmly lined with grasses, horsehair, or feathers; and the nest is then ready for the eggs, which are usally from three to five in number, 0·75 inch in length by 0·5 in breadth. The ground colour is of a delicate white, having numerous dots and freckles of yellowish-brown and faint lilac sprinkled over the whole surface, but more thickly at the larger end. The nests are 4 to 6 inches (wide by 2·5 inches deep."

"In 1858, while fishing off a small steamer, which having been out of use for some months, was moored a few hundred yards from the North Shore, in the Sydney Harbour, I observed a pair of these Swallows fly round the boat, and frequently dive underneath the paddle-box. After a long search I discovered their nest, which was composed of black pitchy mud, lined with seaweed and feathers. It was placed upon one of the horizontal beams of the paddle-box, and contained three young ones about half fledged. The man in charge informed me that the nest had been made when the steamer was lying lower down the harbour, and upon its being tugged to where it then lay the birds flew round and round it the whole time, evidently in a great state of excitement." (*Ramsay*, *Ibis*, 1868, Vol. iv., New Series, p. 275.)

A set of the eggs of this species in the Australian Museum Collection measures as follows :—length (A) 0·73 x 0·5 inch ; (B) 0·74 x 0·59 inch ; (C) 0·72 x 0·58 inch ; (D) 0·76 x 0·6 inch ; (E) 0·73 x 0·58 inch.

Hab. Gulf of Carpentaria, Cape York, Rockingham Bay, Port Denison, Wide Bay District, Dawson River, Richmond and Clarence Rivers Districts, New South Wales, Interior, Victoria and South Australia, Tasmania, West and South-West Australia. (*Ramsay.*)

Genus PETROCHELIDON, *Cabanis.*

ɔ ## PETROCHELIDON NIGRICANS, *Vieillot.*

Tree Swallow.

Gould, Handbk. Bds. Aust, Vol. i., sp. 55, p. 111. ~~XIII~~. *14.*

This species is to be found throughout the whole of Australia, Tasmania, and southern portions of New Guinea. It arrives in New South Wales and Victoria in August, and leaves again at the latter end of February. It deposits its eggs three in number, on the decayed wood, in a hollow branch, or hole in a tree ; the ground colour being of a pinky-white, covered with minute freckles of light rusty-brown, particularly towards the larger end, where in some instances, intermingled with lilac spots they form a zone, others again are of a pure white, with a few fine dots of light red at the larger end.

A set taken by Mr. K. H. Bennett, at Mossgiel on the 16th of September, 1885, measure as follows :—(A) 0·73 x 0·55 inch ; (B) 0·72 x 0·54 inch ; (C) 0·68 x 0·54 inch.

Hab. Gulf of Carpentaria, Cape York, Rockingham Bay, Port Denison, Wide Bay District, Dawson River, Richmond and Clarence River Districts, New South Wales, Interior, Victoria and South Australia, Tasmania, West and South-West Australia, South Coast New Guinea. (*Ramsay.*)

Genus LAGENOPLASTES, *Gould.*

ʒ ## LAGENOPLASTES ARIEL, *Gould.*

Fairy Martin.

Gould, Handbk. Bds. Aust., Vol. i., sp. 56, p. 113.

"About the end of November 1860, I discovered a large batch of nests of this species fastened under an overhanging rock upon the banks of the Bell River. I counted upwards of one hundred nests, all built so closely together, that of many, the entrances were alone visible."

The Martins were flying to and from the nests in great numbers, some carrying in grass for the linings, others busily employed in repairing the old, and building new nests with the mud from the river's bank. Many also I found brooding their eggs, and this gave me a good opportunity of procuring some specimens, which I did not fail to seize. There were usually from three to five eggs, but some nests contained seven. Many of the eggs were altogether white, others were spotted with light brownish-yellow occasionally all over, in other instances only at the larger end. They vary in length from 7 to 8½ lines, and from 6 to 6½ lines in breadth." *(Ramsay, Ibis* 1865, Vol. i., New Series, p. 299.)

On September 29th, 1886, in company with Mr. Geo. Masters we took a number of the nests of this species at Chatsworth, on the Eastern Creek, the eggs varied both in size, shape, and colour, some being white without markings of any kind, others being elongated and heavily marked with yellowish-brown spots, they measure as follows:—length (A) 0·67 x 0·47 inch ; (B) 0·69 x 0·48 inch ; (C) 0·75 x 0·49 inch ; (D) 0·73 x 0·48 inch ; (E) 0·68 x 0·47 inch.

During a visit to Dubbo in August 1887, these birds arrived in great numbers, commencing to build on the 17th, and covering the eaves of the schools, churches, and public buildings with their curiously retort-shaped nests.

Hab. Port Denison, Wide Bay District, Dawson River, Richmond and Clarence Rivers Districts, New South Wales, Interior, Victoria and South Australia. *(Ramsay.)*

GENUS CHERAMŒCA, *Cabanis.*

CHERAMŒCA LEUCOSTERNUM, *Gould.*

White-breasted Swallow.

Gould, Handbk. Bds. Aust., Vol. i., sp. 57 p. 115   *XIII.   15*

I extract the following from Mr. K. H. Bennett's MS. notes on the nidification of this bird :—-

C

"This species of Swallow is the only one with which I am acquainted, that is not migratory, being found in this locality, (Mossgiel New South Wales) and to the same extent as regards numbers all the year round. It is widely distributed throughout the timbered or " back " country, but is never found on the plains, and is generally seen in small flocks of five or six in number. It breeds here during the month of October, in holes in the sides of the entrances to the burrows of either the *Bettongia*, or *Peragales* whether inhabited by these animals or not. I have never yet found more than four eggs in a nest."

A set of three taken by Mr. Bennett on October 19th, 1885, at Mossgiel, are pure white, and measures as follows :—(A) 0·64 x 0·48 inch ; (B) 0·63 x 0·48 inch ; (C) 0·64 x 0·47 inch.

Specimens in my possession taken by Mr. Gardner in South Australia, give the same measurements. .

Hab. New South Wales, Interior, Victoria and South Australia, West and South-West Australia. (*Ramsay.*)

Family MEROPIDÆ.

Genus MEROPS, *Linnæus.*

MEROPS ORNATUS, *Latham.*

Australian Bee-eater.

Gould, Handbk. Bds. Aust., Vol. i., sp. 58, p. 117.

This bird is universally dispersed throughout the whole of Australia and tunnels a narrow hole in the sandy bank of a river or creek at the extremity of which it lays its eggs, five in number. Eggs, beautiful pearly-white ; a set taken at Buldery, measures as follows :—(A) 0·87 x 0·72 inch ; (B) 0·85 x 0·70 inch ; (C) 0·85 x 0·73 inch ; (D) 0·85 x 0·72 inch ; (E) 0·84 x 0·73 inch.

It breeds in New South Wales during October and the two following months.

Hab. Derby, N.W.A., Port Darwin and Port Essington, Gulf of Carpentaria, Cape York, Rockingham Bay, Port Denison, Wide Bay District, Dawson River, Richmond and Clarence River Districts, New South Wales, Interior, Victoria and South Australia, W. and S. W. Australia, South Coast New Guinea. (*Ramsay.*)

Family CORACIDÆ.

Genus EURYSTOMUS, *Vieillot.*

2 - 3 EURYSTOMUS PACIFICUS, *Latham.*

Australian Roller.

Gould, Handbk. Bds. Aust., Vol. i., sp. 59, p. 119. XIV /.

" I found this bird nesting in the hollow Eucalyptus boughs on the Richmond River, in October, 1867 ; they make no nest but lay their eggs on the dust formed by the decayed wood—not unfrequently they fight with, and dispossess the *Dacelo gigas,* and I have seen them take the young of this bird and throw them out of the nest   The eggs are two or three in number, of a dull white, rather glossy, and sometimes variable in form, some being oval and pointed, others almost round. Length (A) 1·45 x 1·05, oblong ; (B) 1·34 x 1·1, roundish." (*Dobr. Mus. P.L.S., N.S.W.,* Vol. vii., p. 46. *Ramsay.*)

Hab. Derby, N.W.A., Cape York, Rockingham Bay, Port Denison, Wide Bay District, Dawson River, Richmond and Clarence River Districts, New South Wales, Victoria and South Australia, South Coast New Guinea. (*Ramsay.*)

Family ALCEDINIDÆ.

GENUS DACELO, *Leach.*

3-4.　　　　　**DACELO GIGAS,** *Bodd.*

Great Brown Kingfisher.

Gould, Handbk. Bds. Aust., Vol. i., sp. 60, p. 122. *II 1.*

This well known bird deposits its eggs on the decaying wood in a hollow branch or hole in a tree, usually a Eucalyptus. Eggs three or four in number for a sitting, of a beautiful pearly-white. Dimensions of a set taken at Cardington, on the Bell River, New South Wales, in October, 1860. Length (A) 1·68 x 1·4 inch; (B) 1·72 x 1·41 inch; (C) 1·8 x 1·45 inch; (D) 1·72 x 1·4 inch.

On one occasion, on Ash Island, a nest of this species was found formed in the side of a clump of Stag-horn fern.

When the young birds are fully fledged they crowd to the entrance of the aperture, in their eagerness to obtain the coveted morsel, which the parent bird procures for them, usually a lizard or field mouse, and their united and incessant clamourings for the same can be heard a considerable distance away.

This species breeds during the months of August, September, and October.

Hab. Wide Bay District, Dawson River, Richmond and Clarence Rivers Districts, New South Wales, Interior, Victoria and South Australia. (*Ramsay.*)

4.　　　　　**DACELO LEACHII,** *Vigors and Horsfield.*

Leach's Kingfisher.

Gould, Handbk. Bds. Aust., Vol. i., sp. 61, p. 124. *II 2.*

"This species takes the place in the North of *D. gigas* in the South. Eggs four in number for a sitting, placed in a hollow bough, or at the end of tunnels excavated in the nest of the Termites. Colour of a pure pearly-white. Length (A) 1·72 x 1·3 inch, oval in shape; (B) 1·6 x 1·34, rather round."

"They breed during the months of September, October, and November." (*Ramsay, P.L.S., N.S.W.*, Vol. vii., p. 45.)

Hab. Port Darwin and Port Essington, Gulf of Carpentaria, Cape York, Rockingham Bay, Port Denison, Wide Bay District, Dawson River, South Coast New Guinea. (*Ramsay.*)

GENUS HALCYON *Swainson.*

HALCYON SANCTUS, *Vigors and Horsfield,*

Sacred Kingfisher.

Gould, Handbk. Bds. Aust., Vol. i., sp. 63, p. 128.

This bird is now known to be universally dispersed throughout the continent of Australia, it having lately been received from North-western Australia, in the collection formed by Mr. Cairn in the vicinity of Derby last year (1886), being the only portion of the continent from which specimens had not been previously received.

The Sacred Kingfisher deposits its eggs which are pearly-white and usually five in number, on the decaying wood in a hollow branch, or hole of a tree, usually a Eucalyptus. A set taken in October 1870, measure as follows : length (A) 1·03 x 0·88 inch ; (B) 1·03 x 0·89 inch ; (C) 1·03 x 0·87 inch ; (D) 1·02 x 0·82 inch ; (E) 1·05 x 0·88 inch.

The breeding season commences in September, and lasts the two following months.

Hab. Derby, N.W. Australia, Port Darwin and Port Essington, Gulf of Carpentaria, Cape York, Rockingham Bay, Port Denison, Wide Bay District, Dawson River, Richmond and Clarence Rivers Districts, New South Wales, Interior, Victoria and South Australia, Tasmania, W. and S.W. Australia, South Coast New Guinea. (*Ramsay.*)

s   # HALCYON PYRRHOPYGIUS, *Gould.*

Red-backed Kingfisher.

Gould, Handbk. Bds. Aust., Vol. i., sp. 64, p. 130.

This species breeds in the Bourke and Cobar districts during October and November, it nests in hollow boughs of trees, but on one occasion Mr. James Ramsay took five eggs from the end of a tunnel in the bank of a recently made dam or tank; these specimens, accompanied by a skin of the parent bird, were sent to me, and measure as follows :—(A) 1·02 x 0·88; (B) 1·02 x 0·88; (C) 1·02 x 0·88; (D) 1·02 x 0·78; (E) 1·04 x 0·87 inch; they are a pure glossy white colour. *(Dobr. Mus. P.L.S., N.S.W.,* Vol. vii., p. 45, *Ramsay.)*

Hab. Derby, N.W. Australia, Port Darwin and Port Essington, Gulf of Carpentaria, Cape York, Rockingham Bay, Port Denison, Wide Bay District, Dawson River, New South Wales, Interior, Victoria and South Australia. (*Ramsay.*)

s   # HALCYON MACLEAYI, *Jardine and Selby.*

Macleay's Kingfisher.

Gould, Handb. Bds. Aust., Vol. i., sp. 66, p. 133.

This beautiful Kingfisher has an extensive range of habitat, being found in the northern provinces of New South Wales, and the whole of Queensland, and has lately been received from North Western Australia, in the collection formed by the late Mr. T. H. Boyer-Bower. It often deposits its eggs in the hollow branch of a tree, at other times, digging a tunnel in the nest of the White Ants, it lays them in a rounded chamber at the extremity. During Dr. Ramsay's visit to Queensland in 1874 a nest was found in the grounds of Mr. Coxen, near Brisbane, on the 6th of January; it was a narrow tunnel about twelve inches or less in length, made in a White Ant's nest on a tree a few yards from the ground, it contained five fresh eggs of a clear pearly white, somewhat transparent shell, and abruptly pointed at the smaller Length (A) 1 inch x 0·83 inch; (B) 1·01 x 0·82 inch; (C) 1·03 end. x 0·83 inch; (D) 1·05 x 0·82 inch; (E) 1·05 x 0·85 inch.

This species breeds during the months of November, December, and January.

Hab. Derby, N.W.A., Port Darwin and Port Essington, Gulf of Carpentaria, Cape York, Rockingham Bay, Port Denison, Wide Bay District, Dawson River, Richmond and Clarence Rivers Districts, New South Wales, South Coast New Guinea. (*Ramsay.*)

Genus TANYSIPTERA, *Vigors*.

TANYSIPTERA SYLVIA, *Gould*.

White-tailed Kingfisher.

Gould, Handbk. Bds. Aust., Vol. i., sp. 68, p. 137.

Mr. J. A. Thorpe of the Australian Museum, informs me that he found this species breeding plentifully during the months of September and October, in the hills or nests of the White Ants *(Termites)* situated in the dense scrubs in the neighbourhood of Cape York. It tunnels a hole a few feet from the ground in one of these hills, about fifteen inches in length, and lays four or five pure white eggs near the extremity, where it is hollowed out in the form of a chamber. A set of these eggs are round in form and measures as follows :—length (A) 0·98 x 0·89 inch; (B) 1 x 0·89 inch; (C) 1 x 0·89 inch; (D) 0·99 x 0·88 inch; (E) 1·03 x 0·98 inch.

Hab. Cape York, Rockingham Bay, Port Denison. (*Ramsay.*)

GENUS ALCYONE, *Swainson.*

6.   ## ALCYONE AZUREA, *Latham.*

Azure Kingfisher.

Gould, Handbk. Bds. Aust., Vol. i., sp. 69, p. 139.

The Azure Kingfisher is always found in the vicinity of water, where it makes a tunnel in the bank of a river or tank; at the extremity of the excavation it deposits its eggs, six in number, on a mass of cast fish bones. Eggs pearly white, and round ; a set taken on September 23rd, 1862, from a tunnel in the side of a waterhole at Dobróyde by Mr. John Ramsay, measure as follows :—length (A) 0·91 x 0·75 inch ; (B) 0·87 x 0·75 inch ; (C) 0 85 x 0·75 inch ; (D) 0·88 x 0·73 inch ; (E) 0·85 x 0·73 inch ; (F) 0·87 x 0·72 inch.

The breeding season commences in September, and continues during the months of October and November.   •

Hab. Rockingham Bay, Port Denison, Wide Bay District, Richmond and Clarence Rivers Districts, Victoria and South Australia. (*Ramsay.*)

ALCYONE DIEMENENSIS, *Gould.*

Tasmanian Kingfisher.

Gould, Handbk. Bds. Aust., Vol. i., sp. 70, p. 141.

The nidification of this species is similar to that of the preceding one.

A single specimen of the egg of this rare bird in the Macleayan Museum Collection is glossy white. Long diameter 0·9 inch, short diameter 0·76 inch.

Hab. Tasmania. (*Ramsay.*)

S.   ALCYONE PULCHRA, *Gould.*

Resplendent Kingfisher.

Gould, Handbk. Bds. Aust., Vol. i., sp. 71, p. 141.

In a collection of birds recently formed on behalf of the Trustees of the Australian Museum, by Messrs. E. J. Cairn and Robert Grant, in the neighbourhood of Mount Bellenden-Ker, Northern Queensland, are two specimens of *Alcyone pulchra.* One of them was obtained with great difficulty, on the Barron River, about thirty miles inland from the coast, by Mr. Cairn, who, having shot the bird, had to swim to procure it. The other was captured on the nest by Mr. Grant. It is worthy of note that after comparing these birds with a large series of *A. pulchra*, from Cape York, Port Essington, and Port Darwin, the flanks are not tinged so deeply with rich lilac as in the specimens from the extreme northern localities; but are similar to others of the same species procured by Mr. Cairn at Derby, North-Western Australia in 1886.

Mr. Grant has kindly supplied me with the following information relative to the taking of the nest :—

"On the 26th December, 1887 at Riverstone, about sixteen miles inland from Cairns, in company with an aboriginal called "Charlie," (native name Euryimba), I saw a Kingfisher fly into a hole in the bank of a creek ; after running forward and placing my hat over the entrance, I with my sheath-knife enlarged the opening, and putting my hand in caught one of the parents; while engaged in securing it, my attention was drawn away from the nest for a moment, when to my surprise another bird flew out, so both of the parent birds were in the hole at the same time. Afterwards, upon dissection, the bird I captured proved to be the male. The nest, if worthy of the name, was placed near the end of the tunnel, which was about sixteen inches in length and inclined upwards ; it was composed of a few cast fish bones, and small pieces of decayed roots, but in all not sufficient to protect the eggs from the sandy soil at the bottom. The nest contained five eggs, three of which were unfortunately broken."

The two remaining eggs are similar to those of the southern representative *A. azurea*, being rounded in form, pearly white, and the texture of the shell fine and very glossy. Length (A) 0·87 x 0·73 inch ; (B) 0·85 x 0·74 inch. (*From the Aust. Mus. Coll. North, P.L.S., N.S.W., 2nd Series*, Vol. iii., pt. i, 1888.)

Hab. Derby, N.W. Australia, Port Darwin and Port Essington, Gulf of Carpentaria, Cape York. (*Ramsay.*)

Family ARTAMIDÆ.

. 3 - 4

ARTAMUS SORDIDUS, *Latham.*
Wood Swallow.

Gould, Handbk. Bds. Aust., Vol. i., sp. 73, p. 143. *VIII. 14.*

This bird makes a round, and rather inclined to cup-shaped nest of roots, lined neatly inside with grasses, three and a-half inches across and two inches in depth, outside measurement. I have found it breeding in all parts of Victoria, during September to January, placing its nest in every conceivable position, sometimes in the dead branches of a fallen tree, on the side of the trunk of a tree, held in position by a projecting piece of bark, or occasionally between bare forked limbs.

Its eggs are three or four in number and are subject to much variation in their markings.

Var. A. Ground colour dull white, spotted with pale brown, dark brown, and brownish black markings, particularly at the larger end. Average length of specimens 0·87 x 0·66 inch. Taken at Dobroyde, September 19th, 1862 (*E.P.R.*)

Var. B. Ground colour almost pure white, with a number of round black spots confined to the larger end where they form a zone. Average length of set of three 0·95 x 0·67 inch. Taken at Cardington, September 1868, by Mr. James Ramsay.

Var. C. A most unusually marked set. Ground colour creamy white with large irregular shaped blotches of reddish- and brownish-black. Average size of a set of three, length 0·87 x 0·68 inch. Taken at Hastings, Victoria, November 23rd 1883. (*A.J.N.*)

This species breeds from September till the middle of January. The young birds are often found on the ground during December, having left the nest before being able to fly, and are unable to get back again.

Hab. Rockingham Bay, Port Denison, Wide Bay District, Dawson River, Richmond and Clarence Rivers Districts, New South Wales, Interior, Victoria and South Australia, Tasmania, W. and S.W. Australia. (*Ramsay.*)

- **3.**   ARTAMUS LEUCOGASTER, *Valenc.*

(A. leucopygialis, Gould.)

White-rumped Wood Swallow.

Gould, Handbk. Bds. Aust., Vol. i., sp. 80, p. 154.

No species of the genus *Artamus* seems to enjoy a wider range than *A. leucogaster*, being found alike at Cape York, and in Victoria and South Australia. It is generally found in the vicinity of watercourses, relining the deserted nest of *Grallina picata* wherein to deposit its eggs. Only on one occasion have I heard of this bird building a nest for itself, and that was found by Mr. K. H. Bennett at Ivanhoe; it was a loosely built structure of dried grasses placed in the horizontal fork of a small tree over hanging water. Eggs three in number for a sitting, of a dull white, spotted and blotched with faint markings of yellowish-brown, reddish-brown and obsolete markings of bluish-grey, the latter appearing as if beneath the surface of the shell; these markings usually assume the form of a zone at the larger end. Dimensions of a set of three taken by Mr. Bennett on the 18th of October, 1885 are as follows, length (A) 0·9 x 0·63 inch ; (B) 0·95 x 0·63 inch ; (C) 0·9 x 0·62 inch.

October and the two following months constitute the breeding season of this species.

Hab. Derby, N.W. Australia, Port Darwin and Port Essington, Gulf of Carpentaria, Cape York, Rockingham Bay, Port Denison,

Wide Bay District, Dawson River, Richmond and Clarence Rivers Districts, New South Wales, Interior, Victoria and South Australia, South Coast New Guinea. (*Ramsay.*)

2 - 3 ARTAMUS PERSONATUS, *Gould.*
Masked Wood Swallow. •

Gould, Handbk. Bds. Aust., Vol. i., sp. 78, p. 150.

The Masked Wood Swallow builds a very frail structure of fine twigs and grasses in some convenient fork of a low tree, or in the hollow in the top of a stump. Eggs two or three in number ; ground colour greyish-white clouded all over with blotches of light brown, a few obsolete spots of grey appearing on the under surface of the shell. Dimensions of a set of three taken by Mr. James Ramsay at Tyndarie, October 1878. Length (A) 0·84 x 0·67 inch ; (B) 0·83 x 0·65 inch ; (C) 0·84 x 0·67 inch. •

This species is to be found breeding in company with *Artamus superciliosus.* On one occasion Mr. George Barnard of Coomooboolaroo, Duaringa, Queensland, found the male of *A. superciliosus* paired with the female of *A. personatus.*

Hab. Wide Bay District, Dawson River, Richmond and Clarence Rivers Districts, New South Wales, Interior, Victoria and South Australia, West and South-West Australia. (*Ramsay.*)

ARTAMUS CINEREUS, *Vieillot.*
Grey-breasted Wood Swallow.

Gould, Handbk. Bds. Aust., Vol. i., sp. 75, p. 147.

I have never seen this bird in any collection, The following description of the nidification of this species is taken from Mr. Gould's Handbook to the Birds of Australia :—"It breeds in October and November, making a round compact nest, in some instances of fibrous roots, lined with fine hair-like grasses ; in others of the stems of grasses and small plants ; it is built either in a scrubby bush or among the grass-like leaves of the *Xanthorrhœa* and is deeper and more cup-shaped than those of the other members of the group. The eggs are subjected to considerable variation in colour and in the character of their markings ; they are usually bluish-white, spotted and blotched with lively reddish-brown, intermingled with obscure spots and dashes of purplish-grey, all the markings being most numerous towards the larger end ; they are about <u>eleven</u> lines long by eight lines broad." (*Gould, Handbk. Bds. Aust.*, Vol. i., p. 148.)

Hab. Western Australia. (*Gould.*)

3   ARTAMUS ALBIVENTRIS, *Gould.*

White-vented Wood Swallow.

Gould, Handbk. Bds. Aust., Vol. i., sp. 76, p. 149.

The White-vented Wood Swallow is an inhabitant of Queensland and the northern portion of New South Wales. The nest of this species is similar to that of other members of the genus, being an open shallow structure composed of fine pliant twigs, and lined inside with fine grasses, and usually placed on the forked branch of a tree. The eggs are three in number for a sitting, in form swollen ovals, some specimens being somewhat sharply pointed at the smaller end, of a dull white ground colour, with irregular shaped blotches, spots, and dots of reddish- and yellowish-brown, intermingled with superimposed markings of greyish-lilac, which are more thickly disposed towards the larger end, where they become confluent, forming a well defined zone ; the markings are very sparingly distributed over the remainder of the surface of the shell. A set before me, taken by Mr. George Barnard of Coomooboolaroo, Queensland, measures as follows :—length (A) 0·87 x 0·69 inch ; (B) 0·85 x 0·69 inch ; (C) 0·87 x 0·68 inch.

Like all the birds of the interior, the time of nesting is greatly influenced by the rainy season; but October and November are the usual months for breeding.

Hab. Gulf of Carpentaria, Rockingham Bay, Interior, Victoria and South Australia. (*Ramsay.*)

7.   ARTAMUS MELANOPS, *Gould.*

Black-faced Wood Swallow.

Gould, Handbk. Bds. Aust., Vol. i., sp. 77, p. 149.   *VIII* /3.

This species is found plentifully dispersed throughout the interior of South Australia, and the Darling and Albert districts of New South Wales. The nest is a round open structure, composed of fibrous roots, lined inside with grasses, and placed in a low bush. Eggs four in number for a sitting, varying considerably in the character of their markings.   •

A set taken by Mr. K. H. Bennett at Mossgiel, on the 17th of October 1886, are of a fleshy-white, thickly freckled and spotted with irregular shaped markings of reddish-brown, and others of a bluish-grey tint appearing as if beneath the surface of the shell. Length (A) 0·87 x 0·69 inch; (B) 0·9 x 0·67; (C) 0·85 x 0·67 inch; (D) 0·89 x 0·68 inch.

Another set of a salmon-white ground colour are heavily blotched all over, but particularly towards the larger end with bright red, and a few indistinct obsolete spots of deep bluish-grey. Length (A) 0·82 x 0·68; (B) 0·87 x 0·67 inch; (C) 0·89 x 0·67 inch; (D) 0·87 x 0·68 inch. Taken by Mr. James Ramsay at Tyndarie, November the 2nd, 1879.

The months of September, October, and November constitute the breeding season of this species. (*North, P.L.S., N.S.W.,* Vol. ii., *2nd Series,* p. 405.)

Hab. Derby, N. W. A., Port Darwin and Port Essington, New South Wales, Interior, Victoria and South Australia, West and South-West Australia. (*Ramsay.*)

2 – 4.   ARTAMUS MINOR, *Vieillot.*

Little Wood Swallow.

Gould, Handbk. Bds. Aust., Vol. i., sp. 74, p. 146.

Mr. George Barnard informs me that this species builds its nests often in the end of a hollow branch, or in hollows in the tops of stumps and broken trees, and posts, sometimes in old mortice holes in fences ; the nest is very frail and scanty structure, merely a few leaves, sticks, and twigs put together so loosely that it will scarcely bare removal. The eggs two, three, or four for a sitting, are of a dull white or cream colour, blotched with yellowish-brown and obsolete markings of slaty-grey, which in some specimens are heavily blotched with these colours, others spotted, blotched, freckled or minutely dotted ; all are more or less zoned at the thicker end, in some the spots are confluent, forming ill-shapen figures, in others round or oval and well defined. Length (1) 0·75 x 0·55—average size ; (2) 0·71 x 0·45 ; (3) 0·76 x 0·55 ; (4) 0·75 x 0·57. (*Ramsay, P.L.S., N.S.W.,* Vol. vii., p. 407.)

Hab. Derby, N.W. Australia, Port Darwin and Port Essington, Gulf of Carpentaria, Rockingham Bay, Port Denison, Wide Bay District, Dawson River, Richmond and Clarence Rivers Districts, New South Wales, West and South-West Australia. (*Ramsay.*)

3   ARTAMUS SUPERCILIOSUS, *Gould.*

White-eyebrowed Wood Swallow.

Gould, Handbk. Bds. Aust., Vol. i., sp. 79, p. 152.

This bird is strictly migratory, arriving in Victoria to breed, about the end of November, and departing again at the commencement of March, sometimes however, three, four, and even five years elapse without seeing a single specimen, and it is remarkable when they visit us in great numbers, as far south as Melbourne, that it is during a period of drought in the interior.

It builds a round, and almost flat, scanty nest of roots and grasses—through which the eggs, in some situations, can be seen from below—in every possible position, both in the indigenous and acclimatized trees of our public parks and gardens. In Albert Park I have found no'less than ten nests, each containing eggs, in a single row of Pines *(Pinus insignus)* of about fifty yards in length ; the trees at that time being of a uniform height of five feet ; at other times the nest is placed in the horizontal fork of the branches of the Eucalyptus or Acacia, the broad flat fronds of the Norfolk Island Pine *(Araucaria excelsa)*, and on two occasions I have found it in the leafy top of a rose bush. The eggs are three in number, usually of a buffy-white ground colour, blotched and freckled all over with light brown, and umber brown markings, particularly towards the larger end, occasionally one egg in a set is found of a dull-white ground colour, with a well defined zone of dark umber round the larger end. The measurements of a set taken at Albert Park in December 1870, are as follows :—length (A) 0·9 x 0·7 inch ; (B) 0·95 x 0·7 inch ; (C) 0·93 x 0·67 inch.

Hab. Wide Bay District, Dawson River, Richmond and Clarence Rivers Districts, New South Wales, Interior, Victoria and South Australia. *(Ramsay.)*

Family PARDALOTINÆ.

Genus PARDALOTUS, *Vieillot*.

4. PARDALOTUS PUNCTATUS, *Temminck*.

Spotted Diamond Bird.

Gould, Handbk. Bds. Aust., Vol. i., sp. 81, p. 157.

The Spotted Pardalote or Diamond-bird is common in all parts of New South Wales, and plentifully dispersed over the whole of eastern and southern portions of the continent.

"Like the Black-headed species *(P. melanocephalus,)* it digs a small narrow burrow in the side of a bank or mound of earth, the end of this it enlarges into a spherical chamber of about four inches in diameter, which it lines all round but more thickly at the bottom, with fine strips of stringy-bark,† or, in the absence of this material, with grass. When the earth is carefully removed and the nest taken out, it is found to be a very loose hollow ball, slightly interwoven and having a small round entrance in the side, opposite the opening of the burrow. Sometimes a small hole in a log of wood is chosen, a crevice in an old wall, a niche under a shelving rock, or the banks of water-holes or creeks, all alike are resorted to ; still I have never known the Spotted Pardalote to breed in the hollow branch of a tree, or take possession of the nests of a Fairy Martin *(Lagenoplastes ariel)* as *P. affinis* and *P. striatus* are wont to do."

The eggs of *Pardalotus punctatus* are four in number, of a beautiful pearly-white after being emptied, but pinkish before, rather roundish, being in length 0·6 inch x 0·5 inch in breadth. The breeding season, which commences sometimes as early as July, lasts until the end of December, during which time three broods are often raised." (*Ramsay, Ibis*, 1868, Vol. iv., New Series, p. 272.)

While collecting with Mr. George Masters at Chatsworth near Mount Druitt, New South Wales, on the 29th September, 1886, a set of these eggs, which are unusally large, were taken, they measure in length (A), 0·77 x 0·58 inch ; (B), 0·71 x 0·55 inch ; (C), 0·7 x 0·54 inch ; (D), 0·71 x 0·57 inch.

Hab. Rockingham Bay, Port Denison, Wide Bay District, Dawson River, Richmond and Clarence Rivers Districts, New South Wales, Interior, Victoria and South Australia, Tasmania, West and South-West Australia. (*Ramsay.*)

---

† The fibrous bark of the *Eucalyptus capitella, E. macrorrhyncha,* and other allied species.

D

4. PARDALOTUS XANTHOPYGIUS, *McCoy.*

Yellow-rumped Diamond-bird.

Gould, Suppl. Bds. Aust., pl. 8.

This beautiful little Pardalote, the last discovered of the family is dispersed over the southern portions of the Australian continent. It gives a decided preference for the Mallee country and scrubby tracts of land; in Victoria it is found rather plentifully in the Whipstick scrub near Sandhurst, and parts of the Wimmera district. It excavates a tunnel about two feet in length, in a bank or in the side of a slight depression of the earth, at the extremity of which it builds a nest of strips of bark and dried grasses; the eggs are four in number for a sitting, pearly-white, rounded in form and slightly pointed at one end, a set taken during November 1883, in the Whipstick scrub, Sandhurst, measure as follows :—length (A) 0·65 x 0·5 inch; (B) 0·66 x 0·52 inch ; (C) 0·65 x 52 inch; (D) 0·64 x 0·51 inch.

Hab. New South Wales, Interior, Victoria and South Australia, West and South-West Australia. (*Ramsay.*)

3-5. PARDALOTUS ORNATUS, *Temminck.*

(P. striatus, Vigors and Horsfield.)

Striated Pardalote.

Gould, Handbk. Birds Aust., Vol. i., sp. 84, p. 161.

"During my first visit to Cardington, on the Bell River, in the Molong district, I was much surprised and delighted at finding this beautiful species of Pardalote in that neighbourhood. My brother, Mr. James Ramsay, informed me at the time, that this bird arrived every year about the beginning of October, and would shortly begin to breed. This I found to be the case. In the course of a few weeks they took possession of their usual breeding places, a batch of old nests of the Fairy Martin (*Lagenoplastes ariel*). These they lined with grass and stringy-bark making a nest similar to that of *Pardalotus punctatus.* The eggs varied from three to five in number. They are very ovate, and of a glossy white; in length $7\frac{1}{2}$ to 8 lines by $6\frac{1}{2}$ to 7 lines in breadth.

"About three weeks after the Pardalotes had taken possession of these nests, the rightful owners returned; but, finding the usurpers unwilling to turn out, the Martins contented themselves by building new nests, and repairing those that had been broken down.

"The nest of this species is usually built in a hole in the dead branch of some tree, and is very compact, being composed of grass strips of bark, and warmly lined with feathers. The breeding season commences in September and lasts during the three following months." (*Ramsay, Ibis,* 1865, Vol. i., New Series, p. 298.)

A set of the eggs of this species taken at Cardington, measure as follows:—length (A) 0·68 x 0·58 inch; (B) 0·64 x 0·55 inch; (C) 0·67 x 0·57 inch; (D) 0·65 x 0·56 inch.

A set in the Macleayan Museum Collection gives the following dimensions:—(A) 0·71 x 0·55 inch; (B) 0·68 x 0·54 inch; (C) 0·71 x 0·55 inch.

Dr. Ramsay and I succeeded in procuring a fine series of these birds on the Bell and Macquarie Rivers, in the neighbourhood of Wellington and Dubbo, during August 1887, and in October of the same year they were found breeding in the nests of the Fairy Martin, in the centre of the town of Orange.

Hab. Port Denison, Dawson River, New South Wales, Interior, Victoria and South Australia, West and South-West Australia. (*Ramsay.*)

4. PARDALOTUS AFFINIS, *Gould.*

Allied Pardalote.

Gould, Handbk., Bds. Aust., Vol. i., sp. 85, p. 163.

This bird, like the preceding species constructs its nest of strips of bark, grasses, and feathers, in a hole of some decayed branch

of a tree. Its eggs usually four in number, are white, and measure
(A) 0·7 x 0·55 inch ; (B) 0·67 x 0·55 inch ; (C) 0·7 x 0·52 inch.

The species breeds during September and the three following
months.

Hab. Dawson River, New South Wales, Victoria and South
Australia, Tasmania. (*Ramsay.*)

4. PARDALOTUS MELANOCEPHALUS, *Gould.*

Black-headed Pardalote.

Gould, Handbk. Bds. Aust., Vol. i., sp. 86, p. 165.

The nidification of this species is similar to that of *P. punctatus*,
a nest of bark and grasses, being constructed at the end of a small
tunnel about eight or ten inches long in a bank. It is a very
common bird at Rockingham Bay, Port Denison, and the Richmond
and Clarence Rivers. Eggs four in number for a sitting, pure
white, rather pointed at the smaller end.

A set taken by Dr. Ramsay at the Salt Water Creek, Richmond
River on the 19th of September 1866 measures as follows :—
length (A) 0·7 x 0·54 inch ; (B) 0·73 x 0·53 inch ; (C) 0·72 x 0·54
inch ; (D) 0·7 x 0·53 inch.

A set taken by Mr. Rainbird at Port Denison measures length
(A) 0·68 x 0·55 inch ; (B) 0·67 x 0·54 inch ; (C) 0·67 x 0·53 inch.

This bird breeds during September and the four following
months.

Hab. Rockingham Bay, Port Denison, Wide Bay District,
Dawson River, Richmond and Clarence Rivers Districts, New
South Wales, Interior. (*Ramsay.*)

4. PARDALOTUS UROPYGIALIS, *Gould.*

Yellow-rumped Pardalote.

Gould, Handbk. Bds. Aust., Vol. i., sp. 87, p. 166.

"This species is an inhabitant of the Gulf of Carpentaria district. I have seen it in collections from the Norman River, and also received the head, wings, and tail, accompanied with eggs from Mr. William E. Armit, taken on the Etheridge River, where this gentleman found it breeding in tunnels dug in the banks of creeks and water-courses, &c., in company with *P. rubricatus.* I can see no difference in the eggs of this and those of the following species *P. rubricatus,* except that they are a trifle smaller. The following remarks on *P. rubricatus* are equally applicable to this species also, Mr. Armit assures me that they breed and nest in the same way, and often accompany each other in small troops, searching for insects among the leafy tops of the trees. Both species seem to be confined to the inland districts. I searched diligently for them at Rockingham Bay, but found only the common species, *P. melanocephalus.* Eggs four in number, length 0·7 x 0·55 inch in breadth, and, like the eggs of all the other species, of a pearly-white colour."

"As I remarked above, *Pardalotus uropygialis* belongs to the same section as *P. melanocephalus, P. rubricatus, P. xanthopygius,* and *P. punctatus,* all digging tunnels in the soft banks of creeks, watercourses, &c., to nest in. On the other hand, *P. affinis,* and *P. striatus* (and, according to Mr. Gould, *P. quadrigintus* also), select holes in hollow branches of lofty trees, where they construct a dome shaped nest of grasses, just as the other species do at the end of their tunnels. The eggs in all instances are white, oval, and rather pointed." (*Ramsay, P.L.S., N.S.W.,* Vol. ii., p. 110, 1877.)

Hab. Derby, N.W. Australia, Port Darwin and Port Essington, Gulf of Carpentaria, Dawson River, and recently ·found at Rockingham Bay. (*Ramsay.*)

⚡ PARDALOTUS RUBRICATUS, *Gould.*

Red-lored Diamond-bird.

Gould, Handbk. Bds. Aust., Vol. i., sp. 82, p. 158.

"From letters received from Mr. William E. Armit, I learn that this species is by no means rare on the Norman River, and is also found rather plentiful on the Etheridge River. It comes as far south as Georgetown, where Mr. Armit obtained the nest and eggs. Like *Pardalotus punctatus, P. melanocephalus,* and *P. uropygialis,* this species digs holes or tunnels in the banks of creeks, &c., making a long narrow tunnel from two to three feet in length, at the end of which it excavates a chamber large enough to contain the nest, which is about four inches in diameter. This round chamber is lined on all sides both above and below, with fine grasses, except a small hole for exit opposite the tunnel. The eggs are four in number, pearly white, 0·8 inch in length by 0·6 inch in width towards the thicker end ; those at present under consideration are rather pyriform, and more pointed than those of any other species I have seen. Some specimens are a little larger than others." (*Ramsay, P.L.S., N.S.W.,* Vol. ii., p. 110.)

Hab. Derby, N.W. Australia, Port Darwin and Port Essington, Gulf of Carpentaria, Rockingham Bay, Dawson River, New South Wales, Interior. (*Ramsay.*)

⚡ PARDALOTUS QUADRAGINTUS, *Gould.*

Forty-spotted Pardalote.

Gould, Handbk. Bds. Aust., Vol. i., sp. 83, p. 160.

This species is confined to Tasmania, the nidification being similar to that of *P. affinis* and *P. striatus.* A set of four eggs which were taken from a hole in the hollow limb of a tree near Hobart, in October 1885 are white, rounded in form, and slightly pointed at one end ; they measure as follows :—length (A) 0·65 x 0·5 inch ; (B) 0·65 x 0·51 inch ; (C) 0·63 x 0·52 inch ; (D) 0·66 x 0·5 inch.

Hab. Tasmania. (*Ramsay.*)

Family LANIADÆ.

Genus STREPERA, *Lesson.*

ɔ-ɏ. STREPERA GRACULINA, *White.*

Pied Crow-shrike.

Gould, Handbk. Bds. Aust, Vol. i., sp. 88, p. 168.

This species is found throughout New South Wales, Victoria, and parts of Queensland. It constructs its nest in the forked branch of a tree, usually a Eucalyptus or Casuarina, it is a large, open, bowl-shaped structure, outwardly composed of sticks, and lined with strips of bark and grasses. Eggs three or four in number for a sitting, of a pale chocolate-brown, with faint blotchings and markings of reddish-brown, in some instances a few obsolete irregular shaped spots of lilac appear as beneath the surface of the shell. Length (A) 1·65 x 1·12 inch; (B) 1·7 x 1·15 inch; (C) 1·63 x 1·2 inch.

The breeding season commences in August and lasts during the three following months.

Hab. Rockingham Bay, Port Denison, Wide Bay District, Richmond and Clarence Rivers Districts, New South Wales, Interior, Victoria and South Australia. (*Ramsay.*)

3. STREPERA CUNEICAUDATA, *Vieillot.*

(S. anaphonensis, Gould.)

Grey Crow-shrike.

Gould, Handbk. Bds. Aust., Vol. i., sp. 91, p. 173. *VIII.* /.

This species has a very wide range, being found at Rockingham Bay in the North and extending to Victoria in the South; the opinion originally held by Mr. Gould that the Western Australia species is different, has been upheld by Mr. Sharpe, therefore the name of *S. plumbea* given to it by Mr. Gould holds good for the West Australian species.

The nest is a large bowl-shaped structure, composed outwardly of sticks, lined with finer twigs, and placed in the small upright branches or fork of a tree, usually a Eucalyptus. The eggs are three in number for a sitting, of a pale reddish-buff or buffy-brown tint, spotted and blotched all over with reddish-brown markings, intermingled with others of a pale slaty-brown, appearing as if beneath the surface of the shell. Length (A) 1·75 x 1·2 inch; (B) 1·73 x 1·18 inch; (C) 1·76 x 1·19 inch.

The breeding season comprises the months of August, September, and October.

Hab. Wide Bay District, Richmond and Clarence Rivers Districts, New South Wales, Interior, Victoria and South Australia. (*Ramsay.*)

STREPERA INTERMEDIA, *Sharpe.*

Brit. Mus. Cat. Bds, Vol. iii., p. 59.

A single egg of this species in the Dobroyde Collection, taken at Mount Gawler, South Australia in 1860, is similar in colour and markings to the egg of the Tasmanian species *S. arguta.* Long diameter 1·77 inch, short diameter 1·17. (*North, P.L.S., N.S.W.,* Vol. ii., 2nd Series, p. 405.)

Hab. Victoria and South Australia. (*Ramsay.*)

STREPERA MELANOPTERA, *Gould.*

Black-winged Crow-shrike.

Gould, P.Z.S., 1846, p. 20. *Brit. Mus. Cat. Bds.,* Vol. iii., p. 61.

"This bird is found breeding in South Australia. It constructs a large open nest of sticks and twigs, lined inside with fibrous roots and grasses, and usually placed in the topmost branches of a Eucalyptus. Two eggs of this species in the Dobroyde Collection taken by Mr. Gardner in 1863, are similar in form to those of

S. arguta ; they are of a light purple or rich vinous brown ground colour, with large irregularly- shaped markings of slaty-brown evenly dispersed over the surface of the shell. Length (A) 1·6 x 1·18 inch ; (B) 1·65 x 1·19 inch." (*North, P.L.S., N.S.W.,* Vol. ii., 2nd Series, p. 406.)

Hab. Victoria and South Australia. (*Ramsay.*)

3-5   STREPERA FULIGINOSA, *Gould.*
Sooty Crow-shrike.
Gould, Handbk. Bds. Aust., Vol. i., sp. 89, p. 170.   *VIII . 2.*

The nidification of this bird is similar to the preceding species, and the same may be said of all members of the genus. The eggs in a number of sets of five species of the genus now before me, show very little variation either in form or colour, and it would be difficult, if possible at all, to single out those of one species from another. Specimens that have been recently taken are of a much brighter and deeper colour.

The Sooty Crow-shrike lays from three to four eggs for a sitting, varying in shape from a short to a long oval, of a pale chocolate or vinous-brown ground colour, and marked all over with blotches of light-brown ; in one specimen (A) the markings are very small and confined to the larger end. Length (A) 1·63 x 1·2 inch ; (B) 1·75 x 1·18 inch ; (C) 1·73 x 1·22 inch.

This species breeds during the months of August, September, and October.

Hab. Victoria and South Australia, Tasmania. (*Ramsay.*)

3.   STREPERA ARGUTA, *Gould.*
Hill Crow-shrike.
Gould, Handbk. Bds. Aust., Vol. i., sp. 90, p. 171.

The nest of this species is similar to that of *S. melanoptera.* Eggs three for a sitting, of a light reddish or buffy-brown ground colour, spotted or blotched with markings of a darker tint, one specimen B, is a rounded-oval in form, and the markings are clouded and not so well defined. Length (A) 1·78 x 1·18 inch ; (B) 1·63 x 1·21 inch ; (C) 1·64 x 1·22 inch.

The above eggs are described from a set taken in Tasmania.

Mr. Sharpe has separated the South Australian examples of this species, and described them under the name of *S. melanoptera* previously given to it by Mr. Gould.

The months of September and October constitute the breeding season of this species.

Hab. Tasmania. (*Ramsay.*)

Genus GYMNORHINA,

GYMNORHINA TIBICEN, *Latham.* •

Piping Crow-shrike.

Gould, Handbk. Bds. Aust, Vol. i., sp. 92, p. 175. *VII*. *4.5.6*.

" None of the Australian birds I have hitherto met with lay eggs that are subject to greater variety than those of the present species. Out of twenty specimens now before me, there are twelve very distinctly marked varieties ; and I will endeavour to describe those of them which are from the neighbourhood of Sydney.

Var. A. In this, which is perhaps the most common variety, the ground colour is of a very pale sky-blue or bluish-white, with spots of lilac and numerous irregular markings of light brown equally distributed over the whole surface of the egg. Length from 16 to 20 lines ; breadth from 13 to 15 lines.

Var. B. Ground colour pale bluish or greenish-white, with long curved markings, smears, and dashes of reddish-brown. Length 18 lines ; breadth 13 lines.

Var. C. When first emptied of the contents, the ground-colour of this variety is a beautiful light green, with deep rust-red blotches over the whole surface, but run together so as to form one large .patch on the thicker end. Length 19 lines; breadth 13 lines. This variety is usually seen in a very long egg.

Var. D. Ground colour bright light green or sky-blue when first taken, but fading when kept, having irregular markings of light wood-brown very sparingly dispersed over the whole surface. Length 18 lines breadth 14 lines.

Var. E. Ground colour very pale sky-blue with distinct oval spots of reddish-brown and obsolete spots of lilac. In some specimens the spots are of a dark deep lilac, having a penumbra. Length from 18 to 20 lines; breadth from 13 to 15 lines.

Var. F. The ground colour a uniform dull dark brown, with numerous minute dots and spots of a deeper hue over the whole surface. Length 20 lines; breadth 13 lines.

Var. G. Ground colour brownish-white, with spots and dashes of wood-brown tinged with lilac, and obsolete lilac spots at the larger end. Length 17 lines; breadth 14 lines.

The nest of the Piping Crow-shrike is a large open structure composed of sticks and twigs, lined with grass and hair. It is usually placed in the fork of a tree, or among the bushy boughs of a species of *Angophora*. The eggs are usually three, but sometimes four in number. They breed during the months of August, September, and October." (*Ramsay, Ibis*, 1865, Vol. i., New Series, p. 300.)

Hab. Rockingham Bay, Port Denison, Wide Bay District, Dawson River, Richmond and Clarence Rivers Districts, New South Wales, Interior, Victoria and South Australia. (*Ramsay.*)

3   GYMNORHINA LEUCONOTA, *Gray.*
White-backed Crow-shrike.
Gould, Handbk. Bds. Aust., Vol. i., sp. 93, p. 176.

This is a very common species in certain districts in Victoria, South Australia and New South Wales. I have found it breeding in the open forest-lands of the former colony; both contiguous to the Dividing Range, and the Sea coast. It constructs a large cup-shaped nest, outwardly composed of sticks, and lined with twigs, grasses, and cow-hair, and placed in the fork of any suitable tree, sometimes in a Eucalyptus at a great height from the ground, at other times in a sapling about twenty feet high, and on several occasions I have taken its nest in the *Melaleuca* within ten feet of the ground. Eggs three in number for a sitting, elongated in form, and varying as much in the colour and disposition of their markings as the preceding species. I give the description of the varieties most frequently found.

Var. A. Ground colour pale blue, marked and streaked all over with irregular shaped lines of reddish-brown, in some specimens the latter colour is smudged and clouded on some parts of the shell. Taken at Yendon, Victoria, September, 1878.

Var. B. Ground colour bluish-white, streaked all over with very narrow faint lines of wood-brown. Taken near Adelaide, South-Australia 1863.

Var. C. Ground colour bright apple-green with minute freckles of light red evenly distributed over the whole surface. Taken at Woodstock, Victoria, August 1870.

Dimensions of Var. A which is an average sized set—length (a) 1·65 x 1·12 inch; (b) 1·67 x 1·1 inch; (c) 1·67 x 1·13 inch.

The months of August, September, and October constitute the breeding season of this species.

Hab. New South Wales, Interior, Victoria and South Australia, West and South-West Australia. (*Ramsay.*)

GYMNORHINA ORGANICUM, *Gould.*
(*G. hyperleucus*, Gould.)
Tasmanian Crow-shrike.

Gould, Handbk. Bds. Aust., Vol. i., sp. 94, p. 178. *VII. 9.*

This species represents in Tasmania the *G. leuconota* of the mainland, its nidification is similar to that of the preceding species, differing only in the nest being more compactly made and warmly lined. A set of eggs, three in number, taken near Hobart in September 1885, are of a greenish-grey ground colour, two of the specimens, being blotched and marked as if with a pen with light umber, while appearing as if beneath the surface of the shell are irregular shaped spots of bluish-grey, becoming confluent near the larger end, the other specimen (C) is rounded in form, ground colour light green, thickly covered with blotches of umber brown, and entirely free from the bluish-grey markings which appear in the other specimens. Length (A) 1·5 x 1·1 inch; (B) 1·45 x 1·9 inch; (C) 1·37 x 1·1 inch.

This species commences to breed in September, and continues during the two following months.

Hab. Tasmania. (*Ramsay.*)

Genus CRACTICUS, *Vieillot.*

3-7   CRACTICUS TORQUATUS, *Latham.*

Collared Crow-shrike.

Gould, Handbk. Bds. Aust., Vol. i., sp. 99, p. 184.

I have seen this species in nearly every part of Victoria and New South Wales. It builds an open nest, outwardly composed of fine sticks, lined with roots, and is placed in any suitable tree, according to the locality which it inhabits; it is usually placed in the top of a gum-sapling or a musk (*Olearia argophylla*), and when frequenting the edges of creeks and water-courses, it resorts to the *Melaleuca.*

The eggs are from three to four in number for a sitting, oval in form, and very variable in their tint and markings. I give the description of three of the varieties most usually found :

Var. A. Ground colour light olive-brown with a well defined zone of reddish-brown spots, intermingled with others of bluish-grey appearing as if beneath the surface of the shell. Length (A) 1·2 x 0·85 inch; (B) 1·2 x 0·85 inch; (C) 1·17 x 0·87 inch. Taken by Mr. James Ramsay, at Tyndarie, August 20th, 1879.

Var. B. Ground colour apple-green, with a few minute dots of reddish-black evenly distributed over the whole surface; these eggs are similar in colour and markings to the eggs of one of the varieties of *Gymnorhina leuconota.* Length (A) 1·21 x 0·87 inch ; (B) 1·16 x 0·9 inch; (C) 1·25 x 0·89 inch. Taken at Childers, South Gippsland, September 1878.

Var. C. Ground colour light yellowish-brown, with a clouded coalesced patch on the larger apex of reddish-brown markings. Taken at Yendon, Victoria, September 1883.

The breeding season commences in August and continues the three following months.

Hab. Rockingham Bay, Port Denison, Wide Bay District, Dawson River, Richmond and Clarence Rivers Districts, New South Wales, Interior, Victoria and South Australia. *(Ramsay.)*

3   CRACTICUS ROBUSTUS, *Latham.*

(C. nigrogularis, Gould.)

Black-throated Crow-shrike.

Gould, Handbk. Bds. Aust., Vol. i., sp. 95, p. 180. *IX. 8.*

The nest of this species is a rather large saucer-shaped structure ; one now before me, from the Collection in the Australian Museum, is composed on the outside of long twigs of the " Cotton Bush," some of the pieces measuring eight inches in length, and the interior is neatly lined with grasses, and finer twigs ; the exterior measurement of the nest is six and a-half inches in diameter, by three inches in depth ; interior diameter four inches and a-half by one inch and three-quarters in depth ; the nest was placed on the horizontal branch of a tree about ten feet from the ground. Eggs three in number for a sitting, and varying in tint from reddish-brown to apple-green. A fine set of these eggs taken by Mr. James Ramsay at Tyndarie, on the 21st of September 1880, are of a faded apple-green ground colour, minutely freckled, and thickly spotted with light umber, brownish-black, and slaty-grey, the latter colour appearing as if beneath the surface of the shell ; a few foreign-looking black spots are scattered over the larger end, which appear common to all the eggs of this genus ; these spots are easily removed by wetting them. Length (A) 1·28 x 0·97 inch ; (B) 1·26 x 0·93 inch ; (C) 1·27 x 0·97 inch.

The breeding season commences in September and continues the three following months.

Hab. Derby, N.W. Australia, Gulf of Carpentaria, Rockingham Bay, Port Denison, Wide Bay District, Dawson River, Richmond and Clarence River Districts, New South Wales, Interior, Victoria and South Australia. (*Ramsay.*)

CRACTICUS CINEREUS, *Gould.*

Ashy-grey Crow-shrike.

Gould, Handbk. Bds. Aust., Vol. i., sp. 100, p. 186.

This is the Tasmanian representative of the Australian form of *C. destructor*, and both the manner of its nidification and the number of its eggs for a sitting are precisely similar. I give the description of two varieties of eggs now before me, taken from different nests near Hobart, in September 1885.

Var. A. Ground colour dull asparagus-green, with indistinct chestnut spots and markings, particularly towards the larger end, where they form a confluent patch on the apex. Length 1·25 x 0·93 inch.

Var. B. A large specimen, ground colour light reddish-brown, thickly covered all over with markings of a darker tint, and a few spots of black on the larger end ; the latter of which are easily removed by wetting them. Length 1·36 x 0·96 inch.

This species breeds during the months of September, October, and November.

Hab Tasmania. (*Ramsay.*)

Sub-Family PACHYCEPHALINÆ.

Genus PACHYCEPHALA, *Swainson*.

— 3.   PACHYCEPHALA GUTTURALIS, *Latham*.

White-throated Thickhead.

Gould, Handbk. Bds. Aust., Vol. i., sp. 113, p. 207. *XII. 9.*

The nest of this species is an open one, and usually composed of twigs and roots, and lined with grass, but differs according to the locality in which it is found ; in Gippsland the nests are built with the fibrous roots of the *Dicksonia antarctica*, in the dead drooping fronds of which they are artfully concealed, at other times they are composed wholly of the long thin leaves of the Casuarina, and placed in some upright fork of a tree about six feet from the ground.   The eggs are subject to considerable variation both in the ground colour and their markings, the most usual form being of a yellow or brownish-buff, thickly marked with freckles of dark umber and blackish-brown, with a few obsolete blotches and spots of lilac, particulary towards the thicker end, where they become confluent and form an irregular zone.   Length (A) 0·92 x 0·67 inch ; (B) 0·88 x 0·67 inch ; (C) 0·95 x 0·68 inch.

While collecting at Heathcote, on the Illawarra Line, in company with Dr. Hurst on the 29th of October, in 1886, a nest was taken, containing two eggs of this species, the ground colour of which is of a deep yellowish-buff ; one specimen (a) has a broad band of dark umber-brown blotches round the centre of the egg, the other (b) has *two* distinct zones, but narrower than in (a), of the same colour, one in the centre, and the other round the smaller end. Length (A) 0·96 x 0·71 inch ; (B) 0·95 x 0·7 inch.

This is the first occasion that either Dr. Ramsay, Dr. Hurst, or I had seen one similarly marked; it is figured in the P.L.S., N.S.W., 2nd ser. Vol. i., pl. xix., fig. 3.

This species lays three eggs for a sitting, and breeds during September and the three following months.

Hab. Rockingham Bay, Port Denison, Wide Bay District, Richmond and Clarence Rivers Districts, New South Wales, Victoria and South Australia. (*Ramsay.*)

3   PACHYCEPHALA OCCIDENTALIS, *Ramsay.*

Western Thickhead.

Ramsay, P.L.S., N.S.W., Vol. ii., p. 212.

Mr. Gould figured this species as *P. gutturalis* of Latham in his folio edition of the Birds of Australia, see Vol. ii., pl. 64, but Dr. Ramsay separates the Western form under the name of *P. occidentalis.* It may therefore be presumed that when Mr. Gould described the eggs of *P. gutturalis,* they were in reality those of *P. occidentalis.* The following is taken from Mr. Gould's Handbook to the Birds of Australia, Vol. i., p. 208 :—

"Gilbert mentions that this species is sparingly distributed throughout the Swan River colony, but is more abundant in the best watered districts, such as Perth and Freemantle."

"I did not succeed in finding the nest of this species, but was informed that it breeds in September and October, and lays three eggs, ten and a-half lines long by eight lines broad, with a ground colour of brownish-buff, sparingly streaked and spotted with reddish-brown and bluish-grey, the latter colour appearing as if beneath the surface of the shell."

Hab. Western Australia. (*Ramsay.*)

E

3   ## PACHYCEPHALA GLAUCURA, *Gould.*
Grey-tailed Thickhead.

Gould, Handbk. Bds. Aust., Vol. i., sp. 114, p. 209.

The nest of this species is similar to that of *P. gutturalis*. The eggs are three in number for a sitting, and closely resemble in shape and colour those of that species, being of a faint creamy-white, with a zone of confluent brownish-black spots and dots, intermingled with others of a dark slaty-grey, the latter appearing as if beneath the surface of the shell. Dimensions of a set of three taken in October 1885, are as follows, length (A) 0·97 x 0·65 inch ; (B) 0·97 x 0·66 inch ; (C) 0·95 x 0·65 inch.

A set in the Collection of Dr. Cox gives the following measurements :—length (A) 0·98 x 0·7 inch ; (B) 0·95 x 0·66 inch ; (C) 0·93 x 0·67 inch.

Specimens in the Macleayan Museum Collection measure—(A) 0·97 x 0·69 inch ; (B) 0·97 x 0·68 inch ; (C) 0·96 x 0·67 inch.

October and the two following months constitutes the breeding season of this species.

Hab. Tasmania. *(Ramsay)*

3   ## PACHYCEPHALA MELANURA, *Gould.*
Black-tailed Thickhead.

Gould, Handbk. Bds. Aust., Vol. i., sp. 115, p. 211.   *XII*   *12.*

The nest is a cup-shaped, shallow, rather scanty structure of fine roots and twigs lined with rootlets and grass, through the bottom of which the eggs can be seen ; it is about three to four inches in diameter by one and a-half deep. The eggs are three in number, of a pale buff with irregular spots of dark umber sparingly scattered over the surface, but forming a zone near the thicker end. Length (A) 0·85 x 0·64 inch ; (B) 0·84 x 0·62 inch. *(Ramsay, P.L.S., N.S.W.*, Vol. ii., p. 47.

Hab. Port Darwin and Port Essington, Gulf of Carpentaria, Cape York, Dawson River, South Coast New Guinea. *(Ramsay.)*

3   PACHYCEPHALA RUFIVENTRIS, *Latham.*

Rufous-breasted Thickhead.

Gould, Handbk. Bds. Aust. Vol. i., sp. 116, p. 212.   *XII*   *11.*

The nest of this species is a scanty structure of rootlets or twigs, about four inches across, and usually placed on a horizontal branch of some tree or sapling—a favourite place being near the summit of a Casuarina about twelve or fifteen feet from the ground. The eggs are three in number for a sitting, of an olive ground colour, spotted and blotched with dark umber and brown markings, which form a zone on the larger end ; in a few instances they are heavily blotched with sepia. A set taken at Dobroyde in 1864, measure as follows :—(A) 0·92 x 0·7 inch, (unusually large and heavily blotched) ; (B) 0·85 x 0·67 inch ; (C) 0·87 x 0·68 inch.

The breeding season commences at the latter end of August, and continues to the middle of December.

Hab. Cape York, Rockingham Bay, Port Denison, Wide Bay District, Dawson River, Richmond and Clarence Rivers Districts, New South Wales, Interior, Victoria and South Australia, West and South-West Australia. (*Ramsay.*)

3   PACHYCEPHALA GILBERTI, *Gould.*

Gilbert's Thickhead.

Gould, Handbk. Bds. Aust., Vol. i., sp. 120, p. 216.   *VIII*   *10.*

Mr. K. H. Bennett found this bird breeding in the neighbourhood of Ivanhoe, New South Wales. A pair last year took possession of a deserted nest of *Cracticus nigrogularis*, and constructed their nest on the old foundation, lining it with roots, strips of fine bark, and grasses ; inside measurement three inches across, and one and five-eighths of a inch in depth. The eggs are three in number for a sitting, and are not unlike large and well-marked specimens of those of *Artamus sordidus*, the ground colour being of a pale yellowish-buff tint ; in one instance (c) spots of blackish-brown are evenly distributed over the whole surface, while appearing as if beneath the surface of the shell, are spots of lilac and bluish-grey; in (a) and (b) they intermingle and form a zone. Length (A) 0·95 x 0·72 inch ; (B) 0·91 x 0·71 inch ; (C) 0·93 x 0·72 inch.

The breeding season commences from the beginning of September and lasts till the end of December.

Hab. New South Wales, Interior, Victoria and South Australia, West and South-West Australia. (*Ramsay.*)

2 PACHYCEPHALA OLIVACEA, *Vigors and Horsfield.*
Olivaceous Thickhead.

Gould, Handbk. Bds. Aust., Vol. i., sp. 122, p. 218.   VIII . //.

For the first eggs of this species, I am indebted to my friend Mr. Charles Mayo, of Childers, South Gippsland, who obtained them during October 1878. The nest, he informs me, was built in a mountain musk (*Olearia argophylla*), and was similar to that of *P. gutturalis,* (which is common in that district) but larger and more compactly built. The egg is pointed at both ends, swelling heavily near the centre, the ground colour being buffy-white, marked all over, but particularly on its thicker end with dots and spots, of reddish and blackish-brown, intermingled with a few here and there of umber, while appearing as if beneath the surface of the shell are larger spots and blotches of dark lilac, which in some places are confluent, and form an irregular zone. Length 1·07 inch x 1·78 inch.

A pair taken in the New England district, (N.S.W.) have the markings more evenly dispersed over the surface, and measure as follows :—(A) 1·08 inch x 1·76 inch ; (B) 1·09 inch x 1·77 inch.

Two eggs are the usual number for a sitting, and the breeding season commences in September and lasts throughout the three following months.

Hab. New South Wales, Victoria and South Australia, Tasmania (*Ramsay.*)

Genus FALCUNCULUS, *Vieillot.*

3 FALCUNCULUS FRONTATUS, *Latham.*

Frontal Shrike-tit.

Gould, Handb. Bds. Aust., Vol. i., sp. 129, p. 228. *VIII. 9.*

" Although this species breeds freely in the neighbourhood of Sydney, its nest is seldom met with, and its eggs are still rarer. This arises chiefly from the inaccessible places in which the birds build, these being mostly the very tops of the tall Eucalypti, so that even when found they are seldom procurable. The nest is a deep cup-shaped structure of fine shreds of bark strongly woven together, strengthened with cobweb, and lined with grasses. The eggs seldom three in number, resemble those of *Myiagra nitida,* Gould, but are more elongated, white, with a few dots of greyish-lilac and slaty-black sprinkled over the surface, but in some cases crowded on the thicker end, or even confluent, forming spots or irregular short linear markings. Length (A) 0·9 x 0·65 inch (Dr. Hurst's Coll.); (B) 0·85 x 0·63 inch; (C) 0·92 x 0·64; B and C have no irregular markings on the shell merely a few minute dots, almost black." (*Ramsay, P.L.S., N.S.W. 2nd Series,* Vol. i., p. 1146.)

Hab. Rockingham Bay, Wide Bay District, Richmond and Clarence Rivers Districts, New South Wales, Interior, Victoria and South Australia. (*Ramsay.*)

3 - 4. FALCUNCULUS LEUCOGASTER, *Gould.*

White-bellied Shrike-tit.

Gould, Handbk. Bds. Aust., Vol. i., sp. 130, p. 229.

" This bird is a native of Western Australia. Gilbert, while staying in the Toodyay district in the month of October, found the nest of this species among the topmost and weakest perpendicular branches of a Eucalyptus, at a height of fifty feet ; it was of a deep cup-shaped form, composed of the stringy bark of the gum-

tree, and lined with fine grasses, the whole matted together
externally with cobwebs; the eggs, which are three or four in
number, are of a glossy white, with numerous minute speckles of
dark olive, most thickly disposed at the larger end; they are
seven-eighths of an inch long by five-eighths of an inch in breadth."
(*Gould, Handbk. Bds. Aust.*, Vol. i., p. 229.)

Hab. West and South-West Australia. (*Ramsay.*)

Genus OREOICA, *Gould.*

2 – 3   OREOICA CRISTATA, *Lewin.*

Crested Oreoica

Gould, Handbk. Bds. Aust., Vol. i., sp. 131, p. 231. VIII. 6.

This bird appears to be almost universally dispersed over the
continent of Australia, specimens having been recently received
from the North Western Coast and the Gulf District; and it is also
found in the most southern parts of Victoria and South Australia.
The powers of ventriloquism of this bird are truly wonderful,
when perched on the thick branch of some lofty gum-tree,
commencing with its singularly low, mournful, and plaintive note,
it appears a long way off, and it is not until it has reached its
fullest and highest bell-like tones, that you are aware it is in the
tree perhaps underneath which you are standing, and even after
ascertaining which tree it is in, it remains so motionless that it is
most difficult of detection.

The nest of this species is a deep cup-shaped structure, composed
of strips of bark, and Eucalyptus leaves, some pieces of the former
measuring three-quarters of an inch across, neatly lined with finer
strips of bark, fibrous roots, and grasses; outside measurement
four and a-half inches across, depth three inches; inside measurement three inches in diameter by two in depth. Eggs two or
three in number for a sitting, usually the former, and varying a

great deal in the disposition of their markings. I give the description of several sets now before me :—

Var. A. Ground colour faint bluish-white, minutely spotted with black, together with larger round markings of the same colour evenly distributed over the whole surface. Length (A) 1·1 x 0·8 inch ; (B) 1 x 0·8 inch. Taken October 1883, at Kewell, Victoria.

Var. B. Ground colour bluish-white, spotted and blotched with irregular shaped black markings, some of them resembling in form ill-shapen figures, crescents, &c. Length (A) 1·05 x 0·87 inch ; (B) 1·1 x 0·88 inch. Taken by Mr. James Ramsay, at Tyndarie, October 1880.

Var. C. Ground colour deep bluish-white, with remarkably heavy black blotches, uniformly dispersed over the surface, resembling in shape, but not so thickly disposed, as those on *Rhynchœa australis*. Length (A) 1·12 x 0·82 inch ; (B) 1·09 x 0·79 inch. Taken by Mr. K. H. Bennett, September 1886.

The breeding season of this species is usually the months of September, October, and November, but Mr. Bennett found several nests of this species, amongst others, containing fresh eggs, on March 19th, 1887, at Ivanhoe, New South Wales.

Hab. Derby, N.W.A., Gulf of Carpentaria, Rockingham Bay, Port Denison, Wide Bay District, Dawson River, Richmond and Clarence Rivers Districts, New South Wales, Interior, Victoria and South Australia. (*Ramsay.*)

Genus SPHENOSTOMA, *Gould*.

2 SPHENOSTOMA CRISTATA, *Gould*.

Crested Wedge-bill.

Gould, Handbk. Bds. Aust., Vol. i., sp. 184, p. 316. 𝐕𝐼𝐼𝐼 ♀

The habitat of this species is the Interior of Australia. For the eggs of this bird I am indebted to my highly esteemed and deeply regretted friend, the late Mr. W. Liscombe, who obtained them at

Wilcannia in New South Wales, in October 1882, and later on in the same year at Adavale, Queensland.

In both instances the nests were built in the *Polygonum* bushes and were cup-shaped structures composed of thin twigs, lined with grasses. Eggs two in number for a sitting; those taken at Wilcannia are elongate in form, of a pale blue ground colour, minutely spotted and blotched with irregular shaped black markings; length (A) 1·03 x 0·68 inch; (B) 0·97 x 0·67 inch. The specimens taken at Adavale, Queensland, are more rounded in form, of a deep blue ground colour, with a few minute dots and nearly round spots of black; closely resembling those of the Song Thrush *(Turdus musicus),* from which they differ only in size; length (A) 0·93 x 0·7 inch; (B) 0·93 x 0·71 inch.

An egg of this species in the Dobroyde Collection, taken by one of Mitchell's party during his expedition to Central Australia in 1835, is similar in size and markings to the last described; another pair taken by Mr. James Ramsay at Tyndarie in 1876, is of a delicate blue ground colour minutely spotted and blotched with irregular shaped purplish-brown markings.

The breeding season of this species commences in September, and lasts the two following months.

Hab. Gulf of Carpentaria, New South Wales, Interior, Victoria and South Australia. *(Ramsay.)*

Genus PSOPHODES, *Vigors and Horsfield.*

? PSOPHODES CREPITANS, *Vigors and Horsfield.*
Coach-whip-bird.

Gould, Handbk. Bds. Aust., Vol. i., sp. 182, p. 312. *VIII* ?

This bird is found in the dense thickets and scrubs that clothe the sides of mountain ranges, rivers, and creeks. I found it very plentiful in the Strzelecki Ranges in South Gippsland, Victoria, making the fern gullies and musk scrubs echo with its peculiar note terminating like the cracking of a stockman's whip, although from its restless habits, while traversing the thick undergrowth, it is always a difficult specimen to procure. A nest of this species in the Australian Museum Collection, presented by Dr. Hurst, and taken by that gentleman in August 1886, at Newington, on the banks of the Parramatta River, is an open, shallow structure, composed of long twigs bent round, and lined inside with the leaves of the *Casuarina;* it measures externally five and a-half inches in diameter, depth two and a-half inches ; internal diameter three inches, depth one inch and a-half. The nest was placed near a fallen log in a tangled mass of vines *(Dioscorea)* about two feet from the ground. Eggs two in number for a sitting, of a bluish-white ground colour, spotted and blotched all over with irregular shaped black markings. Length (A) 1·07 x 0·77 inch ; (B) 1·06 x 0·78 inch. Taken by Dr. Hurst.

Two eggs taken by Dr. Ramsay at Lismore in the Richmond River District, during November 1869, are of a pale bluish-white ground colour, minutely spotted and boldly marked all over with curiously shaped dashes of black, resembling crescents, figures, letters &c., while on the larger end are short wavy linear markings of dark lilac interlacing one another, which together with the spots and dashes of black form a nearly perfect zone. Length (A) 1·07 x 0·82 inch ; (B) 1·04 x 0·84 inch.

Specimens in my own collection, have the markings nearly round, but the above described are the most usual varieties found.

Nests containing eggs of this species have been found as early as July, and as late as November.

Dr. Hurst and myself were successful in obtaining the eggs of this species at Newington on the 30th of July 1887, also two new nests a week after in the same locality, one of the latter being prettily esconced in the centre of a *Zamia spiralis.*

Hab. Port Denison, Wide Bay District, Richmond and Clarence River Districts, New South Wales, Victoria and South Australia. *(Ramsay.)*

Family CAMPOPHAGIDÆ.

Genus GRAUCALUS, *Cuvier*.

—7-3   GRAUCALUS MELANOPS, *Latham*.

Black-faced Graucalus.

Gould, Handbk. Bds. Aust., Vol. i., sp. 103, p. 192. ♂. ♀.

A nest of this species now before me, taken from the Collection in the Australian Museum, is composed entirely of the dried leaves of the Casuarina securely held together on the outside by cobwebs; it is neatly fitted into the angle of a forked horizontal branch, the rim of the nest being but slightly above the level of the top of the branch ; it is a very flat structure and measures externally four and a-half inches across by one and a-quarter inches in depth, and has but the slightest depression in the centre viz., seven-sixteenths of an inch, just sufficient to keep the eggs from rolling out, this nest was taken from a Stringy-bark (*Eucalyptus*). Eggs two to three in number for a sitting, varying in tint from olive-brown to apple-green.

A set taken by Dr. Ramsay at Dobroyde, November 6th, 1860, are two in number, of a light olive ground colour thickly blotched and spotted with dark umber, chestnut-brown, and deep reddish-lilac the latter colour appearing as if beneath the surface of the shell. Length (A) 1·23 x 0·9 inch ; (B) 1·3 x 0·87 inch.

A set of three taken at Hastings, Western Port, Victoria, on October 23rd, 1881, are of a beautiful apple-green ground colour, uniformly spotted and blotched with light chestnut, yellowish-brown, and deep umber-brown, the latter colour in one specimen partaking the form of large clouded blotches. Length (A) 1·25 x 0·9 inch ; (B) 1·3 x 0·91 inch ; (C) 1·26 x 0·87 inch.

They begin to breed in September and continue during the three following months.

Hab. Derby, N.W. Australia, Port Darwin and Port Essington, Gulf of Carpentaria, Cape York, Rockingham Bay, Port Denison, Wide Bay District, Dawson River, Richmond and Clarence

Rivers Districts, New South Wales, Interior, Victoria and South Australia, West and South-West Australia, South Coast New Guinea. (*Ramsay.*)

3. GRAUCALUS MENTALIS, *Vigors and Horsfield,*

Varied Graucalus.

Gould, Handbk. Bds. Aust., Vol. i., sp. 105, p. 195. *IX. 3.*

The nest and eggs of this species is somewhat similar to that of *G. melanops.* A set of three eggs taken by Dr. Ramsay at Dobroyde on the 31st of October 1864, are of a dull asparagus green, uniformly spotted and blotched with pale red markings, and a few obsolete spots of bluish-grey, appearing as if beneath the surface of the shell. Length (A) 1·27 x 0·87 inch ; (B) 1·25 x 0·86 inch ; (C) 1·26 x 0·86 inch.

Another set, elongate in form, are of a dull olive-brown, thickly freckled and spotted all over with dark reddish-brown markings. Length (A) 1·3 x 0·83 inch ; (B) 1·3 x 0·85 inch ; (C) 1·27 x 0·86 inch.

This species breeds during September and the three following months.

Hab. Rockingham Bay, Port Denison, Wide Bay District, Dawson River, Richmond and Clarence Rivers Districts, New South Wales, Victoria and South Australia. (*Ramsay.*)

3 GRAUCALUS PARVIROSTRIS, *Gould.*

Short-billed Graucalus.

Gould, Handbk. Bds. Aust., Vol. i., sp. 104, p. 194.

This is the Tasmanian form of the *G. melanops* of the Australian Continent, and its nidification and the number and colour of its eggs is precisely similar. A set of two, taken near Hobart in November 1885, are of a dull olive-green, thickly freckled, spotted, and blotched with chestnut and wood-brown markings, a few obsolete spots of the latter colour appearing as if beneath the surface of the shell Length (A) 1·25 x 0·87 inch; (B) 1·33 x 0·9 inch.

The months of October, November, and December, constitute the breeding season of this species.

Hab. Tasmania. (*Ramsay.*)

2　　　GRAUCALUS HYPERLEUCUS, *Gould.*
White-bellied Graucalus.

Gould, Handbk. Bds. Aust., Vol. i., sp. 106; p. 196. *IX. l.*

"The nest of this species, like all those of the genus, is a rather flat structure of wiry grasses, securely fastened together by cobwebs, and placed on a horizontal bough, usually over a forked branch; it is very shallow, having but a slight depression, just sufficient to hold the eggs in the centre, round, and about four inches in diameter outside; the eggs in the present instance are two in number, of a pale, and bright asparagus green with a few reddish-brown spots confluent on the thicker end, and others sprinkled over the rest of the surface; length 1·1 x 0·8 inch; some have no confluent markings, but the spots are more evenly distributed, oval or round, but sometimes closer together on the thicker end. *From Mr. Barnard's Collection.*" (*P.L.S., N.S.W.,* Vol. vii., p. 408, *Ramsay.*)

Hab. Port Darwin and Port Essington, Gulf of Carpentaria, Cape York, Rockingham Bay, Port Denison, Dawson River, South Coast New Guinea.? (*Ramsay.*)

Genus PTEROPODOCYS, *Gould.*

3　　PTEROPODOCYS PHASIANELLA, *Gould.*
Ground Graucalus.

Gould, Handbk. Bds. Aust., Vol. i., sp. 108, p. 199. *IX. 2.*

"The nest greatly resembles that of *Graucalus melanops,* and is placed in similar situations on horizontal boughs; it is composed of grasses and the stalks of various herbs slightly interwoven and fastened together by spiders webs, &c., and is lined with finer grasses &c.; inside diameter four inches, the depth 1·4, the height of the rim above the branch on which it is placed is one inch. The eggs are three in number, oblong in form, the shell of a delicate thin texture, the ground colour pale asparagus green with a dull brownish patch of confluent markings at the thicker end, or with freckles of the same tint thinly distributed over the surface, and a few black irregular markings at the thick end. Length (A) 1·3 x 0·87 ; (B) 1·35 x 0·95 ; (C) 1·33 x 0·92 ; (D) 1·3 x 0·87 ; (E) 1·35 x 0·88." (*Ramsay, P.L.S., N.S.W.*, Vol. vii., p. 47.)

Hab. Wide Bay District, New South Wales, Interior, Victoria and South Australia, W. and S.W. Australia. (*Ramsay.*)

Genus LALAGE, *Boie.*

LALAGE LEUCOMELÆNA, *Vigors and Horsfield.*

Black and White Lalage.

Gould, Handbk. Bds. Aust., Vol. i., sp. 111, p. 203.

Respecting the nidification of this species, Mr. R. D. Fitzgerald writes as follows :—

" A nest of this species, taken at Ballina near the mouth of the Richmond River, on November 4th, 1887, is composed of the wiry and pliant stems of herbs and grasses entwined and matted together with cobweb, and a few pieces of lichen felted together, making the outside resemble the branch, in a fork of which it is placed ; the nest is about the size of that of *L. tricolor*, being comparatively small for the size of the bird ; the one at present under consideration, was placed between a fork in a small branch of a Ti-tree (*Melaleuca* sp.); it is a small and shallow structure, being only 2·1 inches outside diameter by 1·35 inches inside, and without any special lining. It contained but one egg, which I believe is all that is laid for a sitting, for on shooting and dissecting the female no other egg was found in any degree of maturity.

" The egg is of a bright apple-green colour, with a well defined zone of reddish-brown spots near the thicker end ; the rest of the surface is thickly sprinkled with dots, freckles, and small spots of the same, or of a slightly brighter tint, which are less close together on the thin end ; it is rather elongated in form, measuring in length 0·98 inch, its short diameter being 0·68 inch."(*Fitzgerald*, ² *P.L.S.*, *N.S.W.*, Vol. ii., 2nd Series, 1887. p. 971.)

Hab. Port Darwin and Port Essington, Gulf of Carpentaria, Cape York, Rockingham Bay, Port Denison, Wide Bay District, Richmond and Clarence Rivers Districts, New South Wales, South Coast New Guinea. (*Ramsay*.)

2 - 3   LALAGE TRICOLOR, *Swainson*.

(*Campephaga humeralis*, Gould.)

White-shouldered Lalage.   •

Gould, Handbk. Bds. Aust., Vol. i., sp. 112, p. 204.

This species is universally dispersed throughout Australia, and its sweet and pleasant note, which can be heard from a great distance, together with the striking contrast of its plumage, makes it one of the most conspicuous of the smaller birds to be met with in the Australian bush.   It usually constructs its nest on a dead horizontal branch, but in the neighbourhood of Mount Buninyong, Victoria, I have found it built in the upright forks, at the top of the stem of a sapling ; it is a shallow structure, composed of grasses, loosely interwoven and held together with cobwebs, and small fragments of bark attached to the outside and rim of the nest ; it measures two and a-quarter inches inside diameter, and one-half inch in depth.   The eggs are two or three in number for a sitting, and are usually of a light green ground colour, blotched uniformly all over with longitudinal markings of reddish-brown and olive-brown, length (A) 0·82 x 0·65 inch ; (B) 0·83 x 0·67 inch.

This species breeds during the months of September, October, and November.

Hab. Derby, N.W. Australia, Port Darwin and Port Essington, Gulf of Carpentaria, Cape York. Rockingham Bay, Port Denison, Wide Bay District, Dawson River, Richmond and Clarence Rivers Districts, New South Wales, Interior, Victoria and South Australia South Coast New Guinea. (*Ramsay.*)

Family PRIONOPIDÆ.

Sub-Family PRIONOPINÆ.

GENUS GRALLINA, *Vicillot.*

GRALLINA PICATA, *Latham.*

Gould, Handbk. Bds. Aust., Vol. i., sp. 102, p. 188. *VIII. 12.*

The nest of this species is usually placed upon a bare horizontal branch, but not unfrequently selecting one where a few green twigs and leaves are growing out, which partially hide the structure. The nest is cup-shaped, composed of mud lined inside with grasses, measuring five and a-half inches external diameter, and four in depth, the inside measurements being four and a-half inches across, and two inches in depth. The eggs are three or four in number, and vary considerably in the tint and disposition of their markings, the most usual variety found being of a pinky-white ground colour, spotted and blotched with reddish-brown and light red, with a few obsolete spots of purple and slaty-lilac appearing as if beneath the surface of the shell, the latter colours predominating on the larger end, where in some instances an irregular zone is formed, and the remainder of the surface almost without spots; in New South Wales this latter variety is the rule. Length (A) 1·18 x 0·81 inch ; (B) 1·17 x 0·81 inch ; (C) 1·17 x 0·8 inch ; (D) 1·19 x 0·82 inch.

This bird breeds during October, November, and December.

Hab. Port Darwin and Port Essington, Gulf of Carpentaria, Cape York, Rockingham Bay, Port Denison, Wide Bay District,

Dawson River, Richmond and Clarence Rivers Districts, New South Wales, Interior, Victoria and South Australia, W. and S.W. Australia. (*Ramsay.*)

Genus COLLYRIOCINCLA, *Vigors and Horsfield.*

- 3   COLLYRIOCINCLA HARMONICA, *Latham.*

Harmonious Thrush.

Gould, Handbk. Bds. Aust., Vol. i., sp. 123, p. 220. *VIII ?.3.¼.*

The Harmonious Thrush may be met with in nearly every part of New South Wales and Victoria, where its rich clear note echoing through the bush is one of the earliest signs of approaching spring.   The nest is a round and rather loosely formed structure, composed of long strips of bark with a lining of fibrous roots ; its external diameter is five inches, and three and three quarter inches in depth, the inside measurement being three and a-half inches across by two and a-half inches in depth.   The nest is usually placed in the hollow top of a stump, at other times in the thick fork of a tree.   The eggs are three in number, and are subject to considerable variation.

Var. A. Ground colour buffy-white, covered evenly all over with bran-like markings of light umber-brown, intermingled with others of a darker tint, while appearing underneath the surface of the shell are freckles of bluish-grey.   Length (A) 1·18 x 0·87 inch ; (B) 1·21 x 0·87 inch ; (C) 1·2 x 0·87 inch.

Var. B. Ground colour beautiful pearly-white, with large blotches of brownish-black, particularly on the larger end, where with superimposed spots of bluish-black, they appear confluent and form an irregular zone.   Length (A) 1·2 x 0·9 inch ; (B) 1·2 x 0·87 inch ; (C) 1·18 x 0·87 inch.

Var. C. Egg in shape elongate, ground colour pearly-white, with a few large dashes of black, confined exclusively to the larger apex of the egg.   Length 1·24 x 0·87 inch.

This bird breeds from August till the end of December.

Hab. Port Denison, Wide Bay District, Dawson River, Richmond and Clarence Rivers Districts, New South Wales, Interior, Victoria and South Australia. (*Ramsay.*)

COLLYRIOCINCLA PARVULA, *Gould.*

Little Grey Thrush.

Gould, Handbk. Bds. Aust., Vol. i., sp. 127, p. 225.

The mode of nidification of the Little Thrush is similar to that of the preceding species. Eggs of this bird in the Dobroyde Collection, taken at Port Darwin, are nearly oval in form, pearly-white, spotted and blotched with olive-brown, blackish-brown, and obsolete bluish-grey markings, which predominate towards the thicker end. Length (A) 1·05 x 0·75 inch; (B) 1·03 x 0·72 inch.

Hab. Port Darwin and Port Essington. (*Ramsay.*)

3 COLLYRIOCINCLA BRUNNEA, *Gould.*

Brown Thrush.

Gould, Handbk. Bds. Aust., Vol. i., sp. 125, p. 223.

The nest of this species is similarly constructed to that of other members of the genus, being composed outwardly of bark, and lined internally with twigs and grasses, and usually placed in the hollow top of a high stump. Eggs three in number for a sitting, pearly-white, spotted and blotched with olive-brown markings, intermingled with superimposed spots of bluish-grey which become confluent towards the larger end, where they form a well defined zone. An average specimen taken by the late Mr. T. H. Boyer-Bower at Derby, North-western Australia, in October, 1886, measures 1·18 inch long diameter, by 0·77 inch short diameter.

F

Specimens in the Macleayan Museum Collection, taken by Mr.
E. Spalding at Port Darwin, have the markings smaller and more
evenly dispersed over the suface of the shell.   Length (A) 1·2 x
0·8 inch ; (B) 1·18 x 0·78 inch.

Hab.   Derby, N.W. Australia, Port Darwin and Port Essington,
Gulf of Carpentaria, Cape York, South Coast New Guinea,
(*Ramsay.*)

3   COLLYRIOCINCLA RECTIROSTRIS, *Jardine and Selby.*

(*C. selbii*, Jardine.)

Selby's Thrush.

Gould, Handbk. Bds. Aust., Vol. i., sp. 126, p. 224.

In Tasmania this bird takes the place of *C. harmonica*, of the
Australian Continent, and in its mode of nidification, and the
number and colour of its eggs, it is so precisely similar to that species
that a second description is hardly required.   The eggs vary in
form from an oval to a lengthened oval, and are of a pearly-white,
spotted and blotched with chestnut-brown, blackish-brown, and
bluish-grey, the latter colour appearing as if beneath the surface
of the shell.   Length (A) 1·2 x 0.86 inch ; (B) 1·18 x 0·84 inch ;
(C) 1·21 x 0·87 inch.

Five eggs of this bird in the Collection of Dr. J. C. Cox, give the
following measurements, :—length (A) 1·2 x 0·87 inch ; (B) 1·17
x 0·85 ; (C) 1·25 x 0·9 inch ; (D) 1·14 x 0·87 inch ; (E) 1.15 x
0·85 inch.

Hab.   Tasmania.

2-3   COLLYRIOCINCLA RUFIVENTRIS, *Gould.*

Buff-bellied Thrush.

Gould, Handbk. Bds. Aust., Vol. i., sp. 124, p. 222.

" This species is a native of Western Australia where it is to
be found in all thickly wooded places feeding as much on the
ground as upon the trees and scrubs.   It breeds in the latter

part of September and the beginning of October, and the nest, which is generally placed in the hollow part of a high tree, is formed of dried strips of gum-tree bark very closely packed; it is deep, and is sometimes lined with soft grasses. The eggs, which are two or three in number, are of a beautiful bluish or pearly white, with large blotches of reddish olive-brown and dark grey, the latter appearing as if beneath the surface of the shell; the medium length of the eggs is one inch and one line, by ten lines in breadth." *(Gould, Handbk. Bds. Aust.,* Vol. i., p. 222.)

Hab. Derby, N.W. Australia, West and South-west Australia. (*Ramsay.*)

COLLYRIOCINCLA RUFIGASTER, *Gould.*
Rufous-breasted Thrush.
Gould, Handbk. Bds. Aust., Vol. i., sp. 128, p. 226.

Eggs of this species taken by Mr. George Barnard, on the Dawson River, Queensland, are of a beautiful pearly-white, spotted and blotched all over the surface with dark olive-brown, and obsolete markings of slaty-grey, which are more thickly disposed towards the larger end, where they become confluent. Length (A) 1·03 x 0·74 inch; (B) 1·01 x 0·73 inch.

Hab. Derby, Wide Bay District, Dawson River, Richmond and Clarence Rivers Districts, New South Wales. (*Ramsay.*)

2   COLLYRIOCINCLA PARVISSIMA, *Gould.*
Smaller Rufous-breasted Thrush.
Ann & Mag. Nat. Hist., Vol. x., p. 114.   VIII /.

Specimens of this bird in the Australian and Macleayan Museums, and in the Dobroyde Collection, differ only from the southern representative *C. rufigaster* in the deeper tints of its plumage and smaller admeasurements. The average length of a number of specimens of *C. rufigaster* obtained at Wide Bay and the Richmond River, is 7·5 inches; wing, 3·9 inches; tail, 3·5 inches; bill from forehead 1 inch. A specimen of *C. parvissima* from Cape York measures 6·5 inches in length; wing, 3·7 inches; tail, 3 inches; bill from forehead, 0·87 inch.

A nest of this species, from the same district, is similar to those of the other members of the genus, being an open cup-shaped structure composed of strips of bark, lined with grasses; the eggs, two in number for a sitting, are pearly-white, spotted all over with slate and umber-brown, with obsolete markings of deep bluish-grey. Length (A) 0·97 x 0·7 inch; (B) 0·96 x 0·7 inch.

Hab. Gulf of Carpentaria, Cape York, Rockingham Bay, South Coast New Guinea. (*Ramsay.*)

Family MUSCICAPIDÆ.

Genus RHIPIDURA, *Vigors and Horsfield.*

RHIPIDURA ALBISCAPA, *Gould.*

White-shafted Fantail.

Gould, Handbk. Bds. Aust., Vol. i., sp. 134, p. 238.

This lively and interesting little bird is plentifully distributed throughout Queensland, New South Wales, Victoria and South Australia, and although it has nothing to recommend it in the sombre tints of its plumage, the remarkable shape of its nest attaches to it an interest that renders it one of the most conspicuous of the smaller birds of the Australian bush.

A nest of this species now before me is funnel-shaped, or like a wine glass with the base broken off; it is composed of strips of very fine bark closely interwoven and securely held together on the outside with spider's web, which is neatly wound round the exterior portion of the nest proper, the thin branch on which it is placed and the upper portion of the stem-like appendage which extends below the branch from the bottom of the nest; the lower portion of the stem is ragged at the end, and just sufficient web placed around it to hold together the fine shreds of bark of which it is composed. The interior of the nest is lined entirely with fine fibrous roots, and the rim of the nest is very thin; external diameter one and seven-eighths of an inch, depth one and three-quarters of an inch, length of stem below the nest, two inches, thickness near the end one quarter of an inch; internal diameter one and seven-eighths of an inch, depth one and one-eighth of an inch.

The nest of this bird is placed in a variety of situations, some times on the thin branch of a *Melaleuca* within a few feet of the ground, but not unfrequently on one of the topmost branches of an Acacia or tall gum-sapling twenty feet from the ground. Eggs two or three in number for a sitting, of a dull, and in some instances creamy-white ground colour, thickly spotted with brown markings, intermingled with a few obsolete spots of bluish-grey towards the larger end, where they become confluent and form a well defined zone. Dimensions of a set taken at Macquarie Fields, length (A) 0·62 x 0·5 inch ; (B) 0·61 x 0·48 inch ; (C) 0·6 x 0·5 inch.

A set taken at the mouth of the Yarra, near Melbourne, are of a dull white, with the brown markings more clouded and evenly dispersed over the whole surface ; length (A) 0·66 x 0·48 inch ; (B) 0·64 x 0·5 inch ; (C) 0·61 x 0·48 inch.

This species commences to breed in October and continues the two following months.

Hab. Rockingham Bay, Port Denison, Wide Bay District, Richmond and Clarence Rivers Districts, New South Wales, Interior, Victoria, and South Australia. (*Ramsay.*)

RHIPIDURA DIEMENENSIS, *Sharpe.*

Tasmanian Fantail.

Sharpe, Brit. Mus. Cat. Bds., Vol. iv., p. 311.

"Two eggs taken near Hobart, in October, 1885, are of a dull white colour, thickly freckled all over with creamy-brown markings

but more particularly towards the larger end.   Length (A) 0·61
x 0·47 inch ; (B) 0·6 x 0·47 inch." *(North, P.L.S., N.S.W., Vol.
ii., 2nd Series, p. 406.)*

Hab. Tasmania. *(Ramsay.)*

?   RHIPIDURA PREISSI, *Cabanis.*

Preiss's Fantail.

Gould, Handbk. Bds. Aust., Vol. i., sp. 135, p. 240.

When collecting in North-western Australia, Mr. W. Froggatt
found a nest of this species built in a climbing plant growing on
the banks of the Fitzroy River, about twenty-five miles inland
from Derby, on the 25th September, 1887 ; it is very likely that
of its near ally *R. albiscapa,* of the eastern and southern portions
of the continent, but much smaller, resembling in shape a miniature
wine glass with the base broken off.   The nest is composed of
shreds of thin fibrous bark and fine grasses, held together on the
outside with spiders' webs, which are neatly wound round the
exterior surface of the nest, also the thin branch on which it is
placed, the stem of the nest having a somewhat ragged appearance
at the extremity.   Exterior diameter 1·7 inch, depth 1·67 inch,
length of stem from the bottom of the nest proper and branch on
which it is placed 2·25 inches.   Interior diameter 1·58 inch, depth
1·2 inch.   The rim of the nest is very thin.   Eggs two in number
for a sitting, of a creamy-white ground colour, spotted and blotched
with dull wood-brown, intermingled with obsolete markings of
slaty-grey which are more thickly disposed towards the larger end
where an ill-defined zone is formed.   Length (A) 0·6 x 0·48 inch ;
(B) 0·63 x 0·5 inch.   *From the Macleyan Museum Collection.*
(North, Proc. Linn. Soc., N.S.W., Vol. iii., 2nd Series, part ii.,
April, 1888.)

Hab. Derby, N.W. Australia, West and South-west Australia.
(Ramsay.)

2   RHIPIDURA RUFIFRONS, *Latham.*

Rufous-fronted Fantail.

Gould, Handbk. Bds. Aust., Vol. i., sp. 136, p. 240.

" This species although a constant visitor to Sydney and the neighbourhood, seldom breeds except in the thick brushes of Illawarra, or such like localities.   The eggs are two for a sitting, of a pale cream colour, or creamy-white, with a zone of spots and dots of light wood-brown, and a few dots of lilac, the markings being confined to the zone, with the exception of one or two large dots on the remainder of the surface.   Length 0·7 x 0·52 inch." *(From Mr. Ralph Hargrave's Coll. Ramsay, P.L.S., N.S.W., 2nd Ser.*, Vol. i., p. 1143.)

Hab. Port Denison, Wide Bay District, Richmond and Clarence Rivers Districts, New South Wales, Victoria and South Australia.   (*Ramsay.*)

2.   RHIPIDURA SETOSA, *Quoy et Gaimard.*

(*R. isura*, Gould.)

Northern Fantail.

Gould, Handbk. Bds. Aus., Vol. i., sp. 138, p. 242.

A nest of this species in the Australian Museum Collection, taken at Port Darwin in 1879, is similar in shape to that of *R. albiscapa*, but slightly larger, the nest is composed entirely of very fine thin strips of bark woven together with the webs of spiders, but not so compactly made as that of the former species, it is also without any special lining, the nest is placed at the junction of a fine three-pronged branch, and the stem-like appendage at the bottom of the nest is built round a twig growing below the fork on which the nest is placed.

" Egg similar to that of *S. motacilloides*, but much smaller.   It is of a light cream colour, with dull wood-brown spots forming a zone at the larger end. Length 0·68 x 0·55 inch. Other specimens similar but with larger and more defined markings have recently been received from the late Mr. T. H. Boyer-Bower. Two is the usual number of eggs laid for a sitting, and the above nest and eggs were taken in the month of October, near Derby, North-Western Australia. (*Ramsay, P.L.S., N.S.W.*, 2nd Ser., Vol. i., p. 411.)

Hab. Derby, N.W. Australia, Port Darwin and Port Essington, Gulf of Carpentaria, Cape York, Rockingham Bay, South Coast New Guinea. (*Ramsay.*)

GENUS SAULOPROCTA, *Cabanis.*

SAULOPROCTA MOTACILLOIDES, *Vigors and Horsfield.*

Black and White Fantail.

Gould, Handbk. Bds. Aust., Vol. i., sp. 139, p. 244.

The nest of this bird is usually placed on the branch of a tree overhanging the water, it is cup-shaped, and is composed of grasses, fine strips of bark &c., neatly woven together with cobwebs, of which latter material the outer surface is entirely covered ; and is lined inside with fibrous roots &c., it measures three inches across in external diameter, and two in depth, the inside measurement being two and three-eighths of an inch by one and a-quarter in depth. It does not always confine itself to building over, or near water, having found its nest on several occasions at a great distance from any stream. The eggs are three or four in number, the ground colour yellowish- or brownish-white, becoming lighter at the smaller end, with a zone at the larger end of minute dots of yellowish-brown, ashy-grey, and bluish-black ; the latter colour appearing on the under surface of the shell. Length (A) 0·8 x 0·6 inch ; (B) 0·78 x 0·57 inch ; (C) 0·8 x 0·6 inch.

This species breeds during September, and the three following months.

Hab. Rockingham Bay, Port Denison, Wide Bay, District, Richmond and Clarence Rivers Districts, New South Wales, Interior, Victoria and South Australia, West and South-West Australia, South Coast New Guinea. *(Ramsay.)*

GENUS SEISURA, *Vigors and Horsfield.*

2-3

SEISURA INQUIETA, *Latham.*

Restless Flycatcher.

Gould, Handbk. Bds. Aust., Vol. i., sp. 141, p. 246.

"The nest of this Flycatcher, like those of most of the family is round and cup-shaped, two and a-half to three inches across by one and a-quarter deep, and placed upon a horizontal bough over a fork, or by the side of an upright twig; it is chiefly composed of bark and grass neatly interwoven; the lining is of grass, hair, or roots, and the edges often ornamented with lichen fastened on by cobweb. It is usually placed at a considerable distance from the ground, and often near the end of a dead bough. The eggs are two or three in number, from nine to ten and a-half lines in length by seven and a-half in breadth, rather rounded in form, having the ground colour of a dull white, stained with spots and blotches of dull chestnut-brown and greyish-lilac, the latter appearing as is beneath the surface. In most of the specimens the spots form only a distinct zone nearer the larger end, but in some are sprinkled over the whole surface. The birds are for the most part found breeding in October, November, and December, but sometimes earlier or later. They have two broods in the year." *(Ramsay, Proc. Phil. Soc., Sydney,* 1865, p. 325, pl. i., fig. 6.)

A set of the eggs of this species now before me measure as follows:—length (A) 0·75 x 0·6 inch; (B) 0·77 x 0·6 inch; (C) 0·76 x 0·52 inch.

Hab. Rockingham Bay, Port Denison, Wide Bay District, Dawson River, Richmond and Clarence Rivers Districts, New

South Wales, Interior, Victoria and South Australia, W. and
S.W. Australia. (*Ramsay*.)

Genus PIEZORHYNCHUS, *Gould*.

2   PIEZORHYNCHUS NITIDUS, *Gould*.
Shining Flycatcher.
Gould, Handbk. Bds. Aust., Vol. i., sp. 142, p. 249.

Mr. J. A. Boyd found this bird breeding early in January 1888,
on the Herbert River, Queensland ; the nest was built on the dead
branch of a Ti-tree, (*Melaleuca*) that had fallen into a waterhole.

This species usually builds its nest near the ground on a
horizontal branch or forked limb ; it is a deep cup-shaped structure
composed of strips of bark and grasses, held together with cobwebs,
and ornamented on the outside with small scales of bark and
lichens, the interior being neatly lined with wiry rootlets.   The
eggs, two in number for a sitting, vary in form from swollen to
elongated ovals, they are of a bluish-white ground colour, minutely
spotted and blotched with olive-brown intermingled with obsolete
spots of greyish-lilac, all of which predominate towards the larger
end, where they assume the form of a zone, but are not confluent.

The dimensions of two sets in the Australian Museum Collection
are as follows :—length No. 1 (A) 0·85 x 0·65 ;   (B) 0·85 x 0·65 ;
No. 2 (C) 0·9 x 0·61 ; (D) 0·9 x 0·62 inch.

Hab. Port Darwin and Port Essington, Cape York, Rockingham
Bay, Port Denison, South Coast New Guinea. (*Ramsay*.)

2   PIEZORHYNCHUS GOULDII, *G. R. Gray*.
(*M. trivirgata*, Gould.)
Black-fronted Flycatcher.
Gould, Handbk. Bds. Aust., Vol. i., sp. 153, p. 262.

"The nest and eggs of this very interesting species were
forwarded to me in 1865, from South Grafton, by the late Mr. J.

Macgillivray, who procured them from one of the neighbouring brushes. Mr. Macgillivray also sent me a skin of one of the parent birds, proving that this rare species is to be found much nearer Sydney than was expected; for until specimens had been received at the Australian Museum from Mr. Rainbird, who had procured them at Port Denison, *Monarcha trivirgata** was looked upon as a bird of the greatest rarity.

The nest is very similar and similarly situated to that of *M. carinata*, but differs in being smaller and composed of finer material; in length it is 3·5 inches, by 2·5 inches diameter at the thickest part, and 1·25 inch deep. In this instance the nest was placed in the upright fork of a small tree, about six feet from the ground, and is composed of very fine fibrous roots, long strings of green moss (*Hypnum; sp.?*) shreds of bark, and soft silky down from the seed pods of some native trees. The whole is closely interwoven and made into a neat cup-shaped structure, lined solely with fine black hair-like roots; the edges and parts of the outside are ornamented with a beautiful *Hypnum* and white cobwebs. Upon the whole the nest and eggs bear a close resemblance to those of *M. carinata* † but, unlike all I have ever seen of this latter species, the nest of *M. trivirgata* is not so entirely enveloped in green moss. The eggs I believe, were only two in number; they are in length .833 inch by .583 in breadth, having a pure white ground thickly sprinkled with dots of bright reddish-brown crowded upon the thicker end, where they form a blotch approaching more to salmon colour." (*Ramsay, Ibis*, 1868, New Series, Vol. iv., p. 271.)

Two eggs of this species in the Australian Museum Collection, taken at Wide Bay, Queensland, are similar to those described above, and measure as follows :—length (A) 0·8 x 0·58 inch ; (B) 0·78 x 0·58 inch.

Hab. Cape York, Rockingham Bay, Port Denison, Wide Bay, District, Richmond and Clarence Rivers Districts, New South Wales. (*Ramsay.*)

---

* = M. gouldii, *G. R. Gray.*   † = M. melanopsis, *Vieillot.*

PIEZORHYNCHUS ALBIVENTRIS, *Gould.*

(Monarcha albiventris, Gould.)

White-bellied Flycatcher.

Gould, Suppl. Bds. Aust., pl. 13.

" The nest of *P. albiventris* is similar in every respect to that of *Monarcha melanopsis,* Vieillot, but slightly smaller, and the eggs of the several species are scarcely to be distinguished from one another excepting by their size. Those of *P. albiventris,* Gould, measure 0·8 inch in length x 0·56 inch in breadth. The ground colour is white, the whole surface being sprinkled with freckles and dots of bright red, which becoming confluent near the thicker end, there form a zone." (*Ramsay, P.L.S., N.S.W.,* Vol. i., 2nd Series, p. 1144.)

Hab. Gulf of Carpentaria Cape York, and Rockingham Bay. (*Ramsay.*)

Genus MYIAGRA, *Vigors and Horsfield.*

2 - 3   MYIAGRA RUBECULA, *Latham.*

· *(M. plumbea,* Vigors and Horsfield.)

Lead-coloured Flycatcher.

Gould, Handbk. Bds. Aust., Vol. i., sp. 144, p. 252.

" This pretty Flycatcher arrives in the vicinity of Sydney about the same time as *Monarcha carinata,** or perhaps a little earlier. It is however, much more regular in its visits than that bird, coming every year, whereas the other is not so regular, nor in such constant numbers. *Myiagra plumbea* † is a pleasing, active little bird, ever on the move, and even when perched continues

---

* = Monarcha melanopsis.   † = Myiagra rubecula.

to pour out its guttural squeaking note, which is always accompanied by a tremulous motion of the wings, as if it were anxious to be off again. It has another melancholy but pleasing note, which, when heard far off in the bush, is never to be forgotten, and at once warns you of its return. Although it is not so numerous during the months of November and December as when it first arrives in September, still many remain and breed with us, pairing off and beginning to build sometimes as early as October, but more usually during the two following months. They then leave the closely wooded sides of the creeks and watercourses, and show a decided preference to the more open or half-cleared land, choosing as sites for their nests the horizontal boughs of the larger trees, upon which they build neat round open nests, two inches in diameter by one and a-half deep, and composed of the inner stringy bark of the *Eucalyptus obliqua*, bound and fastened together with cobwebs, the outside being ornamented with scales of bark glued on with cobwebs, and made to resemble, as much as possible, the boughs to which they are fastened. They are lined with grass and thinner strips of bark. The eggs, which are two or three in number, have the ground colour bluish-white with a zone of slate-blue and lilac dots near the larger end. In some the markings are of a wood-brown tint, or consist of lilac spots alone, with a dot of deeper tint, in the centre of each spot. Their length is from eight to eight and a-half lines, and the breadth from six to seven lines." *(Ramsay, Ibis,* 1865, Vol. i., New Series, p. 401.)

Two sets of the eggs of this species taken at Dobroyde, give the following measurements, length (A) 0·73 x 0·57 inch ; (B) 0·74 x 0·55 inch ; (C) 0·74 x 0·56 inch ; (D) 0·74 x 0·55 inch ; (E) 0·76 x 0·57 inch ; (F) 0·73 x 0·55 inch.

Hab. Derby, N.W. Australia, Port Darwin and Port Essington, Gulf of Carpentaria, Cape York, Rockingham Bay, Port Denison, Wide Bay District, Dawson River, Richmond and Clarence Rivers Districts, New South Wales, Victoria and South Australia, Tasmania, South Coast New Guinea. (*Ramsay.*)

3   MYIAGRA CONCINNA, *Gould.*

Pretty Flycatcher.

Gould, Handbk. Bds. Aust., Vol. i., sp. 145, p. 254.

The nest is a neat cup-shaped structure of bark with a few fine grasses neatly interwoven, and placed on a horizontal bough usually over a fork or junction of two branches ; the whole is cemented together with cobweb and scales of lichens, &c. The eggs, three in number, of a delicate bluish-white when fresh, with a strongly defined band of spots and dots of wood-brown to sienna or yellowish umber, here and there a dot of slaty-blue appearing as if beneath the surface. Length (A) 0·64 x 0·53 inch ; (B) 0·65 x 0·52 inch. *(Ramsay, P.L.S., N.S.W.,* Vol. vii., p. 48.)

Hab. Derby, N.W. Australia, Port Darwin and Port Essington, Gulf of Carpentaria, Rockingham Bay, Dawson River. *(Ramsay.)*

MYIAGRA NITIDA, *Gould.*

Shining Flycatcher.

Gould, Handbk. Bds. Aust., Vol. i., sp. 146, p. 255.

The nest of this species is similar to those of other members of the genus, and is placed in like situations. Specimens of these eggs just received from Mr. E. D. Atkinson, of Table Cape, Tasmania, are in form swollen ovals, of a faint greenish-white, with a well defined zone of irregularly shaped spots and blotches of yellowish-umber, umber-brown, and obsolete markings of purplish-grey on the thicker end, a few dots of the same colour being scattered over the remainder of the surface of the shell. Length (A) 0·77 x 0·58 inch ; (B) 0·78 x 0·58 inch.

Hab. Cape York, Rockingham Bay, Port Denison, Wide Bay District, Richmond and Clarence Rivers Districts, New South Wales, Interior, Victoria, South Australia, and Tasmania. *(Ramsay.)*

GENUS MICRŒCA, *Gould.*

_2-3_ MICRŒCA FASCINANS, *Latham.*

Brown Flycatcher.

Gould, Handbk. Bds. Aust., Vol. i., sp⋅ 149, p. 258.

" This bird, although one of our most common and most sombre-coloured, is one of our sweetest songsters. At day-break it may be seen perched upon the dead top of some lofty Eucalyptus, pouring forth a song of the most cheerful and pleasing strain ; its notes are varied and may be heard at a considerable distance. Mr. Gould remarks that they resemble those of the Chaffinch *(Fringilla cœlebs).* They have a decided preference for perching' while singing, upon the very topmost boughs of the most lofty trees from whence they will dart off to capture some insect on the wing, and then return to complete their song. They are very tame and fearless of man, and will frequently come and perch beside you when walking in the fields or bush, wagging their tails from side to side—as if perfectly sure they were either privieged birds, or, on account of their dull plumage, not worth shooting. The nest is small, but very neat and compact, one inch and three-quarters across by half an inch deep, composed of grasses sunk in the fork of a horizontal bough ; the edge is even with or slightly raised above the branches, and ornamented with small scales of bark securely fastened on with cobwebs and rendered so like the bark of the tree, that it is no easy task for one who is unacquainted with its habits to discover it. The eggs are two in number, but I remember two instances in which we found three in a nest : this is however rarely the case. In length they are from eight and a-half to ten lines by six to seven lines in breadth. They vary considerably in colour, some being of a beautiful bluish-green with a zone of brownish-purple and greyish-lilac blotches round the centre, and a few dots over the rest of the surface ; in others the spots are dispersed equally over the whole. As the eggs fade the ground colour becomes very pale, and the markings turn to dull reddish-brown. This species has two, and sometimes three broods in the year. The peculiar instinct which birds have of ornamenting the outside of their nests with small scales of the bark and lichen which grow upon the same trees, is beautifully illustrated in not only the nest of the present species, but also in those of many other Australian birds : as in that of the genus *Eöpsaltria*, and more particularly in those of the genus *Sittella*, which are not only ornamented on the outside with scales of bark from the same or similar branches to which they are fastened, but the inside is carefully lined with small pieces of the mouse-eared lichen, so arranged as to bear a very close resemblence to the eggs." *(Ramsay, Proc. Phil. Soc., Sydney,* 1865, p. 329, pl. i., figs. 9 and 10.)

Dimensions of a set in the Australian Museum Collection :— length (A) 0·73 x 0·59 inch ; (B) 0·75 x 0·59 inch.

. A set in the Dobroyde Collection measures as follows :—length (A) 0·71 x 0·56 inch ; (B) 0·72 x 0·54 inch.

Hab. Port Denison, Wide Bay District, Dawson River, Richmond and Clarence Rivers Districts, New South Wales, Interior, Victoria and South Australia. *(Ramsay)*

Genus MONARCHA, *Vigors and Horsfield*.

MONARCHA MELANOPSIS, *Vieillot*.

(M. carinata, Swainson.)

Carinated Flycatcher.

Gould, Handbk. Bds. Aust., Vol. i., sp. 152, p. 262. *IX* ·*9.*

Dr. Ramsay writes as follows regarding the nidification of this species :—

"I have never myself had the pleasure of finding the nest of this beautiful species, but perhaps the fact that very few breed about Sydney may be sufficient for this seeming neglect.

For the nest and eggs which at present grace my collection I am indebted to Mr. George Masters, of Petersham, who procured them during a visit to Kiama in January 1864.

The only instance I know of this bird's breeding in the vicinity of Sydney was in December 1860, when I observed a pair, accompanied by two young ones scarcely able to fly. The first specimen I obtained last year was during September, about the 25th. Mr. Masters had also shot some a few days before at Petersham, about three miles distant from Sydney. They seldom remain long, but disappear as miraculously as they come, only a few pairs remaining to breed.

The nest procured by Mr. Masters was placed between the upright forks of a small tree, about eight feet from the ground. It is a neat structure, cup-shaped and open above, composed of grass and fine rootlets closely interwoven; the outside is ornamented with green moss, *Hypnum*, &c., which gives it a very beautiful and pleasing appearance. It is four inches in length by three across, and about an inch and a-half deep inside. The eggs are two in number, their ground colour pinkish-white with numerous bright red or pinkish salmon-coloured spots and markings sprinkled all over the surface, but more numerously towards the thicker end. They measure ten lines in length by eight in breadth. In this bird the plumage of both sexes is alike." (*Ramsay, Ibis*, 1865, Vol. i., New Series, p. 302.)

Another set measures, (A) 0·9 x 0·7 inch ; (B) 0·91 x 0·69 inch.

Hab. Port Darwin and Port Essington, Gulf of Carpentaria, Port Denison, Wide Bay District, Dawson River, Richmond and Clarence Rivers Districts, New South Wales. (*Ramsay.*)

Genus GERYGONE, *Gould.*

GERYGONE ALBIGULARIS, *Gould.*
White-throated Gerygone.

Gould, Handbk. Bds. Aust., Vol. i., sp. 155, p. 266.

"This delicate little bird is only a summer visitant to the neighbourhood of Sydney, arriving regularly in tolerable numbers every year during September, and remaining to breed, taking its

G

departure again in March and April. Its arrival is at once made known by its soft and varied strain of considerable melody. From its song (not that it resembles the notes of any other bird), and partly on account of its yellow breast it has gained the local name of the "Native Canary" Upon its arrival, it betakes itself to the smaller trees and saplings, and almost at once commences to build, selecting some strong twig among the innermost boughs of a bushy tree, to which it suspends its oblong dome-shaped nest, the extremity of which terminates in a well formed tail of about three inches in length, which is extremely characteristic. The body of the nest is in length from six to eight inches, and four in breadth; it is composed of fine pieces of stringy bark and grasses closely interwoven and matted together with cobwebs, being lined with the silky down of the cotton-tree or with opossum fur; the entrance which is about two inches and a-half down the side, is one inch in diameter, and completely hidden from view in front by a neatly woven hood of one inch and a-half in length. The nests are often placed in trees infested with ants, which insects are often found on the nests themselves, but do not, as far as I am aware, cause the bird any anxiety. The eggs, which are laid from October to December, and sometimes even as late as January, are three in number. Their ground colour is of a delicate white, but almost hidden by numerous spots, dots, blotches and freckles of dull red; in some the markings are thicker upon the larger end, where they form a well-defined zone or circular blotch; others are minutely dotted. Upon the whole, both in shape and colour they closely resemble those of *Malurus cyaneus*, but may be distinguished by being more thickly and strongly marked; they are also slightly larger and more lengthened in form.

This species shows a decided preference for the more open parts of the forest, with thickly foliaged trees and young saplings of Eucalyptus; its actions among the leaves, where it searches for insects, their larvæ, &c., are very pleasing and graceful, stopping in its search every now and then to pour forth its curious and varied song, in which it will sometimes stop abruptly and fly off without finishing, as if something had startled it or suddenly attracted its attention. Although well suited for the purpose, *Chalcites plagosus*, and *C. basalis*, seldom lay their eggs in the nests of this species. Still it must be numbered among the foster-parents of these birds, although such is rarely the case." (*Ramsay, P.Z.S.*, 1866, p. 576.)

Two sets in the Dobroyde Collection measure as follows, length No. 1 (A) 0·7 x 0·48 inch ; (B) 0·72 x 0·5 inch ; (C) 0·7 x 0·5 inch. No. 2 (elongated) (A) 0·8 x 0·5 inch ; (B) 0·79 x 0·51 inch; (C) 0·8 x 0·5 inch.

Hab. Derby, N.W. Australia, Cape York, Rockingham Bay, Port Denison, Wide Bay District, Dawson River, Richmond and Clarence River Districts, New South Wales, Interior. (*Ramsay.*)

3   GERYGONE FUSCA, *Gould.*

Brown Gerygone.

Gould, Handbk. Bds. Aust., Vol. i., sp. 156, p. 267.

This bird is to be found in the neighbourhood of Sydney ; the manner of its nidification is similar to that of the preceding species, although when available, it often constructs its nest in a bunch of green moss hanging from the end of a branch. Dr. Ramsay found several nests of this species near Grafton in September 1866; the eggs in every instance were three in number for a sitting, and precisely similar in colour and markings to those of *G. albigularis*, but smaller. I give the measurements of two sets :—No. 1, (A) 0·65 x 0·45 inch ; (B) 0·62 x 0·45 inch ; (C) 0·63 x 0·45 inch. No. 2, (A) 0·61 x 0·44 inch ; (B) 0·63 x 0·45 inch ; (C) 0·66 x 0·46 inch.

The months of September, October, November and December, constitute the breeding season of this species.

Hab. Wide Bay District, Richmond and Clarence Rivers Districts, New South Wales, Victoria and South Australia. (*Ramsay.*)

GERYGONE FLAVIDA, *Ramsay.*

Ramsay, P.L.S., N.S.W., Vol. ii., p. 53.

"This species is common among the dense belts of mangroves near Cardwell; we found several of its nests containing eggs and young birds on February 26th 1874, when my young friend Master J. Sheridan, an enthusiastic young naturalist, kindly waded nearly up to his knees in black mud to secure them for me, one nest contained the egg of a Cuckoo, exactly the same as that of *C. plagosus,* but smaller than any eggs of that bird I have hitherto met with; it is probably the egg of *C. minutillus.* The nest is a somewhat bulky structure, and resembles closely a lump of débris left by the floods hanging to the end of some leafy twig, it is composed of shreds of bark, dried water-weeds, and withered grasses, selected, I have no doubt, from the débris of the floods, plentiful on every side. It is oval oblong, with a small side entrance, and suspended by the top to the end of some hanging branch, often a considerable distance from the shore. The eggs are white, with a few dots of reddish-brown at the larger end; some are altogether white without any markings." Length (A) 0·7 x 0·51 inch. (*Ramsay, P.Z.S.,* 1875, p. 587.)

Hab. Rockingham Bay. (*Ramsay.*)

GENUS SMICRORNIS, *Gould.*

3   SMICRORNIS BREVIROSTRIS, *Gould.*

Short-billed Smicrornis.

Gould, Handbk. Bds. Aust., Vol. i., sp. 161, p. 273.

This little bird is distributed over the greater part of the Eastern and Southern portions of Australia. In the belts of *Melaleuca* at the mouth of the river Yarra, near Melbourne, I have often found the nests of this species. A nest now before me taken from the Australian Museum Collection, is of a swollen pear-shaped form and is constructed of mosses, grasses, spiders' cocoons &c., all securely woven together, with an entrance near the top, and lined inside with feathers and a few dried flowers, it is attached to the fine drooping leafy twigs of a Eucalyptus, and is four and a-half inches in length, by two inches and a-quarter in breadth at the thickest part; aperture one inch long, by six-eighths of an inch in width.

Eggs three in number for a sitting, a set taken at the mouth of the river Yarra in September 1879, are of a creamy-brown ground colour, minutely freckled at the larger end with buffy-brown; length (A) 0·57 x 0·44 inch; (B) 0·57 x 0·46; (C) 0·58 x 0·45 inch. A set taken at Macquarie Fields, in October 1860, are of a dull buffy-white, minutely freckled all over with slaty-brown markings, but particularly towards the larger end where they become confluent, and form a well defined zone; length (A) 0·62 x 0·43 inch; (B) 0·63 x 0·43 inch; (C) 0·63 x 0·44 inch. The latter is the most usual variety found.

This species commences to breed in September and continues during the two following months.

Hab. Wide Bay District, Dawson River, Richmond and Clarence Rivers Districts, New South Wales, Interior, Victoria and South Australia. (*Ramsay.*)

Genus ERYTHRODRYAS, *Gould.*

3. ERYTHRODRYAS RHODINOGASTER, *Drapiez.*

(*E. rhodinogaster*, Gould.)

Pink-breasted Wood Robin.

Gould, Handbk. Bds. Aust., Vol. i., sp. 163, p. 276.

" The nest of this species is formed of narrow strips of soft bark, soft fibres of decaying wood, and fine roots matted and woven together with vegetable fibres, and old black nests of spiders. The eggs are three in number, of a greenish-white, thickly sprinkled with light chestnut and purplish-brown ; eight lines and a-half long by six lines and a-half broad." (*Gould, Handbk. Bds. Aust.,* Vol. i., p. 276.)

Hab. Victoria and South Australia, Tasmania. *(Ramsay.)*

PETRŒCA LEGGII, *Sharpe.*

3-4

(P. multicolor, Vigors and Horsfield.)

Scarlet-breasted Robin.

Gould, Handbk. Bds. Aust., Vol. i., sp. 165, p. 279.

This bird is universally dispersed throughout Queensland, New South Wales, Victoria, South Australia, and Tasmania. It arrives in Victoria about the latter end of June, and commences to breed early in September and continues to the end of December.

On the partially cleared land in the dense forests of South Gippsland I have often found the nest of this species by seeing the bird fly into one of the huge blackened hollow trunks of the Eucalyptus that has been destroyed by fire ; the nest is placed about six or seven feet from the ground on a projecting piece of roughened and charred wood, it is composed of strips of bark, grasses, and mosses, securely held together with cobweb, and lined with hair, fur, feathers &c., and sometimes with the soft downy fibre of the inner bark of the tree-fern *(Dicksonia antarctica);* it is also placed on the horizontal branch, or bole of a tree after the manner of the *Artamus* family. Eggs three or four in number for a sitting.

Var. A. Ground colour buffy-white thickly freckled and spotted all over the surface of the shell with purplish-brown, wood-brown, and bluish-grey markings. Length (A) 0·72 x 0·6 inch ; (B) 0·74 x 0·91 inch ; (C) 0·72 x 0·62 inch. Taken by Dr. Ramsay on Ash Island, October 1860.

Var. B. Ground colour, when just taken greenish-white, fading to a dull white after being emptied of their contents and kept for some time, minutely freckled and spotted with pale umber, purplish-brown and lilac markings, intermingled with others of a deep bluish-grey, the latter colour predominating and appearing as if beneath the surface of the shell; towards the larger end these markings become thicker and form an irregular shaped zone. Length (A) 0·73 x 0·61 inch; (B) 0·7 x 0·6 inch; (C) 0·72 x 0·61 inch; (D) 0·72 x 0·6 inch. Taken at Childers, South Gippsland, November 23rd, 1884.

Hab. Wide Bay District, Dawson River, Richmond and Clarence Rivers Districts, New South Wales, Victoria and South Australia, Tasmania. (*Ramsay.*)

4.   PETRŒCA GOODENOVII, *Vigors and Horsfield.*

Red-capped Robin.

Gould, Handb. Bds. Aust., Vol. i., sp. 166, p. 280.

The nest of the Red-capped Robin is one of the most beautiful belonging to our Australian birds, it is usually placed on a thick branch or upright fork of a Casuarina, near the ground; and is very difficult of detection as it is made to assimilate so closely to its surroundings, and like the nests of the *Sittellæ* it is only by the actions of the birds that its whereabouts is betrayed; it is composed of fine strips of bark, neatly held together with fragments of wool, the inside being lined with cow hair, opossum fur, and a few feathers; the edge of the nest is thick and rounded, and the whole outer surface is beautifully ornamented with a mouse eared lichen which when new gives it a very pretty appearance; external diameter two inches and three-quarters, depth one inch and three-quarters, rim five-eighths of an inch in thickness; internal diameter one inch and a-half, depth one inch. The eggs are four in number, ground colour when fresh greyish-green, thickly covered all over the surface with light purplish-brown markings, but particularly towards the centre where they become larger and confluent, and intermingling with a few spots of lilac form an irregular zone.

A set of four taken at Tyndarie by Mr. James Ramsay in 1879, measure as follows :—length (A) 0·65 x 0·5 inch ; (B) 0‘63 x 0·5 inch ; (C) 0·62 x 0·48 inch ; (D) 0·63 x 0·5 inch.

This bird breeds during August and the two following months.

Hab. Dawson River, Richmond and Clarence River Districts, New South Wales, Interior, Victoria and South Australia, West and South-West Australia. *(Ramsay)*

3   PETRŒCA PHŒNICEA, *Gould.*

Flame-breasted Robin.   •

Gould, Handbk. Bds. Aust., Vol. i., sp. 167, p. 282.

For the nest and several sets of eggs of this bird I am indebted to Mr. Joseph Hill of Pine Rise, Kewell, Victoria, who procured them together with the birds on the borders of the Mallee Scrub, in the Wimmera District, during September 1884. The nest is similar to that of *P. leggii,* and the eggs in every instance were three in number for a sitting. The ground colour of the eggs is greenish-white, freckled and spotted with yellowish and chestnut-brown markings ; in two specimens, A and B, towards the larger end is a well defined zone of delicate lilac spots ; in the other specimen C, the spots are uniformly distributed over the whole surface. Length (A) 0·73 x 0·58 inch ; (B) 0·74 x 0·57 inch ; (C) 0·72 x 0·56 inch.

The breeding season commences in August and lasts during the three following months.

Hab. New South Wales, Victoria and South Australia, and Tasmania *(Ramsay.)*

Genus MELANODRYAS, *Gould.*

3 MELANODRYAS BICOLOR, *Vigors and Horsfield.*

(M. cucullata, Latham.)

Hooded Robin.

Gould, Handbk. Bds. Aust., Vol. i., sp. 168, p. 283.

The nest of this species is usually placed on the thick branch or fork of a tree close to the ground; one before me 'taken by Mr. K. H. Bennett at Mossgiel, September 1886, is a round, open, shallow structure, composed of strips of bark, held together by cobwebs, and lined inside with fine grasses and rootlets, measuring two inches and a-quarter across, by five-eighths of an inch in depth, inside measurement. The eggs three in number for a sitting, are olive-green with a very faint tint of reddish-brown on the larger end. Length (A) 0·83 x 0·65 inch; (B) 0·85 x 0·65 inch; (C) 0·84 x 0·65 inch. The eggs of this species vary in tint; a set I have in my collection taken in the Wimmera District, Victoria, being of a bright apple green.

The breeding season commences in August and lasts through the four following months.

Hab. Wide Bay District, Dawson River, Richmond and Clarence River Districts, New South Wales, Victoria and South Australia, West and South-West Australia. (*Ramsay.*)

2 MELANODRYAS PICATA, *Gould.*

Pied Robin.

Gould, Handbk. Birds Aust., Vol. i., sp. 169, p. 285.

This bird has a wide range over the continent of Australia, specimens having been procured together with the nest and eggs by Mr. James Ramsay, in October 1876, near Bourke, New South Wales; and last year both Mr. Cairn and the late Mr. T. H. Boyer-Bower obtained several specimens about eighty miles inland from Derby, North-western Australia.

The nest is a small shallow structure, composed of strips of bark, grasses, roots, &c., held together on the outside with cobwebs, and placed on the dead branch of a tree within a few feet of the ground. Eggs two in number, one specimen (A), being a dark asparagus-green, faintly tinged with brown on the larger end, the other (B), with the exception of the smaller end which shows the asparagus-green ground colour, is shaded all over with rich brown, more particularly towards the larger end, which is entirely capped with a darker tint of the same colour. Length (A) 0·8 x 0·59 inch ; (B) 0·78 x 0·6 inch. (*North, P.L.S., N.S. W.,* Vol. ii., 2nd Series, p. 554, 1887.)

Hab. Derby, N. W. A., Port Darwin and Port Essington, Gulf of Carpentaria, Interior. (*Ramsay.*)

3-4

Genus AMAURODRYAS, *Gould.*

AMAURODRYAS VITTATA, *Quoy et Gaimard.*

Gould, Handbk. Bds. Aust., Vol. i., sp. 170, p. 286.

The nest of this species is similar to that of *M. bicolor,* but somewhat larger. For the first eggs of this species I am indebted to Mr. Gordon, who obtained them among others for me during his residence on King Island, in Bass's Straits, when superintending the building of the lighthouse at that place. Eggs three or four in number for a sitting of a pale apple-green, faintly tinged with brown on the larger end forming a perfect, but nearly invisible, zone. Length (A) 0·87 x 0·63 inch ; (B) 0·84 x 0·63 inch ; (C) 0·88 x 0·67 inch. Among several sets taken in October 1885, near Mount Wellington, Hobart, Tasmania, are some with a distinct zone of brown spots ; others again appear as if smudged in places with the same colour. Dimensions of one of the sets are as follows :—length (A) 0·87 x 0·66 inch ; (B) 0·84 x 0·64 inch ; (C) 0·85 x 0·63 inch.

The breeding season of this species commences at the latter end of August, and continues till the end of December.

Hab. Victoria and South Australia, Tasmania. (*Ramsay.*)

Genus PŒCILODRYAS, *Gould.*

2. PŒCILODRYAS SUPERCILIOSA, *Gould.*
White Eyebrowed Robin.

Gould, Handbk. Bds. Aust., Vol. i., sp. 172, p. 289.

"The nest of this species somewhat resembles that of an *Eöpsaltria* The eggs also resemble those of *E. australis,* Lath., but are much smaller. The ground colour is of a rich apple-green, but in some of a bluish tint; some are zoned and sprinkled with spots, others have irregular or confluent blotches of reddish-brown. Two eggs are considered by this bird sufficient for a sitting. Length (A) 0·78 x 0·57 inch; (B) 0·8 x 0·55 inch; (C) 0·77 x 0·57 inch; (D) 0·9 x 0·55 inch, this last being an elongated abnormal specimen. They were taken, and the birds shot by Mr. Ed. Spalding, at Rockingham Bay in 1868." (*Ramsay, P.L.S., N.S.W.,* 2nd Series, Vol. i., p. 1145.)

Hab. Port Darwin and Port Essington, Gulf of Carpentaria, Cape York, Rockingham Bay, Port Denison. (*Ramsay.*)

Genus DRYMODES, *Gould.*

2. DRYMODES SUPERCILIARIS, *Gould.*
Eastern Scrub Robin.

Gould, Handbk. Bds. Aust., Vol. i., sp. 174, p. 291.

"Mr. Macgillivray, while traversing on the 17th of November 1849 a thin open scrub of small saplings growing in a stony ground thickly covered with dead leaves, about five or six miles inland from Cape York, observed a nest of this species placed on the earth at the foot of a small tree ; its internal diameter was four inches and a-half ; it was outwardly composed of small sticks, with finer ones inside, and lined with grass-like fibres, and was moreover surrounded with dead leaves heaped up to a level with its upper surface ; it contained two eggs an inch lon; by seven-tenths of an inch broad, of a regular oval shape, and of a very light stone-grey thickly covered with small umber blotches, which increased in size and were more thickly placed at the larger end : they were placed side by side, with the large end of one opposite the small end of the other." (*Gould, Handbk. Bds. Aust.*, Vol. i., p. 291.)

Hab. Gulf of Carpentaria, Cape York. (*Ramsay.*)

Genus EÖPSALTRIA, *Gould.*

2-3

EÖPSALTRIA AUSTRALIS, *Latham.*

Yellow-breasted Robin.

Gould, Handbk., Bds. Aust., Vol. i., sp. 175, p. 293.

" The nest of this species much resembles in form those of the true Australian Robins of the genus *Petrœca*, to which the birds also closely assimilate in their movements and habits, with the exception that the *Eöpsaltriæ* are lovers of the more unfrequented parts of the bush, while nearly all the members of the genus *Petrœca* prefer the open and half cleared patches of land. The nests of the Yellow-breasted Robin are either placed in the upright fork of some small tree, or built upon some horizontal bough, often within two or three feet of the ground. It is a beautifully round and cup-shaped structure, three inches high by two inches across and one inch and a-half deep, composed of strips of bark, and lined most frequently, with the narrow thread-like leaves of the native oak *(Casuarina)* and a few dry leaves of the Eucalypti. The edges and parts of the outside are studded with small pieces of the mouse-ear lichen, and hanging from the sides are long chips of bark, some of them four inches or more in length and half an inch wide, fastened one above the other with cobweb, the lowest of them reaching several inches below the bottom of the nest. The eggs, which are two or three in number, are of an apple-green, or light greenish-blue colour, spotted, blotched, or minutely dotted with deep brownish-red, yellowish-brown, and obsolete spots of faint lilac. Some are thickly speckled all over so as almost to hide the ground colour, and in these the yellowish-brown markings predominate; others are distinctly spotted, or have a zone of markings. They are in length ten and a-half to eleven lines by seven to seven and a-half in breadth, and are usually found in September and the three following months." (*Ramsay, Proc. Phil. Soc., Sydney,* 1865, p. 326, pl. i., figs. 7 and 8.)

A set of the eggs of this species taken at Dobroyde in 1866 measure as follows :—length (A) 0·82 x 0·63 inch ; (B) 0·8 x 0·59 inch ; (C) 0·81 x 0·61 inch.

A set taken in South Gippsland give the following dimensions: length (A) 0·84 x 0·65 inch ; (B) 0·81 x 0·62 inch ; (C) 0·82 x 0·63 inch.

Hab. Dawson River, Richmond and Clarence Rivers Districts, New South Wales, Victoria and South Australia. (*Ramsay.*)

2 EÖPSALTRIA CAPITO, *Gould.*

Large-headed Robin.

Gould, Handbk. Bds. Aust., Vol. i., sp. 178, p. 297.

The localities which this bird frequents are the rich brushes that clothe the sides of the rivers on the eastern coast of Australia, extending from Rockingham Bay in the north to the Clarence River in the south. A nest of this species now before me taken from the low fork of a tree near Ballina, on the Richmond River, is a deep cup-shaped structure composed of portions of the dead leaves of the "lawyer-vine" *(Calamus australis),* held together

with a few wiry grass stems, the whole exterior being covered
with mosses, and ornamented in a few places with large pieces of
lichens. Exterior diameter two and three-quarters of an inch by
two inches and a-half in depth. Eggs two in number for a sitting,
oval in form slightly tapering at one end, of a very faint dull
greenish-white, covered with indistinct markings of yellowish and
reddish-brown, which at the larger end becomes more boldly
defined, where, intermingled with superimposed blotches of wood-
brown, they form an irregularly shaped zone. Length 0·82 x 0·6
inch. The eggs of this bird are entirely devoid of the rich apple-
green colour of those of the southern representative *E. australis.*
*From Mr. R. D. Fitzgerald's Collection. (North, Proc. Linn. Soc.
N.S.W.,* Vol. ii., Second Series, p. 146, 1888.)

Hab. Rockingham Bay, Port Denison, Wide Bay District,
Richmond and Clarence Rivers Districts. (*Ramsay.*)

2 - 3   EÖPSALTRIA GULARIS, *Quoy et Gaimard.*
Grey-breasted Robin.
Gould, Handbk. Bds. Aust., Vol. i., sp. 176, p. 294.

A nest of this species taken by Mr. George Masters at King
George's Sound, on the 24th of September 1868, and now in the
Australian Museum Collection, is a deep cup-shaped structure,
composed entirely of strips of bark held together with a very fine
white fibre resembling cobwebs ; at the bottom of the nest are a
few dried Eucalyptus leaves ; on the outside of the nest are long
strips of bark attached all round, some pieces measuring over three
and a-half inches in length and three-eighths of an inch in width ;
outside measurement of nest three inches, depth two inches,
internal diameter two inches and a-half, depth one inch and a-half.

"Eggs two or three for a sitting; the ground colour is of a pale
apple-green, with a zone of dots and spots round the larger end of
a light reddish-brown, they approach in tint faded eggs of *E. nana,*
Ramsay, but are much larger."   Length (A) 0·83 x 0·6 inch ; (B)

0·84 x 0·62 inch." (*Ramsay, P.L.S., N.S.W.*, Vol. i., 2nd Series, p. 1145.

Hab. West and South-West Australia. (*Ramsay.*)

2 - 3   EÖPSALTRIA NANA, *Ramsay.*

P.L.S. of N.S.W., Vol. ii., p. 372.

" I first noticed this species on the Lower Herbert, and afterwards obtained it in the dense scrubs at Dalrymple's Gap, about fourteen miles from Cardwell; but it was not until Mr. Broadbent had forwarded to me adult specimens, shot from the nest, that I became aware of its being a distinct species; and although very closely allied to *Eöpsaltria capito* (Gould) of our New South Wales brushes it may at once be distinguished by the rufous tint on the lores and round the eye.

In habits *E. nana* resembles all others of the genus, building a similar nest and laying eggs closely resembling those of *E. capito* but smaller. The nest is placed in the fork of a vine or horizontal bough of a tree, and is a remarkably neat structure; one before me is perfectly round, open above, about 1·8 inch inside diameter, 2·8 inches outside; depth inside 1·4 inch; to bottom of nest outside 1·6 inch to 2·5 inches; it is built in the angle formed by a loaf of a species of *Calamus* and the upright cane, and supported by the branching leaflets, narrow strips of withered palm leaves &c., and ornamented on the outside with green mosses and scales of the bark of moss-grown scrub trees.

The eggs are two or three in number, of a dull greenish-yellow, greenish-buff, or greenish grey-brown, blotched and spotted with yellowish-umber, buff, and reddish-brown, with freckles of a slaty-grey tint; the larger spots and blotches forming a zone at the thicker end. Length 0·85 inch; breadth 0·56 inch. (*Ramsay, P.L.S., N.S.W.*, Vol. ii., p. 474.)

Hab. Rockingham Bay. (*Ramsay.*)

Genus MALURUS, *Vieillot.*

MALURUS CYANEUS, *Ellis.*

Superb Warbler.

Gould, Handbk. Bds. Aust, Vol. i., sp. 185, p. 317.

This species is found breeding almost everywhere in New South Wales, even in the public parks about Sydney, a pair successfully rearing a brood of five young ones in the Museum grounds last year. The nest is dome shaped, having a small entrance at the side, and is constructed of grasses, warmly lined with feathers or other soft materials. It is usually placed in some low, thick bush ; when breeding in any of the public gardens, the prickly *Acaciæ* is a favourite situation. Eggs five in number for a sitting, fleshy-white, with rich reddish dots, spots, and blotches, in some these are confluent, forming a broken zone on the larger end, while in others the markings are in a coalesced patch on the apex. A set of five taken at Dobroyde, August 9th, 1862, measure :—(A) 0·68 x 0·5 inch ; (B) 0·61 x 0·5 inch ; (C) 0·62 x 0·48 inch ; (D) 0·66 x 0·47 inch ; (E) 0·68 x 0·5 inch.

They breed from August to January.

Hab. Richmond and Clarence River Districts, New South Wales, Victoria and South Australia. (*Ramsay.*)

MALURUS CYANOCHLAMYS, *Sharpe.*

P.Z.S., 1881, p. 788.

This species is closely allied to *M. cyaneus*, from which it may be known by the much paler cobalt tint of the head, ear-coverts, and mantle. Specimens of this bird were obtained on the Herbert River, Queensland, in November 1868, together with the nest and eggs. The nest is a dome-shaped structure, with an entrance in the side, constructed of dried grasses intermingled with spiders' webs, and lined inside with feathers, hair, &c., it was placed in a thick bush close to the ground. Eggs four in number for a sitting, fleshy-white, sprinkled all over with pale reddish-brown markings, in one specimen (A.) forming a coalesced patch on one end. Length (A) 0·68 x 0·5 inch ; (B) 0·68 x 0·5 inch ; (C) 0·66 x 0·51 inch ; (D) 0·67 x 0·48 inch. *(North, P.L.S., N.S.W.,* Vol. ii., 2nd Series, p. 406.)

Hab. Rockingham Bay, Port Denison, Wide Bay District, Dawson River. *(Ramsay.)*

MALURUS GOULDII, *Sharpe.*
(M. longicaudus, Gould.)
Long-tailed Superb Warbler.

Gould, Handbk. Bds. Aust , Vol. i., sp. 186, p. 320.

The nest of this species is similarly constructed to that of *M. cyaneus,* but rather larger, and is built in some low bush or tuft of long grass. Eggs usually five for a sitting, and the largest of all the genus *Malurus;* they are fleshy-white with blotches and spots of rich red scattered all over the surface, but particularly towards the larger end, where in most instances they form a zone. A set taken near Hobart, Tasmania, last year measures:— (A) 0·75 x 0·55 inch ; (B) 0·7 x 0·52 inch ; (C) 0·72 x 0·54 inch ;

They breed during August and the four following months.

Hab. Victoria and South Australia, Tasmania. *(Ramsay.)*

MALURUS LAMBERTI, *Vigors and Horsfield.*
Lambert's Superb Warbler.

Gould, Handbk. Bds. Aust., Vol. i., sp. 191, p. 327.

The nest of this species is similar to others of the genus. It is still to be found breeding in the neighbourhood of Sydney. The eggs of this species cannot be distinguished from those of *M. cyaneus,* or many others of the genus, two in the Dobroyde

H

Collection taken by Dr. Ramsay on the 16th September 1860, are heavily blotched with red, forming a zone on the thicker end, another has the spots smaller and sprinkled over the whole surface. Length (A) 0·64 x 0·48 inch ; (B) 0·65 x 0·47 inch.

A set taken by Mr. James Ramsay at Tyndarie, are not so heavily blotched as the former, and have nearly all the markings confined to the larger end.   Length (A) 0·62 x 0·5 inch ; (B) 0·66 x 0·47 inch ; (C) 0·65 x 0·47 inch.

They breed from the beginning of September until the end of December, during which time two or more broods are reared.

Hab. Gulf of Carpentaria, Rockingham Bay, Dawson River, Richmond and Clarence Rivers Districts, New South Wales, Interior, Victoria and South Australia.   (*Ramsay.*)

MALURUS MELANOTUS, *Gould.*

Black-backed Superb Warbler.

Gould, Handbk. Bds. Aust., Vol. i., sp. 187, p. 322.   *XIII. 19.*

Mr. K. H. Bennett procured several nests and eggs of this species from the neighbourhood of Ivanhoe and Mossgiel, in the interior of New South Wales, during the months of October and November 1885 and 1886.   The nest is of the usual form, composed throughout of strips of bark, grasses, and wool, with a few feathers inside, rather loosely interwoven, being five inches in length by three in breadth, and placed in a low bush.

" Eggs like those of *M. cyaneus*, from which they are not to be distinguished ; white, with rich red dots, spots, and in some blotches, scattered over the surface, crowded on one end, or forming a broken zone near the thicker end ; the size of an average specimen is—long axis 0·63 inch ; short axis 0·48 inch ; of a heavily blotched specimen 0·65 x 0·45 inch." *Dobr. Mus. Coll.* (*Ramsay, P.L.S., N.S.W.*, 2nd Series, Vol. i., p. 1145.)

Hab. Interior, South Australia. (*Ramsay.*)

H—2

MALURUS SPLENDENS, *Quoy et Gaimard.*

Banded Superb Warbler.

Gould, Handbk. Bds. Aust., Vol. i., sp. 188, p. 323.

Mr. K. H. Bennett found a single nest of this species in the interior of Australia, close to the South Australian border. The nest as usual of this species is dome-shaped and constructed of grasses, wool, and moss, lined with a few feathers ; the outside is ornamented with some bright yellow and white wild " everlasting flowers " ; it is six inches in length by three in breadth. Unfortunately the eggs were too far incubated to be blown, but were similar to other eggs of the genus. " According to Mr. Gould the eggs are four in number, of flesh-white, thickly blotched and freckled with reddish-brown, especially at the larger end ; eight and a quarter lines long by six and a quarter lines broad, and they breed during September and the three following months."

Hab. West and South-West Australia. (*Ramsay.*)

MALURUS ELEGANS, *Gould.*

Graceful Superb Warbler.

Gould, Handbk. Bds. Aust, Vol. i., sp. 189, p. 324.

" The nest of this species is dome-shaped with a hole in the side for an entrance, and is generally formed of the thin paper-like bark of the Ti-tree *(Melaleuca)*, and lined with feathers ; it is usually suspended to the foliage of this tree, and occasionally to that of other shrubs which grow in its favourite localities. The eggs are four in number, of a delicate flesh-white, freckled with spots of reddish-brown, which are much thicker at the larger end ; they are about eight lines long and six lines broad. The breeding season commences in September, and continues during the three following months." (*Gould, Handbk. Bds. Aust.*, Vol. i., p. 324.)

Hab. West and South-West Australia. (*Ramsay.*)

MALURUS PULCHERRIMUS, *Gould*.

Blue-breasted Superb Warbler.

Gould, Handbk. Bds. Aust., Vol. i., sp. 190, p. 326.

A nest of this species found by Mr. Froggatt on the 25th of January 1888, in a low bush on the Napier Range, about one hundred miles inland from Derby, North-western Australia, is similar to those of other members of the genus, being a dome-shaped structure, outwardly composed of long thin strips of bark, matted together with spiders' cocoons and lined inside with the soft downy seeds of a composite plant. It contained three eggs in an advanced state of incubation, oval in form, white, sprinkled all over the larger end with reddish-brown markings. Length (A) 0·68 x 0·48 inch ; (B) 0·68 x 0·45 inch. *From the Macleayan Museum Collection.*

Hab. West and South-West Australia. (*Ramsay.*)

MALURUS LEUCOPTERUS, *Quoy et Gaimard*.

White-winged Superb Warbler.

Gould, Handbk. Bds. Aust., Vol. i., sp. 194, p. 330.

"The nest is from the same district as the following species and composed of the same materials and similarly placed ; it is a little smaller and rather more loosely put together ; the eggs are very similar, only a trifle smaller than those of *M. leuconotus*, and the zone of reddish spots not so distinct ; they nevertheless vary considerably, some having the zone more defined, others which have no zone at all are simply sprinkled all over the thicker end with reddish-brown or light red spots ; they breed during September to December." (*Ramsay, P.L.S., N.S.W.*, Vol. vii., p. 49.)

Dimensions of a set taken by Mr. James Ramsay at Tyndarie, are as follows :—length (A) 0·59 x 0·43 inch ; (B) 0 58 x 0·43 inch ; (C) 0·58 x 0·45 inch ; (D) 0·62 x 0·45 inch.

Hab. New South Wales, Interior, Victoria and South Australia, West and South-West Australia. (*Ramsay.*)

2 ## MALURUS LEUCONOTUS, *Gould.*

White-backed Superb Warbler.

Gould, Handbk. Bds. Aust., Vol. i., sp. 195, p. 332.

"The nest, like that of all other members of the genus is a dome shaped, oblong structure of fine grass, ornamented and mixed with cobweb and wool, and lined inside with the cotton from the native "Cotton Bush" or the silky down from the seed pods of an *Asclepiad.* The length of the nest is 5·5 inch x 2·3 inch, and it was placed in a small tuft of coarse grass near the ground; others were found among the lower branches and grass at the base of "Cotton Bush" shrub. The eggs are three in number, pearly-white with a zone of reddish spots on the thicker end, and a few dots of the same tint sprinkled over the rest of the surface. Length 0·6 x 0·45 inch. They breed during the months of September, October and November." (*Ramsay, P.L.S., N.S.W.,* Vol. vii., p. 49.)

Hab. New South Wales, Interior, Victoria and South Australia. (*Ramsay.*)

3 ## MALURUS MELANOCEPHALUS, *Vigors and Horsfield.*

Black-headed Superb Warbler.

Gould, Handbk. Bds. Aust., Vol. i., sp. 196, p. 333.

While in the Richmond River District Dr. Ramsay found the nest and eggs of this species at Lismore on the 12th of November 1866. The nest is like that of *M. cyaneus,* but smaller, and decorated on the outside with new moss; it was built about three feet from the ground in the top of some high rank grass. Eggs three in number, white, minutely freckled and spotted all over with rich red, particularly on the larger end where they form an irregular zone. Length (A) 0·62 x 0·45 inch; (B) 0·6 x 0·44 inch; (C) 0·61 x 0·44 inch.

This species breeds during September and the three following months.

Hab. Wide Bay District, Dawson River, Richmond and Clarence Rivers Districts, New South Wales, Interior. (*Ramsay*)

MALURUS CALLAINUS, *Gould.*
Turquoisine Superb Warbler.

Gould, Suppl. Bds. Aust., pl. 23.   XIII   18,

"This Wren, one of the latest species described by Mr. Gould is far from rare in the interior, my brother Mr. James Ramsay having no difficulty in obtaining as many specimens as I required during one season, both of its nests and eggs with the birds shot therefrom. Although the eggs appear quite different from those of other species of the genus, still it is difficult to express these differences in a description. Eggs white or pinkish-white with minute dots and small spots of rich red sprinkled over the whole surface, in some forming zones, in others blotches. What I consider the more typical eggs of this species are those with a few dots of dark red sparingly sprinkled over the whole surface of the shell, closer together on the thicker end, but seldom forming a distinct zone; all more or less pointed; (A) 0·67 x 0·48 inch; (B) 0·67 x 0·48 inch; (C) 0·66 x 0·48 inch." *J. R., Dobr. Mus.* (*Ramsay, P. L.S., N.S.W.*, 2nd Series, Vol. i., p. 1145.)

Hab. New South Wales, Interior, Victoria and South Australia. (*Ramsay.*)

MALURUS CRUENTATUS, *Gould.*
Brown's Superb Warbler.
Gould, Handbk. Bds. Aust., Vol. i., sp. 197, p. 334.

"Nest dome shaped with the entrance at the side, slightly protected with a hood, placed among grasses or shrubs near the ground. Eggs four for a sitting, length (1) 0·6 x 0·45 inch; (2) 0·68 x 0·46 inch; the last is an exceptionally large egg of this species, and has the dots crowded into a brownish-red patch on the thicker end, a few specks of the same colour are sprinkled over the rest of the surface ; the ground colour is white in No. 1; it is sprinkled with reddish dots all over the surface, but forming a zone at the thicker end." (*Ramsay*, *P.L.S.*, *N.S.W.*, Vol. vii., p. 408.)

Hab. Port Darwin and Port Essington, Gulf of Carpentaria, Cape York, Rockingham Bay, Port Denison. (*Ramsay.*)

Family MENURIDÆ.

Genus MENURA, *Davies.*

1.

MENURA SUPERBA, *Davies.*

Lyre-bird.

Gould, Handbk. Bds. Aust., Vol. i., sp. 179, p. 298.   𝒳   *4.*

" The nest of this species differs according to the locality frequented by the birds :—some being constructed of rough material, such as large sticks, stringy bark *(Eucalyptus obliqua)* and dead ferns *(Pteris aquilina);* others of very fine rootlets and pieces of *Hymenophyllum tunbridgense*, which makes a remarkably neat nest. Braisher, the most successful of my collectors, who also procured the young birds, called upon me a few days ago with some of the eggs, when I took the opportunity of getting all the particulars respecting the nidification. I find that in no instance did he meet with more than one egg or one young bird in the same nest. The birds commence to build in May, and lay their eggs in June and July. The female is not fed by the male while she is sitting, nor has the male bird ever been observed near the place after she has laid her egg. The female frequently leaves her egg during the middle of the day to search for food. This may account for the length of time taken in the hatching, which sometimes extends over a month. The young do not leave the nest until they are eight or ten weeks old. When one is standing in front of the nest, the egg or the young bird can easily be seen in it. The female enters the nest *head first*, and then turns round and settles herself on the egg, with her tail sometimes over her back, *but more often bent round by her side*. Thus in time the tail becomes quite askew, and is a tolerable guide to the length of time the bird has been sitting.

The nests are for the most part placed on the darker side of the gullies and ravines. They are large, oval, domed structures, with the entrance in the front, and are usually placed on the ground at the foot of some stump or tree, or by the side of a fallen log ; sometimes they are placed on a ledge of rock in the face of the cliff at a considerable height from the ground ; occasionally a nest is found in the end of a log which has been hollowed out by fire and formed in the shape of a scoop. They are always built on some solid foundation, nor do I see how such a bulky and loosely built structure could hold together if placed otherwise ; great care must be taken in moving the nests to prevent their falling to pieces. I have now before me three nests :—No. 1, taken from the hollow end of a log ; No. 2, from a ledge of rock ; while the third was found by the side of a fallen tree. No. 2 is composed of fine roots and *Hymenophyllum tunbridgense*, with pieces of *Hypnum*, and lined with feathers ; this nest is much more neat, smaller than the others, and looked very beautiful while the ferns and moss, which covered the whole of the outside were fresh and green. Nos. 1 and 3 are much the same in appearance and size. being large, oval dome-shaped structures of sticks, twigs, and roots, interwoven loosely with pieces of bark and moss, roots of ferns, and fronds of *Pteris aquilina;* the inside is lined with rootlets, and finally, the long loose feathers from the flanks and backs of the birds. The entrance, which is in the side (or front), is not covered with a hood, nor does its upper edge hang over so as to conceal the egg· The lower edge, if anything, protrudes slightly in all the nests I have examined. The total length of the nest is twenty-six inches, height twelve, and width eighteen inches ; the entrance is five or six inches in diameter, and its lower edge four and a-half in thickness. The whole of the interior is lined with feathers, which being much of the colour as the egg, help to protect it and hide it from view. All the nests and eggs which I possess, with the exception of one, were procured in the Illawarra district, chiefly from the ravines and gullies in the neighbourhood of Appin and Wollongong. Occasionally the same nest is used more than once after being lined afresh with feathers. The eggs are of three varieties at least :—

Var. A, the most common, is of a light stone-grey, with darker coloured blotches and spots, and a few jet-black dots ; length 2·4 to 2·5 inches by 1·6 to 1·7 in breadth. Other specimens are dull brown, stone-brown, or dark blackish-brown, with dull brown spots and blotches when fresh.

Var. B is of a reddish-brown colour, with dark blackish-brown spots, and a beautiful blush of pinkish purple over the whole surface. I have only seen one of this very marked variety, 2·35 inches in length by 1·65 in breadth.

Var. C is a most peculiar looking egg, of a uniform dark metallic blackish-brown, having obscure spots and blotches of a darker tint, almost invisible at a short distance ; length 2·5 by 1·7 inches ; and like many of the other specimens, this variety has jet-black lines and dots dispersed over the surface.

The young which are hatched early in August, but sometimes as late as the end of September, are of a whity-brown colour upon leaving the egg, but become darker as they get older ; the crown of the head is covered with long dusky slate-coloured down, which hangs over the neck (which is quite bare) on to the back ; the wings have a fringe of shorter down round them, which is longest on their lower edge ; the upper part of the rump, centre of the back, and the tail are also covered with down, while two rows of short down grow along the thighs. The bare triangular part of the neck is surrounded by a narrow fringe of very short down, while two ridges still shorter and of a light yellow colour grow on either side of the breast or keel of the sternum. Down on the head from one and a-half to two inches in length, on rump and tail it is two inches long. Bill 0·5 inch in length, blackish-brown at tip ; tarsi 0·8 inch in length." (*Ramsay, P.Z.S.*, 1868, p. 49.)

Hab. New South Wales. (*Ramsay.*)

MENURA ALBERTI, *Gould.*
Prince Albert's Lyre-bird.

Gould, Handbk. Bds. Aust., Vol. i., sp. 181, p. 307.   X. 3.

"Nest similar to that of *M. superba*. I have lately seen a fine specimen of this rare egg in the Macleayan Museum, and another in the Australian Museum Collection, which are all I have met with during the last twenty years ; the egg is oval, almost equal at both ends, the ground colour is of a rich purple-brown, the thicker end of the egg is blotched with large irregular markings of purplish-brown, very dark and almost forming a zone, the remainder of the surface is marked with irregularly shaped spots of the same tint, a few of them inclining to linear, others almost rounded ; length 2·23 x 1·7 inch. The ground colour of the specimen in the Macleayan Museum was a purplish-stone, but has faded to a light slate colour, the markings are irregular, of a dark purplish-brown and sprinkled sparingly over the surface." (*Ramsay, P.L.S., N.S.W.*, Vol. vii., p. 50.)

Hab. Wide Bay District, Richmond and Clarence Rivers Districts. (*Ramsay*)

MENURA VICTORIÆ, *Gould.*
Queen Victoria's Lyre-bird.

Gould, Handbk. Bds. Aust., Vol. i., sp. 108, p. 302.   X. /, 2.

This species differs but slightly from *M. superba*, and is precisely similar in its manner of nidification, and the localities which it frequents. I have often obtained the nests and eggs of this species in various parts of Victoria, but principally in South Gippsland ;

the nest is not always placed in out of the way situations, one being found only a few feet from a well-frequented track, and another within a stone's-throw of the house. The most successful collector I know of, Mr. E. Pakenham, obtained no less than six eggs of this species in one day. The breeding season of this species commences in the month of May and continues the three following months.

" *Menura victoriæ*, Var. A. Ground colour olive-brown, of a rather light tint, with spots of blackish-brown and purple-brown, some confluent, others solitary, rather crowded on the top of the thicker end ; there are also a few obsolete spots of a lilac tint ; length 2·37 x 1·65 inch. Var. B. Ground colour purplish-stone colour or dark brownish-purple, with obsolete spots and irregular markings of blackish, crowded towards the thick end, and forming a dark patch at the top where they overlap, some of the spots on the body of the egg are elongate and interspersed among freckles of the same blackish tint ; length 2·41 x 1·73 inch." (*Ramsay, P.L.S., N.S.W.,* Vol. vii., p. 50.)

Hab. Victoria. (*Ramsay.*)

Family TIMELIIDÆ.

Sub-Family TIMELIINÆ.

Genus AMYTIS, *Lesson.*

3. AMYTIS STRIATUS, *Gould.*

Striated Wren.

Gould, Handbk. Bds. Aust., Vol. i., sp. 199, p. 335. *IX. 10.*

This bird is an inhabitant of the large open grass plains and scrubby portions of the back country of New South Wales, and the interior of Australia. A nest of this bird in the Australian Museum Collection, taken from a tussock of " porcupine grass " by Mr. K. H. Bennett at Mossgiel in 1883, is a rather large open structure, entirely composed of fine shreds of bark fibre and grass, placed upon a foundation of pieces of bark ; it measures four inches and a-half in diameter and two inches and a-half in depth ; inside measurement two inches and a-half in diameter, depth three-quarters of an inch.

" Eggs three in number for a sitting closely resembling those of *Ptenœdus rufescens.* The ground colour is white almost obscured towards the thicker end with freckles and dots forming confluent spots of rich red, in some forming a zone, in others extending in an irregular patch over the end. Length (A) 0·85 x 0·65 inch ; (B) 0·84 x 0·63 ; (C) 8·85 x 0·61 inch." This species breeds during the months of September and October. (*Ramsay, P.L.S., N.S.W..* Vol. i.. p. 1143.)

Hab. New South Wales, Interior, Victoria and South Australia. (*Ramsay.*)

Genus STIPITURUS,

3   STIPITURUS MALACHURUS, *Latham.*
Emu Wren.
Gould, Handbk. Bds. Aust., Vol. i., sp. 201, p. 339.

" I had for many days visited the swamps upon Long Island, where these birds were very plentiful, in hopes of finding them breeding ; but it was not until the 25th of September 1861, that I succeeded in discovering a nest, although I had watched them for hours together for several days. While walking along the edge of the swamp, however, on this day, I was agreeably surprised by disturbing a female which flew from my feet out of an over-hanging tuft of grass growing only a few yards from the water's edge. Upon lifting up the leaves of the grass which had been bent down by the wind, I found its nest carefully concealed near the roots and containing three eggs. As the bird did not fly far, but remained close by in a small Swamp-oak *(Casurina,* sp.), I had a good opportunity of satisfying myself that it was a veritable Emu Wren. The eggs were of course quite warm, and within a few days of being hatched : this may account for the bird being so unwilling to leave the spot; for when I returned about five minutes afterwards, the female was perched upon the same tuft of grass, and within a few inches from where I had taken the nest. The whole nest is of an oval form (but that part which one might term the true nest is perfectly round), placed upon its side ; the mouth very large, taking up the whole of the upper part of the front. It is very shallow—so much so that if tilted slightly, the eggs would roll out, they being almost upon a level with its edge. It is outwardly composed of grass and the young dry shoots of the reeds which are so common in all the swamps near the Hunter River, lined with fine grass, roots, and finally a very fine green moss. It is very loosely put together, and requires to be moved very gently to prevent it from falling to pieces. The eggs were three in number, six and a-half lines long by four and a-half broad, sprinkled all over with minute dots of a light reddish-brown (brighter in tint than those of the *Malurus cyaneus),* but more numerously at the larger end, where they are blotched with the same colour. One of the three had no blotches, but was only minutely freckled all over. The ground colour is a delicate white with a blush of pink before the egg is blown." *(Ramsay, Ibis,* 1863, Vol. v., p. 177.)

Hab. Wide Bay District, Richmond and Clarence Rivers Districts, New South Wales, Victoria and South Australia, Tasmania, West and South-West Australia. *(Ramsay.)*

Genus SPHENURA, *Lichtenstein.*

3   SPHENURA BRACHYPTERA, *Latham.*

Gould, Handbk. Bds. Aust., Vol. i , sp. 202, p. 342.   *VIII* /6.

" The nest is an oval dome shaped structure, composed of grasses and débris; it is placed at the foot of some bushy. shrub and

concealed among the débris and grass which usually accumulates
in such places.   The eggs are three in number, the ground colour
almost white, the whole surface thickly freckled with dots of
blackish-brown and reddish-brown, with a few of a pale lilac tint
here and there, some of the dots very minute, others larger and
roundish in shape ; in one specimen they form a thick crowded
patch on the thicker end, where some are confluent ; the egg
before me is oval, rather swollen, and the shell very thin ; length
1·02 x 0·75 inch, they breed during September to December.
From Mr. Ralph Hargrave's Coll.   (Ramsay, P.L.S., N.S.W.,
Vol. vii., p. 50.)

Hab. Wide Bay District, Richmond and Clarence Rivers
Districts, New South Wales, Victoria and South Australia.
(Ramsay.)

²   SPHENURA LONGIROSTRIS, *Gould.*

Long-billed Bristle Bird.

Gould, Handbk. Bds. Aust, Vol. i., sp. 203, p. 343.   *VIII 15.*

A nest of this species in the Australian Museum Collection,
taken by Mr. George Masters at King George's Sound, Western
Australia, in September 1868, is oval in form with a large entrance
at the side, and is composed entirely of long dried hollow grass-
stalks, with a little grass of a finer description placed inside at
the bottom of the nest; it measures six inches in length, five inches
in width, and four inches in height, and was placed amongst some
dried vegetation close to the ground.   Eggs two in number for a
sitting, of a dull white ground colour minutely dotted, spotted,
and freckled all over with wood-brown and purplish-browń
markings, but particularly towards the larger end, where
intermingled with clouded blotches of dark lilac, appearing as if
beneath the surface of the shell, they become confluent and form an
irregular zone ; length (A) 0·9 x 0·72 inch ; (B) 0·91 x 0·73 inch.

Hab. West and South-west Australia.   *(Ramsay.)*

Genus HYLACOLA, *Gould.*

3   HYLACOLA PYRRHOPYGIA, *Vigors and Horsfield.*

Red-rumped Hylacola.

Gould, Handbk. Bds. Aust., Vol. i., sp. 205, p. 346.

" The nest of this species is usually hidden at the base of a clump of bushes and grass, or in some bushy scrub near the ground; sometimes resting on the ground, and at all times very difficult to find. I first found them breeding at Dobroyde in 1860, where I procured both adults and young. The nest is a loose structure, composed of narrow strips of bark, grasses, and rootlets, (which can be scarcely said to be interwoven), and with which it is chiefly lined with the addition of a few feathers. It is dome shaped in form, and a little larger than that of *Malurus lambertii.* The full number of eggs were in every instance three, the ground colour of a pinkish-salmon tint, fading after being emptied to a dull white, tinged with chocolate pink, in tint not unlike those of a *Sericornis magnirostris* or *S. frontalis.* They are blotched with irregular markings of light chocolate brown at the larger end, and a few dashes and spots of the same tint on the thinner end. The blotches forming a zone near the thick end. Length 0·76 x 0·57 inch. *(Ramsay, P.L.S., N.S.W.,* Vol. ii., p. 108.)

Hab. Wide Bay District, Richmond and Clarence Rivers Districts, New South Wales, Interior, Victoria and South Australia. (*Ramsay.*)

Genus CISTICOLA, *Kaup.*

3 (4)   CISTICOLA RUFICEPS, *Gould.*

Rufous-headed Grass Warbler.

Gould, Handbk. Bds. Aust., Vol. i., sp. 212, p. 353.

" This species is plentifully dispersed over the grass beds; it is common near Sydney, and equally plentiful at Cape York. The nest is a very neat, dome-shaped structure, chiefly composed of fine grasses, thistle down, and cobwebs, or the flowering portions of grasses all matted closely and thickly together, and having the adjacent leaves of the plant in which it is placed neatly sewed on to the side of the nest; sometimes two or three broad leaves are sewed together with cobweb, and the nest made between them. The nest is always placed near the ground where the grass growing through some broad-leaved plant affords it concealment, and is about two inches wide by 1·5 inch deep. The eggs in nearly every instance are three in number, but sometimes four are found. The ground colour is of a delicate pale blue, dotted, spotted, or blotched with brownish-red of various tints and shades. Their length is from 0·5 to 0·65 inch, by from 0·4 to 0·5 inch in breadth. They breed during October and the two following months." (*Ramsay, Ibis*, 1868, Vol. iv., New Series, p. 277.)

Hab. Derby, N.W. Australia, Port Darwin and Port Essington, Gulf of Carpentaria, Cape York, Rockingham Bay, Port Denison, Wide Bay District, Dawson River, Richmond and· Clarence Rivers Districts, New South Wales, Victoria and South Australia, South Coast New Guinea. (*Ramsay.*)

Genus PYCNOPTILUS, *Gould.*

2

PYCNOPTILUS FLOCCOSUS, *Gould.*
Downy Pycnoptilus.

Gould, Handbk. Bds. Aust. Vol. i., sp. 207, p. 348. *IX. 6.*

" This species has hitherto been considered a scarce bird in New South Wales, prior to which it had only been recorded from our more southern provinces. Our taxidermist, Mr. J. A. Thorpe procured some beautiful specimens in the flesh at Cambewarra, about 100 miles south of Port Jackson; and Mr. Yardley of that district has forwarded quite recently the nest and eggs taken by Mr. Sinclair, a timber-getter working in the adjacent scrubs. The nest, I am informed, was placed on or very near the ground among some débris on a bank or slope; it is a rather loose structure, built of shreds of bark chiefly, and lined with feathers of various kinds, among which may be distinguished those of the Lyre-bird, Cat bird, and some of the *Pycnoptilus* itself. In form it is somewhat dome-shaped, placed on its side and with a large, rough, ill-defined opening, which was probably narrowed by the adjacent débris among which it was placed. The eggs, two in number for a sitting, are in tint of a dark rich purplish-brown, like those of *Sericornis citreogularis*, with an indistinct zone at the larger end of a blackish tint, and a few ill-defined obsolete spots of the same on the other parts ; they are smaller and more dot-like nearer the thin end, where the ground colour is slightly lighter in tint ; they measure as follows :—(A) 1 inch x 0·75 inch ; (B) 0·95 x 0·75 inch. They are decidedly swollen and much shorter in proportion but otherwise very like the dark variety of the eggs of *Sericornis citreogularis.* Mr. A. J. North, who took a nest of this species so far back as October 1878, at Childers, in South Gippsland, and exhibited the first specimens I had seen, at the International Exhibition held in Melbourne 1880, informs me that this species was very plentiful in that district up to 1881, but the numerous clearings made by the "selectors" have since driven the birds to other parts. The egg he states shows no difference from those here described, except that some are slightly longer, and not so swollen as others." *(Ramsay, P L.S., N.S.W.,* 2nd Series, Vol. i., p. 1139.)

A set taken at South Gippsland measures, length (A) 0·97 x 0·76 inch ; (B) 0·98 x 0·75 inch.

Hab. New South Wales, Victoria and South Australia. *(Ramsay.)*

Genus SERICORNIS, *Gould.*

3. SERICORNIS CITREOGULARIS, *Gould.*

Yellow-throated Sericornis.

Gould, Handbk. Bds. Aust., Vol. i., sp. 213, p. 354. *IX. 5.*

This species was found breeding plentifully in the Richmond River district in 1866 by Dr. Ramsay, and both nest and eggs

I

have been received lately at the Australian Museum from Cambewarra. The nests which are attached to the end of the stem of some bushy bough are large, bulky, dome-shaped structures, composed of rootlets and moss, often eighteen inches in length and six wide, the opening is midway down the side and completely covered with a hood. The eggs are three in number for a sitting, large for the size of the bird, elongated, and smooth to the touch. Two of a set taken at Taranya Creek, on the 17th October 1866, are of a pale chocolate-brown, becoming lighter towards the thin end, with a well defined zone of dark umber on the larger end, the other (C) is very like the egg of *Pycnoptilus floccosus*, being of a uniform drab ground colour, minutely freckled on the larger end with blackish-brown, forming an indistinct band. On looking closely into the eggs of this species and that of *P. floccosus*, the ground colour in some specimens appears to be cracked, in fine, faint undulating rings, quite encircling the shell. Length (A) 0·96 x 0·67 inch; (B) 0·95 x 0·68 inch; (C) 1 inch x 0·68 inch.

A set taken by Mr. Yardley at Cambewarra measures, length (A) 1·01 x 0·63 inch; (B) 1·01 x 0·67 inch; (C) 1·02 x 0·68 inch.

This species breeds from August till December.

Hab. Rockingham Bay, Port Denison, Wide Bay District, Richmond and Clarence Rivers Districts, New South Wales. (*Ramsay.*)

3   SERICORNIS HUMILIS, *Gould*.

Sombre-coloured Sericornis.

Gould, Handbk. Bds. Aust., Vol. i., sp. 214, p. 356.

This bird is found in Tasmania and the islands of Bass's Straits. The nest is a large dome-shaped structure of roots, grasses, leaves, &c., warmly lined inside with feathers, and usually placed near the ground in a dense mass of vegetation. Eggs three in number for a sitting of a rich fleshy-white, becoming darker towards the arger end, where they are finely freckled with purplish- and

I—2

reddish-brown markings which in some instances assume the form of a zone. Dimensions of a set in the collection of Dr. James C. Cox are as follows:—length (A) 0·87 x 0·66 inch ; (B) 0·9 x 0·67 inch ; (C) 0·87 x 0·65 inch.

Another set gives slightly larger measurements.

Hab. Tasmania.

3.   SERICORNIS OSCULANS, *Gould.*

Allied Sericornis.

Gould, Handbk. Bds. Aust, Vol. i., sp. 215, p. 358.

The nest of this species is dome-shaped, composed of dried grasses, mosses, and wiry rootlets, warmly lined inside with feathers and other soft materials.   In South Gippsland it is usually found at the bottom of a clump of "sword grass," at other times under the shelter of a projecting fern covered bank.   During the period of incubation the female sits very close, and will almost allow itself to be captured before leaving the nest.   Eggs three in number for a sitting, varying in colour from a rich fleshy-white to light purplish-brown, minutely freckled all over with markings of a darker tint, but more particularly towards the larger end where they form a clouded zone or coalesced patch.   Length (A) 0·82 x 0·58 inch ; (B) 0·83 x 0·58 inch ; (C) 0·86 x 0·6 inch.

September and the three following months constitute the breeding season of this species.

Hab. Victoria and South Australia.   (*Ramsay*)

3.   SERICORNIS FRONTALIS, *Vigors and Horsfield.*

White-fronted Sericornis.

Gould, Handb. Bds. Aust., Vol. i., sp. 216, p. 359.   *IX.   16.*

The nest of this species is a dome-shaped structure, with a small entrance in the side, it is outwardly composed of leaves, wiry rootlets and débris, and warmly lined with feathers. It is usually placed near the ground at the bottom of a scrubby bush, or under the shelter of a tuft of grass. Eggs three in number for a sitting. A set taken by Dr. Ramsay on Ash Island in 1861, are of a faint purplish ground colour, with a well defined zone of dark purplish-brown on the larger end. Length (A) 0·78 x 0·62 inch ; (B) 0·8 x 0·6 inch ; (C) 0·76 x 0·6 inch.

Another set taken at Macquarie Fields in 1869, are much lighter in tint, and have the markings more evenly distributed over the surface of the shell ; length (A) 0·8 x 0·6 inch ; (B) 0·78 x 0·6 inch.

Hab. Wide Bay District, Richmond and Clarence Rivers Districts, New South Wales, Interior, Victoria and South Australia. *(Ramsay.)*

SERICORNIS MAGNIROSTRIS, *Gould.*
Large-billed Sericornis.

Gould, Handbk. Bds. Aust., Vol. i., sp. 219, p. 362.

This species is common in the Richmond and Clarence Rivers districts. The nest is similar to that of *S. citreogularis* and placed in like situations, often being slung in the "Lawyer vines" *Calamus australis,* which on account of their long tendrils with saw-like edges affords them ample protection. The eggs are three in number, very thin shelled, the ground colour being of a faint purplish-white, minutely flecked and marked with dark brown, in some instances all over, in others confined to the larger end where they form a well defined zone. A set of three taken by Dr. Ramsay on the 14th November 1858, measure as follows :— length (A) 0·77 x 0·58 inch ; (B) 0·77 x 0·57 inch ; (C) 0·77 x 0·58 inch.

The breeding season commences in August and lasts during the four following months.

Hab. Rockingham Bay, Port Denison, Wide Bay District, Richmond and Clarence Rivers Districts, New South Wales. (*Ramsay.*)

3. SERICORNIS MACULATUS, *Gould.*

Spotted Sericornis.

Gould, Handbk. Bds. Aust., Vol. i., sp. 218, p. 361.

Mr. George Masters procured the nests and eggs of this species while at King George's Sound, Western Australia, in December 1868. The nest is similar to those of other members of the genus being a dome-shaped structure, composed of grasses, rootlets &c., warmly lined inside with feathers, and is placed in the scrubby undergrowth near the ground. Eggs three in number for a sitting of a rich fleshy-white, minutely freckled and spotted with dark purplish- and slaty-grey markings, which are more thickly disposed towards the larger end. Length (A) 0·78 x 0·54 inch ; (B) 0·79 x 0·55 inch ; (C) 0·78 x 0·56 inch.

Hab. New South Wales, Interior, Victoria and South Australia, West and South-west Australia. (*Ramsay.*)

GENUS ACANTHIZA, *Vigors and Horsfield.*

3. ACANTHIZA PUSILLA, *Latham.*

Little Brown Acanthiza.

Gould, Handbk. Bds. Aust., Vol. i., sp. 220, p. 364.

" A lover of the scrubs and thick bushes, this species although plentiful is not so often met with as the other members of its genus. In its habits it seems to be intermediate between *Geobasileus* and the true *Acanthizæ*, being frequently seen on the ground as well as in the trees. I have never noticed it mounting high among the branches, nor does it appear to like thinly wooded districts, showing a decided preference for the brushwood and

edges of the scrubs. Upon every occasion that we have discovered
its nest it has been placed within a few inches of the ground.
One I have at present before me is suspended to the underside of
a fern *(Pteris aquilina)*, it is a closely interwoven dome-shaped
structure, in form closely resembling that of *A. lineata*, but differs ·
from it in the outside being made as rough as possible with coarse
pieces of strong bark and leaves of grasses, which hang down and
stick out from it in various directions; it is composed chiefly of
stringy-bark and the white paper-like bark of the Ti-tree, lined
with cotton-tree down and feathers; length four inches by three
inches in breadth. The eggs three in number have a pure white
ground, zoned at the larger end with freckles of light brown (in
tint duller than those of *A. lineata)*, which in some specimens are
also distributed over the rest of the surface. Its note is much
louder and more varied than that of any other species. Besides
being the foster parent of *Chalcites basalis*, and *C. plagosus*, this
species has frequently the pleasure of rearing the young of *Cuculus
flabelliformis*, three nests out of four lately found of this Acanthiza
having contained an egg of the *C. flabelliformis*." (*Ramsay, P.Z.
S.*, 1866, p. 574.)

The set of eggs described above measure as follows:—length
(A) 0·65 x 0·49 inch; (B) 0·66 x 0·49 inch; (C) 0·64 x 0·46 inch.

I found these birds building in the low ferns *(Pteris aquilina)*
at Narrabeen, New South Wales, on the 20th of June, 1888.

Hab. Wide Bay District, Richmond and Clarence Rivers
Districts, New South Wales, Victoria and South Australia.
(*Ramsay*.)

3 - 4.   ACANTHIZA DIEMENENSIS, *Gould.*
                    Tasmanian Acanthiza.
Gould, Handbk. Bds. Aust, Vol. i., sp., 221, p. 365.

This species is strictly confined to Tasmania, through the
greater part of which it is freely dispersed. The nest is a dome-
shaped structure, similar to those of the Australian species, and
is composed of roots, grasses, spiders' cocoons &c., and lined with

feathers and similar soft substances. Eggs three to four in number for a sitting, white, spotted and freckled with red and reddish-brown markings. Length (A) 0·7 x 0 52 inch ; (B) 0·67 x 0·51 inch ; (C) 0·68 x 0·53 inch. Taken near Hobart, September 1885.

This bird commences to breed in August and continues the four following months.

Hab. Tasmania. (*Ramsay.*)

3.   ACANTHIZA UROPYGIALIS, *Gould.*
Chestnut-rumped Acanthiza.
Gould, Handbk. Bds. Aust., Vol. i., sp. 222, p. 367.

For the eggs of this species I am indebted to Mr. K. H. Bennett, who procured them at Mossgiel on the 15th of October 1886. The nest, he informs me was similar to that of *Acanthiza pyrrhopygia*, and was packed between the upright stems of a thickly foliaged tree about five feet from the ground. Eggs three in number for a sitting, of a delicate fleshy-white minutely freckled all over with light reddish-brown markings, but particularly towards the larger end where they form an ill defined zone. Length (A) 0·65 x 0·5 inch ; (B) 0·65 x 0·48 inch ; (C) 0·66 x 0·48 inch. (*North, P.L.S., N.S.W.*, Vol. ii., 2nd Series, p. 407.)

Hab. New South Wales, Interior, Victoria and South Australia. (*Ramsay.*)

3 - 5.   ACANTHIZA APICALIS, *Gould.*
Western Acanthiza.
Gould, Handbk. Bds. Aust., Vol. i., sp. 223, p. 368.

" This species which is a native of Western Australia is to be met with in all wooded situations. It breeds in September and October. The nest which is usually placed in a thick foliaged bush or in a clump of the Ti-tree is of a domed form with an entrance in the side, and is composed of dried grasses and strips of Ti-tree bark, and lined with feathers. The eggs are from three to five in number, of a flesh-white, thickly freckled with reddish-chestnut, the freckles becoming so numerous at the larger end as to form a complete zone; their medium length is eight lines, and breadth six lines." *(Gould, Handbk. Bds. Aust.,* Vol. i., p. 360.)

Hab. West and South-west Australia. *(Ramsay.)*

3 ACANTHIZA PYRRHOPYGIA, *Gould.*
Red-rumped Acanthiza.
Gould, Handbk. Birds Aust., Vol. i., sp. 224, p. 369.

This bird is to be found plentifully dispersed throughout the Mallee Scrub in the Wimmera District of Victoria. For the nest and eggs of this species together with the bird, I am indebted to Mr. James Hill of "Pine Rise," Kewell, Victoria. The nest is a dome-shaped structure, having a small entrance in the side, outwardly composed of strips of fine bark, grasses, roots, &c., lined inside with feathers and a little wool, it is usually placed in the drooping leaves of a Casuarina or Eucalyptus, about six feet from the ground. Eggs three in number for a sitting, of a fleshy-white thickly freckled with reddish-chestnut and reddish-brown markings chiefly towards the larger end, where they become confluent and assume the form of an ill-defined zone, some have the markings almost confined to a coalesced patch on the larger apex. Length (A) 0·65 x 0·47 inch; (B) 0·63 x 0·48 inch; (C) 0·62 x 0·45 inch.

Hab. Interior, Victoria and South Australia. *(Ramsay.)*

ACANTHIZA INORNATA, *Gould.*
Plain-coloured Acanthiza.
Gould, Handbk. Bds. Aust., Vol. i., sp. 225, p. 370.

This bird is found in the southern portions of Western and South Australia, being particularly abundant in the neighbourhood of King George's Sound in the former colony, where Mr. Masters succeeded in obtaining a number of specimens during 1868, likewise the nest and eggs. A nest of this species now before me taken from the Australian Museum Collection, is a dome-shaped structure composed of the dried wiry stems of a *Drosera*, and the flowering portions of the *Banksia* cones, spiders' webs, &c., all matted up together, and lined inside with the white downy seeds of some composite plant. It measures exteriorly four inches and a quarter in height, by three inches in width, the aperture which is oval and near the top being one inch high by one inch and a-quarter in width. The nest is firmly packed in the upright branches of a *Banksia*, and was placed about five feet from the ground; it contained two eggs of a fleshy-white ground colour, freckled all over with irregular shaped markings of reddish-brown, particularly towards the larger end, where they form a well defined zone. Length (A) 0·7 x 0·52 inch; (B) 0·69 x 0·52 inch. *(North, P.L.S., N.S.W.,* Vol. ii., 2nd Series p. 406.)

Hab. Victoria and South Australia, West and South-West Australia. *(Ramsay.)*

3   ACANTHIZA NANA, *Vigors and Horsfield.*

Little Acanthiza.

Gould, Handbk., Bds. Aust., Vol. i., sp. 226, p. 371. XIII /6.

"The nest of this species is not by any means as neat a structure as that of *A. lineata,* it is moreover placed in situations quite different. It is of an oblong form and placed among the topmost twigs of some bushy scrub, composed of thin shreds of stringy-bark and grasses, and often beautifully decorated with green mosses and lichens, and lined with native cotton-tree down, feathers, or fine grasses. The entrance, which is about one inch and a-half from the top, having its edges but roughly finished off and not covered by any hood, is one inch in width. The Yellow Acanthiza shows a decided preference for the tops of the native Ti-trees, but its nest may also be found in various other trees and shrubs, but always placed among the outside twigs. We have taken nests from a species of Acacia overhanging the creeks and rivers. Sometimes they are wholly composed of fine strips of stringy-bark, which when new give them a reddish-brown appearance. At other times they are composed of dry grass, a great quantity of white cobweb being used in all cases. The total length of the nest of *A. nana* is three inches by two inches and a-half in breadth, being somewhat narrower at the bottom. The eggs are three in number, from 0·6 to 0·7 inch in length, and 0·4 inch in breadth, strongly blotched, dotted, or freckled with dark dull reddish-brown inclining to chocolate in some, to red in others, and having a few dots of dull lilac towards the larger end. In some specimens the markings form a zone on the thick end, in others they are equally dispersed over the whole surface and take the form of irregular blotches. The birds may be found breeding in September and the three following months, and are frequently the foster parents of *C. plagosus* and *C. basalis.*" (*Ramsay, P.Z.S.*, 1866, p. 573.)

A set in the Australian Museum Collection give the following dimensions. Length (A) 0·63 x 0·46 inch ; (B) 0·67 x 0·47 inch ; (C) 0·63 x 0·45 inch.

Hab. Wide Bay District, Richmond and Clarence Rivers Districts, New South Wales, Victoria and South Australia. (*Ramsay.*)

3   ACANTHIZA LINEATA, *Gould.*

Lineated Acanthiza.

Gould, Handbk. Bds. Aust., Vol. i., sp. 227, p. 372.

"The nest of the Lineated Acanthiza is one of the most beautiful of those of our Australian birds. It is a neat, oval, compact, and remarkably strong structure, in length four inches and a-half to five inches by three inches through, composed of fine shreds of stringy bark closely interwoven, and frequently ornamented with pieces of white spiders' nests. It is warmly lined with feathers, opossum-fur or the silky down from the seed-pods of the native cotton-tree. The nest is suspended to a thin twig at the end of some leafy bough by the top, and the small opening about two inches down the side is neatly covered with a hood, which excludes both the sun and the rain. Some of the nests are without any ornament; others are decorated with pieces of white paper-bark, or with green and white spiders' nests. Long streamers of bleached seaweed are also often used; and when the nests are placed in the gullies of the ranges, a beautiful bright green string-like *Hypnum* is employed.

We find this species of Acanthiza usually the first to commence breeding. I have taken its eggs in July, but for the most part find them from August to September. They are three in number rather long and of a beautiful pinky-white zoned at the larger end with minute freckles and irregular markings of a light brownish-red, having also a few minute linear dashes of the same colour over the rest of the surface. The zone at the tip of the larger end is extremely characteristic; a few specimens are found without it, but some which I believe to be the eggs of young birds breeding for the first time, are of a pure white without any markings whatever. The average length is 0·7 inch by 0·5 inch in breadth. This species has two and sometimes three broods in the year, stragglers breeding as late as December and January, and is perhaps more frequently the foster parent of *Chalcites plagosus* and *C. basalis*, than any other species." (*Ramsay, P.Z.S.*, 1866, p. 571, Pl. of nest, p. 572.)

Eggs of this species in the Australian Museum Collection measure as follows :—length (A) 0·62 x 0·47 inch ; (B) 0·67 x 0·47 inch ; (C) 0·65 x 0·46 inch.

Hab. Wide Bay District, Richmond and Clarence Rivers Districts, New South Wales, Victoria and South Australia. (*Ramsay.*)

Genus GEOBASILEUS, *Cabanis.*

¥ GEOBASILEUS REGULOIDES, *Vigors and Horsfield.*

Buff-rumped Geobasileus.

Gould, Handbk. Bds. Aust., Vol. i., sp. 230, p. 376.

" Little or no preference seems to be shown in the selection of a site for the nest of this bird. It is a dome-shaped structure, having a small entrance in the side and composed of grasses and stringy bark &c., lined with feathers, cotton-tree down, or opossum-fur. It is placed in a tuft of grass, or low bushy shrub, but just as often among the loose pieces of bark which having accumulated in the forks of the Eucalypti hide all except the entrance of the nest. A hole morticed in the side of a post and the fork of a Ti-tree where rubbish has accumulated alike serve its purpose, the shape depending upon the position chosen. The nests resemble those of the *Malurus cyaneus* both in size and shape ; they are however, much more bulky, thicker, and have a great quantity of lining, which renders them much warmer and more comfortable. The eggs which may be taken from August to December, are four in number 0·6 to 0·7 inch in length by 0·4 to 0·5 inch broad, having a delicate white ground colour, spotted, freckled, or dashed with markings of reddish-brown of various tints and a few of purplish-lilac-brown, in most forming a zone at the larger end ; the eggs of the young breeding for the first or second time are white without any markings. This species has three broods during the season, and if the nest be taken will frequently build another in the same place." *(Ramsay, P.Z.S.,* 1866, p. 575.)

Three eggs of this bird in the Australian Museum Collection give the following measurements :—length (A) 0·67 x 0·47 inch ; (B) 0·65 x 0·46 inch ; (C) 0·62 x 0·45 inch.

Hab. Wide Bay District, Dawson River, Richmond and Clarence River Districts, New South Wales, Interior, Victoria and South Australia. *(Ramsay.)*

3-4. GEOBASILEUS CHRYSORRHŒA, *Quoy et Gaimard.*

Yellow-rumped Geobasileus.

Gould, Handbk. Bds. Aust., Vol. i., sp. 229, p. 374.

"I found this bird one of the most common upon the banks of the Hunter River, also in the Wellington and Lachlan districts. Its nest is a bulky, rough, oblong structure, composed of grasses and strips of bark interwoven in a loose ragged manner, with a little cobweb and wool ; it is lined with feathers and fine grasses. The entrance is about half-way down the side, with rounded and thickened edges but without any hood. The most peculiar characteristic of the nest is a cup-shaped framework placed upon the top (often a little to the one side), as if formed for the commencement of another nest ; this I found is made when the framework of the true nest is formed ; but I believe it is added to after the nest is lined and while the bird is still laying. The whole structure is eight inches high by four wide, the framework on top being two inches by three wide. The breeding season commences sometimes as early as July and ends in December during which time three broods are often reared ; the most usual months are from August to November. Three or four eggs are the number laid for a sitting : they are of a beautiful pure-white colour, having brownish-red dots, centred with a deeper hue and sprinkled over the surface or forming an indistinct zone upon the larger end. Eggs of this species are often found without any markings whatever.* Length 0·67 inch by 0·5 inch in breadth. Almost any bushy tree or bough affords a safe place for the nest of this species : the ends of mangrove boughs overhanging a stream or even those of the Casuarina, the branches of the Ti-trees as well as orange trees are resorted to. The birds may frequently be found in the gardens and orchards, and not unfrequently hopping over the roofs of the houses." (*Ramsay, P.Z.S.*, 1866, p. 575.)

A set in the Australian Museum Collection measure as follows : length (A) 0·68 x 0·48 inch ; (B) 0·67 x 0·48 inch ; (C) 0·7 x 0·49 inch.

---

* In Victoria the latter variety is the rule.

Hab. Rockingham Bay, Port Denison, Wide Bay District, Dawson River, Richmond and Clarence Rivers Districts, New South Wales, Interior, Victoria and South Australia, Tasmania, West and South-West Australia. (*Ramsay*.)

Genus ORIGMA, *Gould*.

ORIGMA RUBRICATA, *Latham*.

Rock Warbler.

Gould, Handbk. Bds. Aust., Vol. i., sp. 236, p. 385. XIII. 8.

"This bird may always be found in the neighbourhood of gullies and ravines, especially where there is running water. It seems to give preference to the rocky sides of steep gullies, where it may be seen running over the rocks uttering its shrill cry, entering into the crevices under the low shelving rocks, and reappearing again many yards in advance. It is a very pleasing and lively little bird and seems to love solitude. I have never seen it perch upon a tree, although I have spent several evenings in watching it. It runs with rapidity over the ground and over heaps of rubbish left by the floods, where it seems to get a good deal of its food. Sometimes it will remain for a minute on the point of a rock, then as if it were falling over the edge, repeat its shrill cry, and dash off again into some hole in the cliffs. The nest is of an oblong form very large for the size of the bird, with an entrance in the side about two inches wide. It is generally suspended under some overhanging rock and is composed of fibrous roots interwoven with the web of spiders, the birds having a preference for those webs which contain the spiders' eggs, and that are of a greenish colour. The mass does not assume the shape of a nest until a few days before it is completed, when a hole for entrance is made and the inside warmly lined with feathers ; however, even when finished it is a very ragged structure and easily shaken to pieces. The birds take a long time building their nests : one I found on the 6th August 1861, was not finished until the 25th of the same month ; on the 30th we took three eggs from it. This nest was suspended from the roof of a small cave in the gully of George's River, near Macquarie Fields, and was composed of rootlets and spiders' webs warmly lined with feathers and opossum-fur, it contained three eggs of a pure and glossy-white, each egg being eight and a-half lines in length by six and a-half lines in breadth. Sometimes the eggs are nine but more often eight and eight and a-half lines long. They are very similar in appearance to those of Latham's Grass-Finch, *Amadina lathami.* The breeding time lasts from August to December, during which time two broods are raised." *(Ramsay, Ibis,* 1863, Vol. v., p. 445.)

A set of three in the Australian Museum Collection, taken at Middle Harbour, Sydney, measures as follows :—length (A) 0·8 x 0·6 inch ; (B) 0·79 x 0·62 inch ; (C) 0·78 x 0·6 inch.

Hab. Wide Bay District, New South Wales. *(Ramsay)*

Genus EPHTHIANURA, *Gould.*

3 EPHTHIANURA TRICOLOR, *Gould.*

Tri-coloured Ephthianura.

Gould, Handbk. Bds. Aust., Vol. i., sp. 233, p. 380. ✗III. /2

" The nest is of fine grasses lined with fine rootlets and a few hairs, it is cup-shaped, two inches in diameter inside and two inches deep, and was placed in a wind-bent tuft of coarse grass, the sides of the nest were hidden by the tops of grasses stuck in perpendicularly round the rim, hanging over it in some places and forming a more secure framework all round. The eggs were three in number, of a pure white with rich clear red dots sprinkled over the surface a little closer together at the thick end, but not forming a zone there. Length (A) 0·63 x 0·5 inch ; (B) 0·65 x 0·5 inch." *(Ramsay, P.L.S., N.S.W.,* Vol. vii., p. 48.)

A set taken by Mr. K. H. Bennett at Mossgiel, New South Wales, on the 24th October 1883, measures as follows:—length (A) 0·63 x 0·51 inch ; (B) 0·65 x 0·51 inch ; (C) 0·66 x 0·52 inch.

Hab. New South Wales, Interior, Victoria and South Australia. (*Ramsay.*)

EPHTHIANURA AURIFRONS, *Gould.*

Orange-fronted Ephthianura.

Gould, Handbk. Bds. Aust., Vol. i., sp. 232, p. 380.

" The nest similar to that of the last species ; a round open cup-shaped structure made of fine twigs and grasses—the one before me has the feather of an Emu worked into the side, and is lined with fine grass—the inside diameter two inches, depth one inch, and was placed in a low bush. The eggs white with small red dots, sometimes confined to the thicker end ; length 0·7 x 0·52 inch ; 0·6 x 0·5 inch." (*Ramsay, P.L.S., N.S.W.,* Vol. vii., p. 48.)

A set taken by Mr. K. H. Bennett, measures :—length (A) 0·67 x 0·5 inch ; (B) 0·68 x 0·5 inch ; (C) 0·67 x 0·51 inch.

Hab. New South Wales, Interior, Victoria and South Australia. (*Ramsay.*)

EPHTHIANURA ALBIFRONS, *Jardine and Selby.*

White-fronted Ephthianura.

Gould, Handbk. Bds. Aust., Vol. i., sp. 231, p. 377.   XIII. //.

" These birds arrive in the vicinity of Sydney about the beginning of September and October. In the latter month they commence to build ; for this purpose they choose some open land studded with low bushes. The stunted *Bursariæ*, the prickly twigs of which are often used to form the framework of their nests seem their favourite building-places. The nests are usually situated a few inches from the ground and are cup-shaped, and placed upon a strong framework of twigs, and neatly lined with grass, hair, &c. I have frequently found them among the dead leafy tops of a fallen Eucalyptus which has been left by the wood-cutters when clearing a piece of new ground. The eggs of this bird are usually three, but sometimes four in number, from six to seven lines long by five broad, beautifully white, some spotted, and others irregularly marked with bright deep reddish-brown at the larger end, where in some the spots form an indistinct zone. In other specimens the spots are crowded at the top and very sparingly sprinkled on the other parts of the egg. These birds easily betray the position of their nest or young by their anxiety and attempts to draw one from the spot by feigning broken wings and by lying struggling upon the ground as if in a fit. They have two broods (and perhaps more) in the year, after which the young accompany the parent birds to feed, generally on the salt marshy grounds near the water's edge. About Botany and the Parramatta River, they are plentiful." (*Ramsay, Ibis,* 1863, Vol. v., p. 178.)

I have found as many as twenty nests of this species in a day among low ferns *(Pteris aquilina)* growing near the mouth of the Yarra in Victoria. Measurements of two sets of eggs are as follows :—length No. 1 (A) 0·65 x 0·52 inch; (B) 0·67 x 0·5 inch; (C) 0·68 x 0·49 inch. No. 2 (D) 0·66 x 0·33 inch; (E) 0·63 x 0·43 inch; (F) 0·61 x 0·43 inch.

Hab. Wide Bay District, Richmond and Clarence Rivers Districts, New South Wales, Interior, Victoria and South Australia, West and South-West Australia. (*Ramsay.*)

Genus PYRRHOLÆMUS, *Gould.*

PYRRHOLÆMUS BRUNNEUS, *Gould.*
Red-throat.

Gould, Handbk. Bds. Aust., Vol. i., sp. 235, p. 384.

" This is a remarkable species and peculiar in the colour of its eggs, the nest is very similar to that of a *Malurus,* it is composed

J

wholly of grasses loosely thrown together without being interwoven more than is necessary to keep them in their place ; the structure would hardly bear removal ; the lining is of hair or fur of the " Rabbit-rat " *Lagorchestes*, it is five inches in diameter by three and a-quarter across outside, with no hood over the opening ; the structure was placed on its side among the twigs of a small shrub with grass growing through its branches near the ground and hidden by the grass. The eggs are of a dull olive-brown, of a nearly uniform bronze tint, usually without markings, one specimen has an indistinct ring of minute dots on the larger end forming a patch of a darker shade, the eggs are three or four in number, length 0·78 x 0·59 inch ; 0·78 x 0·58 inch ; 0·79 x 0·58 inch." *Dobr. Mus.* (*Ramsay, P.L.S., N.S.W.*, Vol. vii., p. 49.)

Hab. New South Wales, Interior, Victoria and South Australia, West and South-West Australia. (*Ramsay.*)

Genus MEGALURUS, *Horsfield.*

ʲ.   MEGALURUS GRAMINEUS, *Gould.*

Little Grass Bird.

Gould, Handbk. Bds. Aust., Vol. i., sp. 245, p. 400.

This little bird was at one time very common in the vicinity of Melbourne, frequenting and breeding freely in the tufts of rushes that skirt the edges of the sheets of water in the Government Domain, Botanical Gardens, and the southern portion of the Albert Park lake, also in the *Melaleuca* scrub that formerly clothed the sides of the Lower Yarra. The site chosen for the nest is somewhat varied, I have usually taken it from the bottom of a clump of long rushes within eight inches of the water but not unfrequently in the upright pronged fork of a *Melaleuca*, about five feet from the ground, when growing in wet and swampy localities.

J—2

A nest of this species, in the Australian Museum Collection presented by Dr. Hurst together with its eggs, and taken by that gentleman on the 16th of October 1886, at Newington on the Parramatta River, was built in the forked branches of a young mangrove, it is a deep cup-shaped structure, the opening at the top being narrowed, and is outwardly composed of long fine twigs, lined with dried grasses and a few feathers, it measures six inches in height by four inches in breadth, and two inches and a-half in depth. It is a much longer nest than those built in the rushes. Eggs four in number for a sitting, in form lengthened ovals, of a reddish-white ground colour finely freckled all over with purplish-red markings ; length (A) 0·73 x 0·52 inch ; (B) 0·76 x 0·53; (C) 0·75 x 0·54 inch. A set taken by myself on the 8th of October 1876, in the Albert Park near Melbourne, are much darker and more heavily marked, the ground colour at the larger end of the eggs being almost obscured with deep reddish-brown and lilac markings, the latter colour appearing as if beneath the surface of the shell. Length (A) 0·68 x 0·5 inch ; (B) 0·72 x 0·5 inch ; (C) 0·75 x 0·52 inch ; (D) 0·68 x 0·52 inch.

September and the two following months constitute the breeding season of this species.

Hab. Port Denison, Wide Bay District, Richmond and Clarence River Districts, New South Wales, Interior, Victoria and South Australia, Tasmania, West and South-West Australia. (*Ramsay.*)

Genus CALAMANTHUS, *Gould.*

ɔ - ʮ   CALAMANTHUS FULIGINOSUS, *Vigors and Horsfield.*

Striated Calamanthus.

Gould, Handbk. Bds. Aust. Vol. i., sp. 237, p. 388.

This bird is confined to Tasmania, the manner of its nidification is similar in every respect to the following species *C. campestris.* Eggs three or four in number for a sitting, of a light chocolate-brown thickly freckled all over with markings of a darker tint, but particularly towards the larger end ; specimens in my own collection taken near Hobart in August 1882 measure as follows : Length (A) 0·85 x 0·64 inch; (B) 0·87 x 0·63; (C) 0·85 x 0·67 inch.

Four eggs in Dr. James C. Cox's Collection give the following measurements :—Length (A) 0·84 x 0·63 inch ; (B) 0·89 x 0·65 inch ; (C) 0·85 x 0·67 inch ; (D) 0·87 x 0·62 inch.

Hab. Tasmania.

CALAMANTHUS CAMPESTRIS, *Gould.*
Field Calamanthus.

Gould, Handbk. Bds. Aust., Vol. i., sp. 238, p. 389.

This species is found breeding in the neighbourhood of Melbourne, and its nest was one of the first taken by me during my early collecting days. It is without exception the first of all birds to commence breeding in Victoria, starting to build before the winter has commenced, and rearing its young through the coldest months of the year.   June and July are the principal months for obtaining the eggs of this species and I have known them taken as early as the 24th of May ; on the 17th of June 1880, I found four nests of this species each containing three fresh eggs, which is the usual number laid by this bird for a sitting. The situation chosen for the nest is somewhat varied, sometimes being placed underneath a tuft of rank grass but more often have I found it artfully concealed at the bottom of a low, stunted, thick shrub growing in wet and swampy ground at the mouth of the Yarra. The nest is rounded in form, composed of grasses, and lined with feathers, and usually one or two projecting from the entrance, the nests found at the mouth of the Yarra were all composed exteriorly of an aquatic weed ; the bird at all times sits very close, and it is only when the bush is pulled open that the bird will leave it, which is the easiest way of finding the nest of this species. On one occasion, when the nest was built in the grass, the bird allowed itself to be trodden upon before leaving its eggs, which were in an advanced state of incubation. Eggs oval in form varying in tint from a light chestnut to pale chocolate-brown, finely freckled all over with nearly invisible markings of a darker tint, particularly towards the larger end where in some instances they form a perfect zone. Dimensions of a set taken 17th June 1880, length (A) 0·81 x 0·58 inch ; (B) 0·78 x 0·57 inch; (C) 0·8 x 0·58 inch.

A set taken at Albert Park, Melbourne, during July 1875 measure as follows :—length (A) 0·77 x 0·58 inch ; (B) 0·79 x 0·6 inch; (C) 0·8 x 0·6 inch.

Hab. Victoria and South Australia, West and South-west Australia. (*Ramsay.*)

Genus CHTHONICOLA, *Gould.*

2 -4   CHTHONICOLA SAGITTATA, *Latham.*

Little Chthonicola.

Gould, Handbk. Bds. Aust., Vol. i., sp. 239, p. 390.

This species constructs a dome-shaped nest, well concealed underneath a tuft of overhanging grass ; it is built throughout of dried grasses. Eggs in form swollen ovals and four in number for a sitting, of a uniform bright chocolate-red. Dimensions of a set taken at Macquarie Fields in October 1860, by Dr. Ramsay : length (A) 0·74 x 0·6 inch ; (B) 0·76 x 0·6 inch; (C) 0·76 x 0·61 inch; (D) 0·76 x 0·59 inch.

A set taken at Oakleigh, Victoria, November 1879 measure as follows :—length (A) 0·75 x 0·59 inch ; (B) 0·76 x 0·6 inch ; (C) 0·76 x 0·61 inch.

In New South Wales this bird is often the foster parent of *C. flabelliformis.* The breeding season commences in September, and continues the three following months.

Hab. Wide Bay District, Dawson River, Richmond and Clarence Rivers Districts, New South Wales, Victoria and South Australia. (*Ramsay.*)

Genus XEROPHILA, *Gould.*

5.   ## XEROPHILA LEUCOPSIS, *Gould.*
White-faced Xerophila.

Gould, Handbk. Bds. Aust., Vol. i., sp. 234, p. 382. *IX* *14.*

" The eggs of this species have been unfortunately described as being white by Mr. Gould ; that many of our Australian birds lay eggs other than of the normal colour must be well known to all Australian Oologists, who are not unfrequently a little puzzled at getting eggs of the same species totally different from one another, nevertheless I believe the eggs described by Mr. Gould as those of this species really belong to *Geobasileus chrysorrhœa.* The eggs of the present species as shown by numerous authentic examples taken by Mr. James Ramsay and Mr. K. H. Bennett, are of a dull white thickly spotted and freckled all over with reddish-brown, dull chocolate-brown, or dark wood-brown; in some specimens the whole of the ground colour is obscured by reddish-brown freckles, others have a zone of confluent spots of dark blackish-brown on the larger end and only a few dots or freckles on the remaining portion ; average length 0·7 x 0·53 inch. The nests vary in structure according to the situation chosen, some being neat and compact placed among the twigs of some low shrub, others which are more commonly placed in the hollow branches of trees or holes in the sides of dead trees or posts are rather scanty ; all are composed of grasses and lined with feathers, wool, hair, &c." (*Ramsay, P.L.S., N.S.W..* Vol. vii., p. 407.)

Eggs five in number for a sitting. A set taken by Mr. James Ramsay at Tyndarie, measures as follows :—length (A) 0·72 x 0·54 inch ; (B) 0·74 x 0·55 inch ; (C) 0·71 x 0·54 inch ; (D) 0·71 x 0·54 inch ; (E) 0·73 x 0·55 inch.

Hab. New South Wales, Interior, Victoria and South Australia, West and South-West Australia. (*Ramsay.*)

Genus ORTHONYX,

ORTHONYX SPINICAUDUS, *Temminck.*

(*O. temminckii,* Vig. & Horsf.)

Spine-tailed Orthonyx.

Gould, Handbk. Bds. Aust., Vol. i., sp. 372, p. 607.

" Mr. Gould in his Handbook has already described the nest of this species. Nests obtained by my collectors in the Richmond River Scrubs in 1865-66 were all placed on the ground at the base or between the "buttresses" of trees, and composed of mosses and débris of leaves, &c. Eggs white, large comparatively, 1·13 x 0·85 inch." (*Ramsay, P.L.S., N.S.W.,* Vol. i., New Series, p. 1148.)

Nests which I have examined are dome-shaped, having an entrance at the side, composed of fallen leaves and mosses (*Hypnum*), they are usually placed on or near the ground. Eggs of this species have been taken in June, also in the month of December.

Specimens in the Australian Museum Collection give the following dimensions :—length (A) 1·13 x 0·84 inch; (B) 1·16 x 0·86 inch.

Hab. Wide Bay District, Richmond and Clarence Rivers Districts, New South Wales, Victoria and South Australia. (*Ramsay.*)

Genus CINCLOSOMA, *Vigors and Horsfield.*

₂ - ₃ CINCLOSOMA PUNCTATUM, *Latham.*

Spotted Ground Thrush.

Gould, Handbk. Bds. Aust., Vol. i., sp. 271, p. 433. *XI. 10.*

This species is found in scrubby, and lightly timbered country with a slight undergrowth. The nest is a round, open, and rather loosely built structure composed of fine strips of bark, grasses and leaves, placed under the shelter of a bush, or clump of ferns

Pteris aquilina, generally.in close proximity to a fallen log. Eggs two or three in number for a sitting, usually the former, varying considerably in their size and markings. A set of two taken at Hastings, Victoria, in September 1883, are of a dull white, freckled and blotched with wood-brown and umber-brown markings, while appearing underneath the surface of the shell are large clouded blotches and spots of dark lilac, which are more numerous than the markings, on the outer surface.   Length (A) 1·31 x 0·92 inch; (B) 0·27 x 0·9 inch.

A set of three taken by Mr. James Ramsay at Merule Creek, are of a dull white thickly freckled with wood-brown markings intermingled with others of bluish-grey, appearing underneath the surface of shell, more particularly towards the larger end ; one specimen (A) is elongated in form and is almost devoid of markings with the exception of a few very minute spots of umber brown scattered over the surface of the shell.   Length (A) 1·27 x 0·82 inch ; (B) 1·2 x 0·87 inch ; (C) 1·2 x 0·85 inch.

Young specimens of this bird were obtained at Heathcote, on the Illawarra Line on October 30th 1886.   This species commences to breed early in August and continues through the three following months.

Hab. Wide Bay District, Richmond and Clarence Rivers Districts, New South Wales, Interior, Victoria and South Australia. (*Ramsay.*)

Genus CINCLORAMPHUS, *Gould.*

3   CINCLORAMPHUS CRURALIS, *Vigors and Horsfield.*

Brown Cincloramphus.

Megalurus cruralis, Vig. & Horsf. in Linn. Trans., Vol. xv., p. 228.

Cincloramphus cruralis, Gould, P.Z.S., v., p. 150, id Bds. of Aust. Handbk., sp. 241.

Cincloramphus cantillans, Gould, Bds. of Aus., folio Vol. iii., pl. 75.

Gould, Handbk. Bds. Aust., Vol i. sp. 241, p. 395.

" The nest of this species is placed upon the ground beside some tuft of long grass or among the dead leafy tops of trees which have been felled.   It is composed often of stringy-bark alone, at other times with grass, fibrous rootlets and the like.   The eggs of this species have a white ground, spotted, blotched, or minutely freckled with bright reddish-brown or salmon colour : their number is usually three." (*Ramsay, Ibis*, 1866, Vol. ii., New Series, p. 328.)

Eggs of this species taken in Victoria have the ground colour entirely obscured by very fine salmon coloured freckles and spots, in some instances forming a zone at the larger end.   Length (A) 0·93 x 0·71 inch ; (B) 0·97 x 0·7 inch ; (C) 0·9 x 0·7 inch.

A set taken by Mr. James Ramsay at Tyndarie, are evenly freckled all over, length (A) 0·95 x 0·64 inch ; (B) 0·95 x 0·62 inch ; (C) 0·95 x 0·65 inch.

Hab. Rockingham Bay, Port Denison, Wide Bay District, Richmond and Clarence Rivers Districts, New South Wales, Interior, Victoria and South Australia   (*Ramsay.*)

4.   CINCLORAMPHUS RUFESCENS, *Vigors and Horsfield.*

Rufous-tinted Cincloramphus.

Gould, Handbk. Bds. Aust., Vol. i., sp. 243, p. 397.

The nest of this species is placed in a hollow scraped in the earth at the side of a tuft of grass, or amongst the withered leaves of a fallen tree, and on several occasions I have taken it from underneath the overhanging bank of a dry creek.   The nest is cup-shaped, and is composed of dried grasses, lined with hair.   Eggs four in number for a sitting, in form pointed ovals, varying a great deal in their tints, and in the disposition of their markings.

A set taken at Nanama, near Yass, New South Wales, by Mr. James Ramsay, in November 1866, have the entire surface finely but thickly freckled with chestnut and purplish-brown markings. Length (A) 0·85 x 0·6 inch ; (B) 0·84 x 0·6 inch ; (C) 0·84 x 0·61 inch ; (D) 0·87 x 0·63 inch.

A set taken by myself at Toorak, near Melbourne, Victoria, in November 1873, are of a purplish-white ground colour, thickly freckled, spotted, and blotched with deep reddish-brown, but particularly towards the larger end, where a few obsolete spots of lilac appear as if beneath the surface of the shell. Length (A) 0·86 x 0·63 inch ; (B) 0·84 x 0·63 inch ; (C) 0·85 x 0·62 inch ; (D) 0·84 x 0·61 inch.

This species commences to breed at the latter end of September and continues till the middle of January. I have frequently found the young of this species just fledged at the end of December.

Hab. Derby, N.W. Australia, Port Darwin and Port Essington, Gulf of Carpentaria, Cape York, Rockingham Bay, Port Denison, Wide Bay District, Dawson River, Richmond and Clarence Rivers Districts, New South Wales, Interior, Victoria and South Australia, West and South-West Australia. *(Ramsay.)*

Genus POMATOSTOMUS, *Cabanis*.

POMATOSTOMUS TEMPORALIS, *Vigors and Horsfield*.

Temporal Pomatostomus.

Gould, Handbk. Bds. Aust., Vol. i., sp. 292, p. 479. *II. 7.*

"This species breeds chiefly in September, October, and November making a large coarse nest of twigs slightly interwoven, the lower part is much rounded, the upper rather elongated and drawn out into a neck, the back twigs being brought forward so as almost completely to hide the small opening, which has as it were a thatch of twigs over its entrance. Very often too, the twigs from the lower side project upwards, rendering it seemingly, almost impossible for the bird to enter without disarranging them. It is lined with a great quantity of grass or stringy bark, with which the eggs are frequently covered when the birds leave their nests. The top of some bushy tree or the end of some thickly branched bough are the sites chosen for the nests, which when in the former situations are placed nearly upright, but when in the latter upon their sides, being built of course to suit the boughs in which they are placed. Several nests may be found within a few yards of each other in the same clump of trees with birds sitting in each of them. The number of eggs in a nest varies from five to ten. My brother Mr. James Ramsay, informs me that he has taken no less than fourteen from one nest, and in these cases believes them to be the joint property of several birds ; the usual number however is five, which are either much elongated or rounded in form, and not unfrequently have the ends of equal thickness ; the medium size is one inch in length by nine lines in breadth. The ground colour is brownish, yellowish, or purplish-buff covered with a most peculiar network of veins and hair-lines running in various directions, both across and round the surface ; these lines are of a dark purplish-brown. The colouring matter has the peculiarity of being easily rubbed off." (*Ramsay, Proc. Phil. Soc., Sydney*, 1865, p. 316, pl. i., fig. 1.)

A set in the Australian Museum Collection measures as follows: length (A) 1·03 x 0·73 inch ; (B) 1·05 x 0·73 inch ; (C) 1·04 x 0·76 inch ; (D) 1·04 x 0·73 inch ; (E) 1·07 x 0·75 inch.

I found this species breeding in great numbers on the Bell and Macquarie Rivers during August 1887.

Hab. Gulf of Carpentaria, Rockingham Bay, Port Denison, Wide Bay District, Richmond and Clarence Rivers Districts, New South Wales, Interior, Victoria and South Australia. (*Ramsay*.)

3-5. ## POMATOSTOMUS RUBECULUS, *Gould*.

Red-breasted Pomatostomus.

Gould, Handbk. Bds. Aust , Vol. i., sp. 293, p. 481.

"Nest flask-shaped, of thin twigs and sticks interwoven, lined with fine grasses, shreds of bark and sometimes a few feathers ; it is placed at the end of some bushy branch or among thick upright twigs, and is very similar to that of *P. temporalis* as described by

Mr. Gould (Handbk., I., p. 479). The eggs three to five in number, are of a yellowish-brown tint, some with the ground colour of a somewhat saturnine hue almost obscured by hair lines and veins of blackish-sienna or of a blackish-chocolate colour ; they vary considerably in tint, some have fleecy cloud-like markings and but few hair lines, some are pointed in form, others oblong with both ends almost equal. Length 1·05 x 0·75 inch, oblong ; 1·07 x 0·74 inch, pointed ; 1·02 x 0·7 inch, rounded. They breed during September and the three following months. *Mr. Barnard's Collection." (Ramsay, P.L.S., N.S.W.*, Vol. vii., p. 46.)

Hab. Derby, N.W. Australia, Port Darwin and Port Essington, Gulf of Carpentaria, Dawson River. *(Ramsay.)*

POMATOSTOMUS SUPERCILIOSUS, *Vigors and Horsfield.*

White-eyebrowed Pomatostomus.

Gould, Handbk. Bds. Aust., Vol. i., sp. 294, p. 482.

" The nest is similar to that of *P. temporalis* but smaller, and has the entrance more completely covered by a thatch of twigs. The eggs are three to five in number, their usual length is ten and a-half or eleven lines by seven and a-half to eight lines in breadth, some are rounded in form others more elongated. The ground colour is of a brownish-grey tinged with olive, clouded with purplish brown and greyish-olive and sparingly veined with dark bistre. some specimens are of a uniform dull greyish-olive-brown, clouded with a deeper hue and without veins, and have a clouded band round the centre. Like the foregoing species, this is frequently found upon the ground hopping about with the greatest agility under the trees, especially during the early part of the day ; when flushed they fly off to the nearest tree and commence to ascend it by a series of hops and jumps until they reach the end of the boughs from which they fly off in a string. They are very sprightly and quick in their movements, and have the peculiarity of drawing their heads in and puffing out their feathers as they ascend the branches, looking like a number of brown balls bouncing among the limbs. This species has a wide range of habitat, being found equally common on the Darling, Lachlan, Murray, and Bell Rivers, as well as over the whole southern portion of the country and in Western Australia. Upon the Bell River and near the Lachlan, I found them very plentiful in company with the *P. temporalis*, and have frequently found several nests of both species built in the same clump of trees, for which purpose they show a preference to the thick bushy tops of a species of Acacia allied to the ' Myall '." (*Ramsay, Proc. Phil. Soc., Sydney*, 1865, p. 318, pl. i., fig. 2.)

A set in the Australian Museum Collection measure (A) 0·93 x 0·63 inch ; (B) 0·9 x 0·64 inch ; (C) 0·89 x 0·65 inch ; (D) 0·91 x 0·63 inch ; (E) 0·93 x 0·62 inch.

Hab. Port Darwin and Port Essington, Gulf of Carpentaria, Rockingham Bay, Port Denison, Wide Bay District, New South Wales, Interior, Victoria and South Australia, West and South-West Australia. (*Ramsay.*)

POMATOSTOMUS RUFICEPS, *Hartlaub.*

Chestnut-crowned Pomatostomus.

Gould, Handbk. Bds. Aust , Vol. i., sp. 295, p. 484.

" Nest similar to that of *P. temporalis*. Eggs a little smaller, five in number. In several the ground colour has a very faint tinge of green, the blackish hair lines are finer and closer together in some nearly obscuring the ground colour, others have a pinkish-chocolate tinge. Length 0·95 x 0·72. *Dobr. Mus.*" (*Ramsay, P.L.S., N.S.W.,* Vol. vii., p. 46.)

Hab. New South Wales, Interior, Victoria and South Australia. (*Ramsay.*)

Family MOTACILLIDÆ.

Genus ANTHUS, *Bechstein.*

ANTHUS AUSTRALIS, *Vigors and Horsfield.*

Gould, Handbk. Bds. Aust., Vol. i., sp. 240, p. 392.

This bird has been received in collections from all parts of Australia, with the exception of the North-western portions, and it is probable that it may be found there also ; it is such a well known and common species that collectors generally neglect to procure specimens. The nest of this bird is formed in a hollow scraped out in the ground, under some overhanging tuft of grass, or in a furrow, and sometimes amongst low rushes; it is cup-shaped and entirely formed of dried grasses. Eggs varying in shape from short to long ovals, three in number, seldom four for a sitting, of a dull white ground colour, which is almost obscured with freckles of slaty-brown, umber-brown, and ashy-grey. In some instances I have found eggs of this species of a buffy-white ground colour with a zone of creamy-brown markings upon the larger end.

Dimensions of a set in the Australian Museum Collection taken in October 1880 :—length (A) 0·81 x 0·62 inch ; (B) 0·85 x 0·63 inch ; (C) 0·81 x 0·6 inch.

A set taken at Macquarie Fields in October 1860 measure :— length (A) 0·86 x 0·65 inch ; (B) 0·9 x 0·65 inch ; (C) 0·9 x 0·63 inch. The breeding season of this species commences in September and continues during the three following months.

Hab. Port Darwin and Port Essington, Gulf of Carpentaria, Cape York, Rockingham Bay, Port Denison, Wide Bay District, Dawson River, Richmond and Clarence Rivers Districts, New South Wales, Interior, Victoria and South Australia, Tasmania, West and South-West Australia. (*Ramsay.*)

Genus MIRAFRA, *Horsfield.*

3. MIRAFRA HORSFIELDII, *Gould.*

Horsefield's Bush Lark.

Gould, Handbk. Bds. Aust., Vol. i., sp. 248, p. 404.

" The nests of *Mirafra horsfieldii* are usually found during the months of November, December, and often as late as January and February. They are loose ragged structures, and not finished off nicely like those of *Anthus australis.* They are cup-shaped, and are composed wholly of grasses without any particular lining. The situation chosen is a little hollow scraped out by the side of a tuft of grass or straw, or behind a clod of earth ; the front edge of the nest is alone smoothed down—the back part being left ragged and often drawn forward as if to help to conceal the eggs. The nest is about two inches and a-half in diameter by one inch in depth. On the 4th of February 1861, we took a nest from a hay-field at Macquarie Fields containing three eggs, which is the usual number. They are in length from eight to ten lines by from six to seven in breadth, and of a light earthy-brown, thickly marked over the whole surface with freckles of a much darker hue. Some specimens are darker in colour than others, and after a time the ground-colour becomes of a more yellowish tint, and the markings much duller and more indistinct." *(Ramsay, P.Z.S.,* 1865, p. 689.)

This is the only species of Australian bird of which I know, that sings at night ; especially is it to be heard on bright moonlight nights about mid-summer, flying slowly about high in the air, apparently filled with pleasure and delight at the continued sweet and varied notes at its command.

Dimensions of eggs taken by me at Moonee Ponds on the 14th of January 1882 are as follows :—length (A) 0·79 x 0·55 inch ; (B) 0·78 x 0·59 inch ; (C) 0·78 x 0·5 inch.

Hab. Derby, N.W. Australia, Gulf of Carpentaria, Rockingham Bay, Port Denison, Wide Bay District, Dawson River, New South Wales, Interior, Victoria and South Australia. *(Ramsay.)*

Family PLOCEIDÆ.

Genus ZONÆGINTHUS, *Cabanis.*

S.   ZONÆGINTHUS BELLA, *Latham.*

Fire-tailed Finch.

Gould, Handbk. Bds. Aust., Vol. i., sp. 249, p. 406.

Tasmania is the stronghold of this species, but it is also found
on the Australian continent; a nest of this species in the Australian
Museum Collection, taken by Mr. J. A. Thorpe at Hornsby on
the 9th of November 1886, (together with the birds and eggs) is
a large structure, composed exteriorly of long pieces of coarse dried
grasses, lined with others of a finer description, it has a long
narrow neck ten inches in length by two inches and a-half in
width, the nest proper which is globular, measuring seven inches
through external diameter : it was built in a Native Broom about
ten feet from the ground. Eggs five in number for a sitting,
pure white elongated in form, being nearly equal in size at both
ends ; length (A) 0·73 x 0·48 inch ; (B) 0·72 x 0·47 inch ; (C)
0·71 x 0·48 inch ; (D) 0·7 x 0·49 inch ; (E) 0·71 x 0·49 inch.

Specimens from Tasmania give the following measurements :—
length (A) 0·73 x 0·51 inch ; (B) 0·72 x 0·5 inch ; (C) 0·75 x 0·52
inch. Specimens in the Macleayan Museum Collection, measure
as follows :—length (A) 0·78 x 0·51 inch ; (B) 0·73 x 0·5 inch ; (C)
0·73 x 0·51 inch ; (D) 0·75 x 0·52 inch.

The breeding season commences in September and continues
till the end of December.

Hab. Wide Bay District, New South Wales, Interior, Victoria
and South Australia, Tasmania. (*Ramsay.*)

Genus STICTOPTERA, *Reichenbach.*

S.   STICTOPTERA BICHENOVII, *Vigors and Horsfield.*

Bicheno's Finch.

Gould, Handbk., Bds. Aust., Vol. i., sp. 251, p. 409.

The habitat of this Finch is the interior and the northern and
eastern portions of Australia. Like all other members of this

family it constructs a flask-shaped nest of dried grasses, which is usually placed in a low bush or long grass. Eggs five in number for a sitting, pure white. Specimens taken by Mr. Geo. Barnard of Coomooboolaroo, Queensland, measures as follows :—length (A) 0·6 x 0·41 inch ; (B) 0·63 x 0·4 inch ; (C) 0·63 x 0·41 inch.

Hab. Gulf of Carpentaria, Rockingham Bay, Port Denison, Dawson River, New South Wales, Interior, Victoria and South Australia. (*Ramsay.*)

5. STICTOPTERA ANNULOSA, *Gould.*

Ringed Finch.

Gould, Handbk. Bds. Aust., Vol. i., sp. 252, p. 410.

" This pretty little Finch is found frequenting the northern and north-western portions of the Australian Continent, where it takes the place of its near ally *S. bichenovii*, of the eastern coast. Both Mr. E. J. Cairn and the late Mr. T. H. Boyer-Bower obtained a number of specimens of this bird in 1886, at Derby, North-western Australia. For the opportunity of describing the eggs I am indebted to the Hon. William Macleay, who has lately received them from one of his collectors ; they were taken near the head of the Leonard River, North-western Australia, on the 2nd of October, 1887. The nest was a flask-shaped structure of dried grasses, similar to those of other members of the family, and was built in a low bush. In this instance the nest contained three fresh eggs, but five is the usual complement ; in colour they are white, of a uniform size, each of them giving exactly the same measurement, viz. :— 0·55 inch in length by 0·44 inch in width. These are among the smallest of our Australian birds' eggs." *From the Macleayan Museum Collection. (North, Proc. Linn. Soc., N. S. W.*, Vol. iii., Second Series, p. 146.)

Hab. Derby, N.W. Australia, Port Darwin and Port Essington. (*Ramsay.*)

K

Genus ÆGINTHA, *Cabanis*.

- *5*   ÆGINTHA TEMPORALIS, *Latham.*

Red-eyebrowed Finch.

Gould, Handbk. Bds. Aust., Vol. i., sp. 253, p. 411.

This bird is one of the commonest of the *Ploceidæ* in New South Wales and Victoria; it constructs a flask-shaped nest of dried grasses &c., placed in a bush or low tree; a favourite breeding locality of this species is in the *Melaleuca* scrubs that fringe the edges of rivers and creeks. At Heidelberg in Victoria, I have seen upwards of fifty nests of this species while out collecting in a single day. Eggs five in number for a sitting, pure white; dimensions of a set in the Australian Museum Collection taken at Macquarie Fields, October 1860 by Dr. Ramsay; length (A) 0·6 x 0·45 inch; (B) 0·61 x 0·44 inch; (C) 0·62 x 0·44 inch; (D) 0·56 x 0·45 inch; (E) 0·63 x 0·43 inch.

It was from a nest of this species, built in an Acacia opposite the entrance gates to the old Government House at Toorak, Victoria that I first took the egg of *Chalcites plagosus*, and although I have taken it on many occasions since, it has been very rarely that I have found it in the nest of this bird. Eggs of *Æ. temporalis* may be taken in September and all through the season to the latter end of February.

Hab. Rockingham Bay, Port Denison, Wide Bay District, Richmond and Clarence Rivers Districts, New South Wales, Victoria and South Australia. *(Ramsay.)*

Genus AIDEMOSYNE, *Reichenbach*.

AIDEMOSYNE MODESTA, *Gould.*

Plain-coloured Finch.

Gould, Handbk. Bds. Aust., Vol. i., sp. 255, p. 414.

This Finch is an inhabitant of Queensland and the northern portions of New South Wales. Living well in confinement,

K—2

numbers of them are trapped annually and sent to Sydney, and other markets for sale. The nest of this bird is a large dome-shaped structure, composed of dried grasses, thickly lined with feathers, and is usually placed in a low shrub or among long grass, the eggs are five in number for a sitting, pure white, specimens taken by Mr. Geo. Barnard of Coomooboolaroo, Queensland, measure :— (A) 0·64 x 0·44 inch ; (B) 0·62 x 0·45 inch. October and the three following months constitute the usual breeding period of this species, but like many of the birds of central Queensland, the breeding season is greatly influenced by the rains.

Hab. Wide Bay District, Dawson River, Richmond and Clarence Rivers Districts, New South Wales, Interior. (*Ramsay*)

Genus NEOCHMIA, *Hombron et Jacquinot.*

4-5. NEOCHMIA PHAETON, *Hombron et Jacquinot.*

Crimson Finch.

Gould, Handbk. Bds. Aust., Vol. i., sp. 256, p. 315.

" The eggs here described were taken by Mr. J. Rainbird in 1864 from some of the nests at that time common on extensive grass lands near Port Denison. The nest is like all others of the family a flask-shaped structure of grasses, with a long narrow entrance, placed on its side in any convenient place, either in Pandanus trees or adjacent shrubs or among the stronger of the grass stems. The eggs four or five for a sitting are small in comparison with the size of the birds ; length 0·65 x 0·45 inch in breadth." (*Ramsay, P.L.S., N.S.W.,* 2nd Series, Vol. i., p. 1148.)

Mr. J. A. Boyd informs me that a pair of these birds built their nest on the wall-plate in one of the corners of the verandah of his house on the Herbert River, Queensland, utilising the iron roof as a shelter to the nest. In his opinion they were probably induced to do this by some captive compatriots placed there, and the canary seed they picked up near their cage.

Hab. Derby, N. W. A., Port Darwin and Port Essington, Gulf of Carpentaria, Cape York, Rockingham Bay, Port Denison, South Coast New Guinea. (*Ramsay.*)

Genus STAGONOPLEURA, *Cabanis.*

_~ 5 ~6_   STAGONOPLEURA GUTTATA, *Shaw.*

Spotted-sided Finch.

Gould, Handbk. Bds. Aust., Vol. i., sp. 257, p. 417.

This beautiful bird is plentifully dispersed over New South Wales and Victoria, and is still to be found breeding close to Sydney. The nest, like all other members of the genus is composed of dried wiry grasses, &c., spherical in form with an elongated neck, used for ingress and egress, it is usually placed low down in the thick foliage of a *Syncarpia, Eucalyptus,* or *Angophora* in New South Wales. Eggs pure white, five or six in number for a sitting, lengthened in form, being nearly equal in size at both ends; a set taken at Ashfield by Dr. Ramsay in October 1860, measure as follows :—length (A) 0·75 x 0 53 inch ;   (B) 0·71 x 0·52 inch ; (C) 0·75 x 0·05 inch ;   (D) 0·7 x 0·49 inch ;   (E) 0·74 x 0·48 inch ; (F) 0·78 x 0·48 inch.

The dimensions of a set of four eggs taken at Broadmeadows, Victoria, in November 1873, are as follows :—length (A) 0·77 x 0·52 inch ; (B) 0·73 x 0·5 inch ; (C) 0·74 x 0·51 inch ; (D) 0·76 x 0·49 inch.

September and the three following months constitute the breeding season of this species.

Hab. Wide Bay District, Dawson River, Richmond and Clarence Rivers Districts, New South Wales, Interior, Victoria and South Australia. (*Ramsay.*)

Genus TÆNIOPYGIA, *Reichenbach.*

~~ *5-6*   TÆNIOPYGIA CASTANOTIS, *Gould.*

Chestnut-eared Finch.

Gould, Handbk. Bds. Aust., Vol. i., sp. 258, p. 419.

The Chestnut-eared Finch is found breeding in companies in the neighbourhood of the Lachlan and the Darling Rivers, during September and the two following months. It constructs a flask-shaped nest of dried grass stems &c., and is placed in the branches of a low tree or thick bush. Eggs five or six in number for a sitting, in colour faint bluish-white ; a set now before me taken by Mr. James Ramsay at Tyndarie, in October 1879, has one specimen (A) with a distinct and well defined band of blue round the centre of the egg ; this is the only occasion that I have ever seen any variation from the typical egg of this species. Length (A) 0·6 x 0·43 inch ; (B) 0·61 x 0·45 inch ; (C) 0·56 x 0·42 inch ; (D) 0·62 x 0·43 inch ; (E) 0·66 x 0·46 inch ; (F) 0·65 x 0·45 inch.

Sets of these eggs in Mr. K. H. Bennett's and my own collection give the same average measurements.

This bird together with *Poëphila cincta* and *P. acuticauda,* are breeding readily in confinement in Sydney at all times of the year. In Dr. Ramsay's aviary at the Museum, a brood of *T. castanotis* and *P. acuticauda* left the nest on June 3rd, 1887 ; this was the third brood of *T. castanotis* from the same pair of birds since January.

Hab. Derby, N.W. Australia, Port Darwin and Port Essington, Gulf of Carpentaria, Dawson River, New South Wales, Interior, Victoria and South Australia, West and South-West Australia. (*Ramsay.*)

Genus DONACICOLA, *Gould.*

4-5.   DONACICOLA CASTANEOTHORAX, *Gould.*

Chestnut-breasted Finch.

Gould, Handbk. Bds. Aust., Vol. i.. sp. 265, p. 426.

"This species is widely distributed over the whole of the northern parts of New South Wales and Queensland. It breeds plentifully in the extensive grass beds of the Clarence and Richmond River Districts, also at Maryborough, Queensland. The nest is a large structure, in shape like a flask or bottle placed on its side, and the entrance which is about an inch and a-half wide, is situated at the end of a long neck, the whole being about fourteen inches in length by six inches in diameter at its widest part. It is usually built near the top of some bushy shrub, or in tangled masses of vines, and composed of grasses and the leaves of reeds, with fine stems of plants *(Goodenia* or *Lobelia,* according to the district its owner frequents), being lined with finer materials— the downy tops of reeds and flags, and occasionally a few feathers. It closely resembles the nest of *Neochima phaeton,* which I have received from Port Denison, and like that is often found placed among the stiff leaves of a grass-like plant growing upon the sides of the trees in and about the edges of the scrubs. The eggs are four or five in number, slightly larger than those of *Ægintha temporalis,* of a dead limy-white colour ; a set taken at Iindah, on the Mary River, Queensland, measure as follows :—length 0·64 x 0·48 inch ; 0·65 x 0·48 inch ; 0·67 x 0·5 inch ; two other eggs from the same nest are slightly smaller." (*Ramsay, Ibis,* 1868, p. 232.)

Hab. Gulf of Carpentaria, Cape York, Rockingham Bay, Wide Bay District, Richmond and Clarence Rivers Districts, New South Wales. (*Ramsay.*)

Genus POËPHILA, *Gould.*

5.

POËPHILA CINCTA, *Gould.*

Banded Grass-finch.

Gould, Handbk. Bds. Aust., Vol. i., sp. 264, p. 425.

"This species was formerly abundant in the neighbourhood of Rockhampton, but during my visit to those parts in 1869-70, not a specimen could be found, the bird having been entirely exterminated by the "trappers," for the European markets. It is thinly dispersed over the country to the north, but is replaced in the Gulf districts by its near ally *P. atropygialis.* It nests in the long grass and Pandanus bushes, and lays five eggs of a bluish-white, elongated in form. Length (A)0·7 x 0·48 inch; (B)0·72 x 0·5 inch. We have a present among others, both *P. cincta* and *P. longicauda* breeding in our aviaries." *(Ramsay, P.L.S., N.S.W.,* Vol. i., 2nd Series, p. 1147.)

Hab. Rockingham Bay, Port Denison, Wide Bay District, Dawson River. *(Ramsay.)*

5.   POËPHILA ACUTICAUDA, *Gould.*
              Long-tailed Grass-Finch.

Gould, Handbk. Bds. Aust., Vol. i., sp. 261, p. 422.

"Of this handsome bird, the late Mr. Boyer-Bower procured a fine series while collecting in North Western Australia. It breeds like its ally *P. cincta* of the eastern coast, in the long grass and low bushes, building a flask-shaped nest of grasses, and laying usually five eggs for a sitting. Eggs white, somewhat lengthened in form, they measure as follows :—length (A) 0·68 x 0·48 inch ; (B) 0·65 x 0·4 inch ; (C) 0·69 x 0·46 inch ; (D) 0·71 x 0·48 inch ; (E) 0·65 x 0·43 inch.

September and the three following months constitute the breeding season of this species." *(North, P.L.S., N.S.W.,* Vol. ii., 2nd Series, p. 408.)

Hab. Derby, N.W. Australia, Port Darwin and Port Essington. *(Ramsay.)*

5-6   POËPHILA ATROPYGIALIS, *Diggles.*
              Black-rumped Grass-Finch.

"This fine species is distributed over the country between the Gulf of Carpentaria and Georgetown and its neighbourhood, where it is said to be common along with *Donacicola pectoralis, Poëphila leucotis* and *P. personata.* Its nest is an oval structure of interwoven

grasses, having an opening at one end partly concealed by long
grasses drawn over the entrance.   It is placed among the stronger
grasses, or small bushes which grow here and there on the grass
flats or among the leaves of the *Pandanus aquaticus.* The eggs
are five or six in number, in length from 0·6 inch to 0·64 inch ;
diameter at the larger end from 0·44 to 0·46 inch.   The shell is
white outside, with a faint greenish tinge inside." (*Ramsay,
P.L.S., N.S.W.,* Vol. ii., p. 111.)

Hab. Gulf of Carpentaria.

5.   POËPHILA MIRABILIS, *Hombron et Jacquinot.*

(P. gouldiæ, Gould ; Black-headed phase.)

(*P. armitiana,* Ramsay ; Yellow-headed phase.)

Crimson-headed Grass-Finch.

Gould, Handbk. Bds. Aust., Vol. i., sp. 260, p. 421.

The nest of this species, like that of other members of the
genus, is a dome-shaped structure, composed entirely of dried
grasses.   It is usually placed in a low tree or bush not far from
the ground.   Eggs white, five in number for a sitting, varying
from oval to pyriform in shape.   A set measures as follows :—
length (A) 0·72 x 0·48 inch ; (B) 0·69 x 0·49 inch ; (C) 0·68 x
0·48 inch ; (D) 0·7 x 0·47 inch ; (E) 0·69 x 0·47 inch.

Var. P. GOULDIÆ, *Gould.*

Gould, Handbk. Bds. Aust., Vol. i., sp. 259, p. 420.

In its nidification and eggs, this variety is precisely similar, in
fact it has been recently proved that it is only a black-headed
phase of *P. mirabilis,* the females of both being very similar to
the males of *P. gouldiæ,* and can only be distinguished outwardly
by their duller colours.   In the aviary at the Museum, both
phases have paired.   Eggs from the black-headed phase are white,
and measure as follows :—length (A) 0·66 x 0·5 inch ; (B) 0·65 x
0·49 inch ; (C) 0·68 x 0·49 inch ; (D) 0·64 x 0·48 inch ; (E) 0·68
x 0·47 inch.

At a meeting of the Linnean Society of New South Wales, (March 1889) the following notes were read relative to the breeding of these Finches :—" It may be interesting to know that several of the Gouldian Finches have bred in Dr. Ramsay's aviary at the Museum. A pair, male and female, of the black-headed phase hatched out on May 13th last (1888), three young ones, one of which, although having a dull coloured breast, has developed the crimson head of *P. mirabilis*. There can be now no doubt whatever, that *P. gouldiæ* the black-headed phase, and *P. armitiana* the yellow-headed phase, are merely varieties of *P. mirabilis* originally described by Hombron and Jacquinot in the " Voy. au Pôle Sud." Many specimens recently brought to Sydney show the various stages of plumage above mentioned, bearing out Dr. Ramsay's previous statement respecting the various phases of plumage exhibited in this species.

Hab. Derby, N.W. Australia, Port Darwin and Port Essington, Gulf of Carpentaria. (*Ramsay*.)

Family TURDIDÆ.

Sub-Family SYLVIIANÆ.

Genus ACROCEPHALUS, *Naum*.

3 -4.   ACROCEPHALUS AUSTRALIS, *Gould*.

Reed Warbler.

Gould, Handbk. Bds. Aust., Vol. i., sp. 246, p. 402.

This bird is found frequenting the reedy edges of rivers, creeks, and lagoons. In Victoria I found their nests in great numbers along the banks of the Yarra and Saltwater Rivers. A nest of this species now before me from the Australian Museum Collection, is built between three upright reeds ; it is a deep cup-shaped structure, outwardly composed of the soft paper-like strips of bark and the partially decayed fibre of reeds, lined inside with dried grasses, and has two feathers worked into the bottom; external diameter three inches, depth three inches and three-quarters, internal diameter two inches, depth two inches. This nest was placed as usual, about two or three feet from the surface of the water. Occasionally a frog is found in snug possession of one of these nests. Eggs three or four in number for a sitting, but usually four; they vary considerably in the tints and disposition of their markings.

A set taken by Dr. Ramsay at Macquarie Fields in October 1860, are of a yellowish-brown ground colour, spotted and blotched all over with markings of umber, blackish-brown, and nearly obscure patches of light olive-brown. Length (A) 0·84 x 0·59 inch; (B) 0·78 x 0·58 inch; (C) 0·79 x 0·6 inch.

A set taken on the Lower Yarra, in November 1878, are of a pale bluish-white ground colour, heavily blotched with purplish-black, olive-brown and bluish-grey, the latter colour appearing as if beneath the surface of the shell. Length (A) 0·81 x 0·58 inch; (B) 0·79 x 0·55 inch : (C) 0·79 x 0·59 inch ; (D) 0·8 x 0·61 inch.

The breeding season commences in October and lasts during the three following months.

Hab. Rockingham Bay, Port Denison, Wide Bay District, Richmond and Clarence Rivers Districts, New South Wales, Interior, Victoria and South Australia. (*Ramsay.*)

4   ACROCEPHALUS LONGIROSTRIS, *Gould.*

Long-billed Reed Warbler.

Gould, Handbk. Bds. Aust, Vol. i., sp. 247, p. 403.

According to Mr. Gould the nest of this species "is placed on four or five upright reeds growing in the water, at about two feet from the surface. It is of a deep cup-shaped from, and is composed of the soft skins of reeds and dried rushes. The breeding-season comprises the months of August and September. The eggs are four in number, of a dull greenish-white, blotched all over, but particularly at the larger end with large and small irregularly shaped patches of olive, some being darker than the others, the lighter coloured ones appearing as if beneath the surface of the shell; they are three-quarters of an inch in length by five-eighths of an inch in breadth." (*Gould, Handbk. Bds. Aust.,* Vol. i., p. 403.)

Hab. Derby, N.W. Australia, West and South-west Australia. (*Ramsay.*)

Sub-Family TURDINÆ.

Genus GEOCICHLA, *Kuhl.*

2   GEOCICHLA LUNULATA, *Latham.*

Mountain Thrush.

Gould, Handbk. Bds. Aust., Vol. i., sp. 275, p. 439.

The home of this species is in the thickly wooded mountain ranges of the coast. It is very plentiful in the Strzelecki Ranges in South Gippsland, frequenting alike the ground and fallen logs, and is a bird easily procured, in fact in shooting the difficulty is to get far away enough from it so as not to damage the specimen, a task which the intervening undergrowth seldom permits, without losing sight of the bird altogether.

The nest is a round open structure, outwardly composed of fine strips of bark and mosses, lined with roots; the rim of the nest is very thick and rounded. The position of the nest varies with the locality in which it is found, in Gippsland it is usually placed between the thick fork of a Eucalyptus about seven or eight feet from the ground; at Cheltenham I have found it placed at the top of a *Melaleuca.* It commencs to breed about the middle of July, and continues the three following months. Eggs two in number for a sitting, elongated in form, in some instances of a pale stone ground colour, in others greenish, thickly freckled with reddish-brown over the entire surface of the shell, occasionally eggs are found with only a few minute markings upon the larger end. Dimensions of a set taken at Cheltenham, July 1878 :— length (A) 1·42 x 0·91 inch ; (B) 1·4 x 0·91 inch.

A set taken at Dobroyde, July 1860, measures as follows :— length (A) 1·4 x 0·89 inch ; (B) 1·38 x 0·87 inch.

Hab. Wide Bay District, Richmond and Clarence Rivers Districts, New South Wales, Victoria and South Australia. *(Ramsay.)*

3   GEOCICHLA MACRORHYNCHA, *Gould.*

Gould, P.Z.S., 1837, p. 145.   *XI.* */.*

" The nest and eggs very much the same as those of *G. lunulata,* Lath., but are larger ; the eggs are three for a sitting, of a greenish-white, strongly freckled all over, but more numerously at the larger end with rich reddish-brown ; some confluent markings take a longitudinal direction or run obliquely with the long axis of the egg. An average specimen measures 1·33 inches in length by 0·95 inch through its short diameter." *(Ramsay, P.L.S., N.S.W.,* 2nd Series, Vol. i., p. 1147.)

Hab. Tasmania.

Family PITTIDÆ.

GENUS PITTA, *Vieillot.*

4.   PITTA STREPITANS, *Temminck.*

Noisy Pitta.

Gould, Handb. Bds. Aust., Vol. i., sp. 269, p. 430.   *XI.* */.*

"This species is found plentifully in the dense " brushes " of the Clarence and Richmond Rivers, and that I believe is its nearest habitat to Sydney ;* while to the north its range extends to the

* A single specimen was shot near Wollongong in 1883. (E.P.R.)

vicinity of Cardwell. It frequents the thickets and densest parts of the scrubs, and, were it not far its loud liquid call, would seldom be found even when searched for. I know of no bird more elegant and which trips over the fallen leaves and logs, or threads its way throug the tangled masses of vegetation, with such grace and ease as *Pitta strepitans*. The nest is a round dome-shaped structure, having a large opening at the side, composed of roots, sticks, and twigs, with a little moss, and lined with rootlets, mosses, and a few feathers. It is usually placed upon the ground, but sometimes a few inches from it, in the angle which the "spurs" make with the stems of the trees, or some other suitable place. The eggs are four in number, in length from 1·2 to 1·3 inch by 0·9 to 1 inch in breadth. Their ground colour is of a delicate white, in some specimens bluish-white, having elongated, irregularly-shaped spots of brown and blackish-brown evenly dispersed over the whole surface, with obsolete spots of bluish-grey, which are usually largest on the thicker end of the egg. A second variety of the egg of this bird, one of which is usually found in a set, is much more elongated in form, and has the whole of the thick end freckled with minute dots of bluish-grey, without any other markings save here and there a small blackish dot. Length 1·6 inch ; breadth 0·9 inch." *(Ramsay, Ibis,* 1867, Vol. iii., New Series, p. 414, pl. viii., fig. 2.)

Two eggs in the Australian Museum Collection measure as follows :—length (A) 1·25 x 1·02 inch ; (B) 1·23 x 0·98 inch.

Hab. Rockingham Bay, Port Denison, Wide Bay District, Richmond and Clarence Rivers Districts, New South Wales. *(Ramsay.)*

3 PITTA SIMILIMA, *Gould. (subsp.)*

Gould, P.Z.S., 1868, p. 76. XI 8

"This northern variety of *Pitta strepitans* I found common enough at the Herbert River, and scrubs near Cardwell. Some of the specimens are deeper coloured and smaller even than any I have seen from Cape York; others again, are not to be distinguished from the New South Wales birds; the white spot on the wing is almost obsolete in many from the ranges near Cardwell. Their notes are exactly the same in all localities. The nest and eggs are the same, and are found to vary in the same way as those described and figured by me in the "Ibis," 1867, p. 414. In size they are slightly smaller. I believe the finely spotted variety of the eggs of this species, taken at Cape York by Cockerell and Thorpe, was at the time mistaken for the eggs of *Pitta macklotii* —which is very probable. One thing is certain, I never knew a nest of either *Pitta strepitans* or *P. similima* to contain more than *three eggs alike;* and often two out of the four (the *number invariably laid for a sitting)* have been of the finely spotted and light-coloured variety, the other two strongly and deeply marked as figured in the Ibis, 1867, p. 414." (*Ramsay, P.Z.S.,* 1875, p. 591.)

Hab. Cape York, Rockingham Bay, South Coast New Guinea. (*Ramsay.*)

PITTA MACKLOTII, *Müller and Schlegel.*

Macklot's Pitta.

Gould, Suppl. Bds. Aust., Pl. 29.   XI. 9.

The habits and nidification of this bird is similar to the preceding species. Two eggs in the Dobroyde Collection are in form swollen ovals, creamy-white, blotched and spotted all over with irregular shaped markings of light purplish-brown and obsolete spots of purplish-lilac and bluish-grey, the latter colour appearing as if beneath the surface of the shell. Length (A) 1·17 x 0·86 inch; (B) 1·18 x 0·87 inch.

Hab. Cape York, South Coast New Guinea. (*Ramsay.*)

Family SCENOPIIDÆ.

Genus PTILONORHYNCHUS, *Kuhl.*

PTILONORHYNCHUS VIOLACEUS, *Vieillot.*

(P. holosericeus, Kuhl.)

Satin Bower-bird.

Gould, Handbk. Bds. Aust., Vol. i., sp. 276, p. 442. *XI. 6.*

"The eggs of the Satin Bower-bird are slightly larger than those of the Spotted Bower-bird, more strongly marked if anything, have the same thin delicate shell and elongated form ; in length they average 1·6 inch, in breadth 1·1 inch, are of a light rich cream-colour, and are marbled all over, more closer at the thicker end, with short wavy irregular lines of deep olive-brown, umber, and sienna. These markings are peculiar in form, some resembling ill-shaped figures of fives, eights, and sevens, others being long and wavy, but few if any encircling the shell altogether. These lines are thick in proportion to their length, and in places are looped, curled, and twisted in various directions, often crossing each other at right angles." *(Ramsay, P.Z.S.,* 1875, p. 112.)

In the Proceedings of the Zoological Society of London for 1875 (March 2nd) p. 112, where I first described the egg of this species, I laid stress on the peculiar *short, wavy,* and *irregular markings,* drawing attention to the somewhat similar characters exhibited on the egg of *Chalamydodera maculata ;* at that time I had only two perfect specimens from nests taken in the Wollongong district. Since then however, I have received two well authenticatod sets, which show that the eggs previously described were not of the normal form, hence the necessity for describing the most common variety, in which irregular blotches and spots form the characteristic markings. The eggs vary in proportionate length, but are usually long ovals, seldom even slightly swollen towards the thicker end ; the ground colour is of a rich cream or light stone-colour, spotted and blotched with irregular patchy markings, and a few dots of umber and sienna-brown of different tints, in some almost approaching blackish-brown, in others of a yellowish colour ; the larger markings are as usual on the thicker end, but a few appear with the small dots on the thin end. In this, the usual form, the irregular short wavy lines previously mentioned, seldom appear except where the larger spots or blotches are confluent ; as if beneath the surface of the shell are a few irregularly shaped faint markings of slaty-grey or pale lilac. The following are the measurements of two normal sets :—1, length (A) 1·75 x 1·15 inch ; (B) 1·7 x 1·16 inch. 2, (C) 1·82 x 1·18 inch; (D) 1·76 x 1·15 inch. Both of the above sets were taken from open nests composed of sticks and twigs, and lined with grass ; by Mr. Ralph Hargrave, at Wattamolla, New South Wales." *(Ramsay, P.L.S., N.S.W., 2nd Series, Vol. i., p. 1059.)*

Hab. Rockingham Bay, Port Denison, Wide Bay District, Richmond and Clarence Rivers Districts, New South Wales, Victoria and South Australia. (*Ramsay.*)

Genus AILURŒDUS, *Cabanis.*

AILURŒDUS VIRIDIS, *Latham.*

(A. crassirostris, Paykul.)

Cat Bird.

Gould, Handbk. Bds. Aust., Vol. i., sp. 277, p. 446.

" The nest of this species is not unlike that of *Oreocincla lunulata*, it is rounded, open above, and placed between upright forks of trees in dense scrubs and thickly wooded parts of the country ; it is composed of rootlets, moss, and shreds of fern bark &c., and ornamented with green mosses, chiefly a species of *Hypnum* found in the dense and damp scrubs; the lining is chiefly composed of fine rootlets. Height two inches ; diameter six inches ; depth inside, one inch and a-half ; diameter inside, three inches and a-quarter. The eggs are three in number, comparatively small for the size of the bird, being in length 1·2 inch by 0·85 inch in breadth ; the ground colour is of a delicate bluish-green, sprinkled all over with light reddish-brown dots and spots, larger and more crowded on the thicker end, and with also a few irregular linear scratchy markings or hair lines. The nest and eggs were taken at Stanwell in the Illawarra district, by Mr. Ralph Hargrave." *(Ramsay, P.L.S., N.S.W.,* Vol. ii., p. 107.)

I had the pleasure of examining the above set of eggs, and the most striking characteristic about them is their unusually small dimensions for the size of the bird.

Hab. Wide Bay District, Richmond and Clarence Rivers Districts, New South Wales, Victoria and South Australia. *(Ramsay)*

2. AILURŒDUS MACULOSUS, *Ramsay.*
The Queensland Cat-bird.

P.Z.S., 1874, p. 601.

" This bird is a native of the dense scrubs that are to be found in the neighbourhood of Rockingham Bay, and the Johnstone, Russell, and Mulgrave Rivers in tropical Queensland. They congregate in small flocks in the palms and fig-trees from which they obtain their food. During a recent excursion to the Bellenden-Ker Ranges, Messrs. E. J. Cairn and Robert Grant, collecting on behalf of the Trustees of the Australian Museum, succeeded in obtaining, among others, a fine series of these birds in different stages of plumage ; and, besides finding several nests with young birds, they were fortunate in obtaining, although very late in the season, a nest containing eggs. The nest and eggs in question were found on December 2nd, 1887, in the fork of a sapling about seven feet from the ground, on the Herberton road at a distance of thirty-two miles from Cairns. The nest is a neat bowl-shaped structure, composed of long twigs and leaves of a *Tristania,* lined inside with twigs and the dried wiry stems of a climbing plant ; on the outside several nearly perfect leaves of the *Tristania* are worked in, and partially obscure one side of the nest. Exterior diameter seven inches, by four inches and

L

a-half in depth ; interior diameter four inches and three-quarters, by two inches and a-half in depth. Eggs two in number for a sitting, nearly true ovals in form, tapering but slightly at one end, of a uniform creamy-white ; the shell is thin, the surface being smooth and slightly glossy. (A) 1·67 x 1·11 ; (B) 1·63 x 1·1 inch. $\frac{4}{4}$ Both parent birds were procured at the time of taking the eggs, which were in a very advanced state of incubation. In addition to finding a great number of other nests, several very young birds of *Macropygia phasianella, Ptilopus superbus,* and *Orthonyx spaldingi* were also obtained in the same locality, showing that the breeding season had just terminated. It is only right to mention that the eggs described above are not altogether, what from analogy, they might be expected to be, being quite different from those of any other species of the family *Scenopiidæ.* Messrs Cairn and Grant, however, state that there can be no doubt as to their authenticity. *(Aust. Mus. Coll.)" (North, P.L.S., N.S.W.,* Vol. iii., 2nd Series, p. 147.)

Hab. Rockingham Bay District and Bellenden-Ker Ranges. (*Ramsay.*)

Genus CHLAMYDODERA, *Gould.*

CHLAMYDODERA MACULATA, *Gould.*

Spotted Bower-bird.

Gould, Handbk. Bds. Aust., Vol. i., sp. 279, p. 450.   𝓧 𝓢.

" Our knowledge of the range of this species has recently been extended to Cape York ; previously Rockingham Bay was considered its northern limit on the coast, and the Murray district in Victoria, and South Australia, its most southern range. The interior provinces are the stronghold of this species, where it is found plentifully dispersed all over the Lachlan and Darling River districts. It also occurs inland about eighty miles west from Rockhampton.

" The nest is an open structure usually placed in a low tree, and is saucer or bowl-shaped, composed of sticks, and lined with grass, about five inches inside diameter by three deep, and four inches high. It is very rarely indeed that *C. maculata*, is found near the coast, although on one occasion Dr. Ramsay procured an egg on Ash Island, near Hexham, on the Hunter River, about ten miles from the sea coast. This was in 1861, and probably the first time that the egg had been found, although this fact appears to have escaped the Doctor's memory, since he described another egg of the same species thirteen years afterwards—(P.Z.S., 1874, p. 605), when Mr. J. B. White was credited with having obtained the first specimen. I give Dr. Ramsay's description, which is that of the typical egg, and of the most usual variety found :—

" ' In form elongate, tapering ; shell thin and delicate, somewhat shining and smooth. Ground colour of a delicate greenish-white tint, surrounded with narrow, wavy, twisted, irregular, thread-like lines of brown, dark umber, light umber-brown, and a few blackish-brown, which cross and recross each other, forming an irregular network round the centre and thicker end ; towards the thinner end they are not so closely interwoven, and light brown lines appear as if beneath the surface of the shell, also a few black irregular shaped linear markings, much broader than the rest, show conspicuously against the pale greenish-white ground ; and here and there, over the whole surface, are scattered ill-shapen figures resembling two's, three's and five's (2, 3, 5) of various tints of colour. Length 1·5 inch ; breadth, 1 inch.'

"In 1875, Mr. James Ramsay obtained several specimens of both birds and eggs at Tyndarie ; and others were received from the Clarence River district. Since then the eggs have become less rare, and are to be found in most collections formed in the interior. The eggs of *C. maculata* vary considerably in the extent of their markings, and sometimes in the tints of colouring ; one I have from the Dawson River district is slightly smaller than usual, and has the ground colour a faint greenish-grey, covered *all over* with a fine network of light brownish linear markings, closer together near the thicker end ; others have their markings confined altogether to the larger end of the egg." *(North, P.L.S., N.S.W.,* Vol. i., 2nd Series, p. 1157.)

A set taken by Mr. John Macgillivray, at Grafton on the Clarence River, on the 7th of September 1864, measures :—length (A) 1·47 x 1·09 inch ; (B) 1·5 x 1·09 inch.

A set taken by Mr. James Ramsay at Tyndarie in 1879, measure as follows :—length (A) 1·5 x 1·1 inch ; (B) 1·53 x 1·09 inch.

Hab. Cape York, Rockingham Bay, Port Denison, Wide Bay District, Dawson River, New South Wales, Interior, Victoria and South Australia. *(Ramsay.)*

CHLAMYDODERA CERVINIVENTRIS, *Gould.*

Fawn-breasted Bower-bird,

Gould, Handbk. Bds. Aust., Vol. i., sp. 281, p. 454.   *XI . 4.*

"This species is found at Cape York, the Islands of Torres Straits, and in the southern portions of New Guinea. This is the only known species of the genus that has not the handsome rose-coloured frill on the nape of the neck. Its bower is larger than that of any of the foregoing, and has the sides nearly parallel with one another, with a very slight curvature at the top. It is not so highly ornamented as the bowers of other members of this genus. The nest is an open one, cup-shaped, and built near the ground ; it is composed of twigs, pieces of bark and moss, and is lined inside with grass, &c. The egg is very like that of *C. maculata* in colour, with the same peculiar linear markings crossing and recrossing each other all round ; but confined more to the larger end of the egg than is usually the case in *C. maculata.* A specimen of this egg in the Australian Museum Collection, taken at Cape York, measures 1·4 inch in length by 1·03 inch in breadth." *(North, P.L.S., N.S.W.,* Vol. i., 2nd Series, p. 1160.)

Hab. Cape York, South Coast New Guinea. *(Ramsay.)*

Genus SERICULUS, *Swainson.*

SERICULUS MELINUS, *Latham.*

Regent Bird.

Gould, Handbk. Bds. Aust., Vol. i., sp. 282, p. 456.

" This, perhaps, the most beautiful of all the Bower-builders, and one of the earliest known species, was described by Latham in 1801, under the name of *Turdus melinus;* since that date, however it has been redescribed many times and under various names, of which that given to it by Swainson, *S. chrysocephalus*, appears the most appropriate, if not the oldest. Dr. Ramsay discovered the bower of this species on Ash Island in 1861, (although at the time he was not aware that it belonged to this species), and again in 1866 while in the Richmond River district, an account of which is given in the Ibis for 1867, p. 456, as follows: ' Allow me to confirm the facts respecting the bower-building habits of the Regent-bird, *(Sericulus melinus).* Several years ago (September 23rd 1861) I found what I thought was the bower of the Satin-bird *(Ptilonorhynchus holosericeus);* but it was a very small one, and in my diary I mentioned that ' the only birds seen near it were two or three Regent-birds.' I thought no more of the matter until I saw some remarks on the subject by Mr. Coxen of Brisbane. [Cf. Gould, Handbook Birds Australia, Vol. i., pp. 458 – 461.] During my visit to the Richmond River I determined to pay close attention to the fact, and was not long before I had an opportunity of making some observations. On the 2nd of October 1866, when returning to our camp, some twenty miles from the township, I stopped to look for an *Atrichia*, which, three days before I had heard calling at a certain log; and while standing, gun in hand, ready to fire as soon as the bird, which was at that moment in a remarkably mocking humour, should show itself, I was somewhat surprised at seeing a male Regent bird fly down and sit within a yard of me. Between the two I hardly knew which choice to take—the *Atrichia*, which was singing close in front of me, or the chance of finding the long-wished-for bower. I decided on the former, and remained motionless for full five minutes, while the Regent-bird hopped round me, and finally on to the ground at my feet, when, looking down I saw the bower scarcely a yard from where I was standing : had I stepped down off the log I must have crushed it. The bird after hopping about and rearranging some of the shells *(Helices)* and berries, with which its centre was filled, took its departure, much to my relief, for I was beginning to feel uncomfortable with standing so long in the same position. Further research was not very successful ; we met with only one other bower. Wishing to obtain a living specimen of so beautiful a bird as the adult male of this species, I determined to leave the structure until the last thing on my final return to Lismore, which was on the 3rd of November following. We then stopped on our way, and setting eight snares round the bower, anxiously awaited the result. It was not long before we heard the harsh scolding cry of the old bird, and knew that he had 'put his foot in it.' Having taken him out and transferred him to a temporary cage, we carefully pushed a board brought for the purpose, underneath the bower and removed it without injury. It is now before me, and is placed upon and supported by a platform of sticks, which, crossing each other in various directions, form a solid foundation, into which the upright twigs are stuck. This platform is about fourteen inches long by ten broad, the upright twigs are some ten or twelve inches high, and the entrances four inches wide. The middle measures four inches across, and is filled with land shells of five or six species, and several kinds of berries of various colours, blue, red, and black, which gave it when fresh, a very pretty appearance. Besides these there were several newly-picked leaves and young shoots of a pinkish tint, the whole showing a decided 'taste for the beautiful ' on the part of this species.'

While in the same locality where the above bower was discovered (Taranya Creek), Dr. Ramsay was successful in finding the nest of this species on the 12th of November of the same year ; it was built in a cluster of " Lawyer vines " *Calamus australis.* In shape the nest was like that of *Collyriocincla harmonica*, and composed of twigs, mosses, leaves &c., about five inches across by three deep.

The only egg known of this species at present, which was taken from the oviduct, is in the Australian Museum Collection, and is of a long oval slightly swollen at one end, the ground colour being of a pale lavender ; upon the larger end and beneath the surface of the shell is a zone of nearly round and oval-shaped spots of a uniform pale lilac colour, which in some places are confluent ; on the outer surface all over the larger end, to the lower edge of the zone, are irregularly shaped but well-defined linear markings of sienna, assuming strange shapes ; two prominent markings being a double loop, and a scroll, others less conspicuous are in the shape of the letter Z and the figure 6, while several of the markings stand at right angles to one another ; from the lower edge of the zone and dispersed over the rest of the surface, are a few bold dashes of the same colour, several lines being straight, but marked obliquely across the egg, others are like the letter V with one side lengthened at a right angle, and the figure 7, while upon the lower apex is a single mark in the shape of the letter M.   The peculiarity of the markings of this egg is, that the spots appear to be on the *under* surface, and the linear markings on the *outer* surface of the shell.   Length 1·35 inch x 0·9 inch." (*North, P.L.S., N.S. W.*, Vol. i., 2nd Series, p. 1160.)

Hab. Wide Bay District, Richmond and Clarence Rivers Districts, New South Wales. (*Ramsay.*)

Family ORIOLIDÆ.

Genus MIMETA, *Vigors and Horsfield.*

2 -> *(⁴⁾*   MIMETA VIRIDIS, *Latham.*

New South Wales Oriole.

Gould, Handbk. Bds. Aust., Vol. i., sp. 283, p. 462.

"The nest of this bird is like that of the Friar-bird(*Tropidorhynchus corniculatus*), differing only in the size, which is a little smaller, being from four to five inches in diameter, three to four inches

wide inside, and about three inches and three-quarters deep. It
is cup-shaped, composed of shreds of bark of the Stringy-bark tree
(Eucalyptus obliqua) strongly interwoven; the inside is made
thick and more compact by addition of the white paper-like bark
of the Ti-tree, or, in its absence, any other material adapted for the
purpose; lastly, it is lined with the narrow leaves of the native
oak, or with grass and hair. The nest is usually suspended between
a fork at the very end of a horizontal bough of *Eucalyptus*,
Melaleuca, Syncarpia, &c., and often in very exposed situations.
The eggs are two or three in number, usually three. In two
instances only did we find four—the first of these being in 1860,
and the second in 1861. In length the eggs are from one inch
two lines to one inch four lines; in breadth from nine lines to
one inch. The ground colour varies from a rich cream to a dull
white or very light brown, minutely dotted and blotched with
umber and blackish-brown, with faint lilac spots which appear
beneath the surface, all over in some; but generally the spots are
more numerous at the larger end, where they form an indistinct
band." *(Ramsay, Ibis,* 1863, Vol. v., p. 179.)

A set taken at Dobroyde in 1860, measures as follows :—length
(A) 1·35 x 0·98 inch ; (B) 1·35 x 0·93 inch ; (C) 1·35 x 0·95 inch.

Hab. Wide Bay District, Richmond and Clarence Rivers
Districts, New South Wales, Interior, Victoria and South
Australia. *(Ramsay.)*

3   MIMETA AFFINIS, *Gould.*
                    Allied Oriole.

Gould, Handbk. Bds. Aust., Vol. i., sp. 284, p. 465.   XI 12.

The nest of this species is similar to that of *M. viridis,* and
is built in like situations. The eggs, three for a sitting, are
of a very light creamy-buff with dark olive-brown spots, and
a few of a pale lilac or slaty tint, appearing as if beneath the
shell ; the spots are sprinkled all over the surface rather widely
apart. Length (A) 1·3 x 0·9 inch ; (B) 1·22 x 0·88 inch. They

breed from the beginning of September to the end of December. (*Ramsay,* P.L.S., N.S.W., Vol. vi., p. 576.)

Hab. Derby, N.W. Australia, Port Darwin and Port Essington, Gulf of Carpentaria, Cape York, Rockingham Bay, Dawson River. (*Ramsay.*)

Genus SPHECOTHERES, *Vieillot.*

3   SPHECOTHERES MAXILLARIS, *Latham.*

Gould, Handbk. Bds. Aust., Vol. i., sp. 286, p. 467. XI. //.

Mr. R. D. Fitzgerald writes as follows regarding the nidification of this bird :—" This remarkable species, which appears to be somewhat gregarious in its habits, I found breeding during the latter part of October and the beginning of November in the brushes of the Richmond River, where the birds are plentiful. The nests, of which several were discovered in adjoining trees, are rather slight and shallow, constructed of small thin twigs interwoven loosely, not unlike a large nest of *Pachycephala gutturalis,* and are usually placed at the extremity of a horizontal branch about twenty feet from the ground ; the tree most favoured being the *Flindersia.* Three nests obtained on the 4th of November 1886, contained each three fresh eggs, which appears to be the regular number for a sitting, all quite fresh. An average-sized pair of these eggs measure as follows :—length (A) 1·25 x 0·88 inch ; (B) 1·25 x 0·9 inch. The ground colour varies from olive-brown to dull apple-green ; the spots sometimes confluent and forming small irregular blotches, are of a reddish-brown, in some brighter and redder, in others very like those on the eggs of *Cracticus destructor;* the markings are distributed over the whole surface, but are usually closer together on the thicker end, where in some they form an irregular zone." (*Fitzgerald,* P.L.S., N.S. W., Vol. ii., Second Series, p. 970.)

Hab. Gulf of Carpentaria, Rockingham Bay, Port Denison, Wide Bay District, Dawson River, Richmond and Clarence Rivers Districts, New South Wales. (*Ramsay.*)

Family CORVIDÆ.
Sub-Family CORVINÆ.

Genus CORVUS, *Linnæus.*

4-5.   CORVUS CORONOIDES, *Vigors and Horsfield.*

(C. australis, Gmelin.)

White-eyed Crow.

Gould, Handbk. Bds. Aust., Vol. i., sp. 290, p. 475.   *VII.* 8

" The nests of this species are large bulky structures of sticks and twigs, some often half-an-inch thick. These form the ground work of the nest, which is usually placed in the most inaccessible trees. Finer materials are used for the inner parts, and it is lastly lined with grasses, stringy-bark, and tufts of hair from various dead animals. The eggs are four or five in number for a sitting, of a bright green, strongly blotched with deep black and brown, with a tinge of yellowish wood-brown in some places ; they are from 19½ to 21 lines in length by 14 or 15 lines in breadth. They usually have two broods a year, beginning to breed in August, and continuing until November, or even later in some instances, according to the locality."

It was in a paper to the Ibis from which the above is extracted, that Dr. Ramsay first drew attention to there being two distinct birds described under the name of *C. coronoides*, a fact since recognized by Mr. Sharpe, who has separated them under the names of *Corone australis,* and *Corvus coronoides.*

C. coronoides can easily be distinguished from *Corone australis* by being the smaller bird of the two, and having the bases of the feathers snow white. *C. australis* has long greenish-black plumes on the throat; bases of feathers dusky-brown or black. The nest of *C. coronoides* is built at the extremity of a long, naked, slender branch terminating in a thick bunch of twigs and leaves, and are in most instances impossible to get at, thus the eggs are very difficult to obtain.

I give the measurements of a set of eggs of *C. coronoides*, taken by Mr. K. H. Bennett, a gentleman who has paid particular attention to this and the following species : a set of four taken at Ivanhoe on the 1st of September 1885, are elongated in form, of a pale greenish ground colour, finely, but thickly freckled with scratches of light umber ; length (A) 1·43 x 1·01 inch ; (B) 1·62 x 1·03 inch ; (C) 1·63 x 1 inch ; (D) 1·63 x 1 inch.

Hab. Derby, N.W. Australia, Port Darwin and Port Essington, Gulf of Carpentaria, Cape York, Rockingham Bay, Port Denison, Wide Bay District, Dawson River, Richmond and Clarence Rivers Districts, New South Wales, Interior, Victoria and South Australia. (*Ramsay.*)

Genus CORONE, *Kaup.*

4-5. CORONE AUSTRALIS, *Gould.*

Australian Raven.

Sharpe, Brit. Mus. Cat. Bds, Vol. iii., p. 37. *VII. 2.*

From a most interesting and exhaustive account by Mr. K. H. Bennett in his MS. notes, of the nidification of this species, I extract the following :—"The breeding season of this species is during the months of September and October ; the nest is composed outwardly of sticks, and lined with soft bark fibre, wool, &c., and is placed in a variety of situations from the highest tree in a clump to bushes of the giant or "old man" saltbush not more than four feet high. Scattered over the plains, singly or in clumps of five or six in number, are numerous small trees from six to twelve feet in height, these are favourite situations, and one or more nests will be found in each of these trees, where possible of construction; the eggs are generally four, sometimes five in number."

A set of four eggs now before me, taken by that gentleman at Mossgiel on the 31st of August 1815, are of a dull green, thickly freckled, spotted, and blotched with umber, wood-brown, and blackish-brown markings, but more particularly towards the larger end; length (A) 1·72 x 1·19 inch; (B) 1·7 x 1·16 inch; (C) 1·68 x 1·22 inch; (D) 1·68 x 1·17 inch.

This species is also found breeding in the low trees in the vicinity of Laverton and the Werribee, near Melbourne, Victoria.

Hab. Derby, N.W. Australia, Port Darwin and Port Essington, Gulf of Carpentaria, Cape York, Rockingham Bay, Port Denison, Wide Bay District, Dawson River, Richmond and Clarence Rivers Districts, New South Wales, Interior, Victoria and South Australia, West and South-West Australia, Tasmania. *(Ramsay.)*

Sub-Family FREGILINÆ.

Genus STRUTHIDEA, *Gould*.

3-4　　STRUTHIDEA CINEREA, *Gould*.

Grey Struthidea.

Gould, Handbk. Bds. Aust., Vol. i., sp. 289, p. 472. *XIII　8*

" The nest is a round cup or basin-shaped structure, composed of mud or clay, about four inches inside diameter; it is lined with grasses, and placed on a horizontal bough, often only a few feet from the ground, but occasionally at a height of about twenty to thirty feet; the eggs are three or four in number, but sometimes five and seven have been taken by Mr. James Ramsay from a single nest. They are of a milky-white, sometimes of a skimmed milk colour, with spots, and here and there a blotch of blackish-umber and blackish-slate colour, or occasionally streaked—some altogether without markings, or with only one or two blackish specks. A set of four measure as follows :—length 1·26 x 0·85 inch; 1·18 x 0·85 inch; 1·27 x 0·88 inch; 1·18 x 0·85 inch." (*Ramsay, P.L.S., N.S.W.,* Vol. vii., p. 406.)

I found these birds in great numbers on the Bell and Macquarie Rivers, during August 1887, congregating in small companies of six or seven in number, making the bush resound with their peculiar harsh grating cries when flying.

Hab. Gulf of Carpentaria, New South Wales, Interior, Victoria and South Australia. (*Ramsay.*)

Genus CORCORAX, *Lesson*.

4. CORCORAX MELANORHAMPHUS, *Vieillot.*

White-winged Corcorax.

Gould, Handbk. Bds. Aust., Vol. i., sp. 288, p. 470.

This bird is plentifully distributed throughout New South Wales, building its large bowl-shaped structure of mud on the horizontal branch of a tree, in any convenient situation. The usual number of eggs for a sitting is four, but as many as eight have been taken from one nest, it would therefore appear that more than one bird lays in a single nest; it is well known that often more than one pair of birds assists in the construction of one nest. Ground colour of eggs dull-white, blotched all over beneath the surface of the shell with deep bluish-black markings ; on the outer surface the blotches become larger, but fewer, and vary from olive-brown to blackish-brown in tint ; the shell of the eggs being rough. Size of a set of four taken at Macquarie Fields ; length (A) 1·56 x 1·16 inch ; (B) 1·5 x 1·18 inch ; (C) 1·51 x 1·18 inch ; (D) 1·53 x 1·13 inch.

The breeding season commences in August, and lasts during the three following months.

I found these birds very plentiful in the neighbourhood of Wellington and Dubbo, in New South Wales.

Hab. Rockingham Bay, Port Denison, Wide Bay District, Dawson River, Richmond and Clarence Rivers Districts, New South Wales, Interior, Victoria and South Australia. (*Ramsay*)

Family STURNIDÆ.

Genus CALORNIS, *G. R. Gray*.

3 - 4   CALORNIS METALLICA, *Temminck*.

Shining Calornis.

Gould, Handbk. Bds. Aust., Vol. i., sp. 291, p. 477.

" This is one of the most common birds in the scrubs of the Herbert River. They breed in companies, seemingly all through the year, making large bulky nests of grass and fine twigs with a side opening, hanging from the ends of the leafy boughs in clusters or singly ; at times the branches break off with the weight of the nests and their contents. On the Herbert River I noticed they gave preference to a small-leaved species of fig resembling *Ficus syringifolia;* and before a colony began to build the twigs on many of the branches were broken and began to wither, and hanging down, at a distance resembled in colour the brown nests of this species. I noticed this on two occasions, and remarked to Inspector Johnstone that the birds were building near his camp. However, when examining the trees through our field-glasses, we found nothing but bunches of dry leaves swinging about with the wind. A few days afterwards we noticed a neighbouring fig-tree in a similar condition, and as both trees were resorted to by these birds, I was under the impression that it was caused by the ravages of some insect which the birds came to feed on ; however, about a month afterwards, Mr. Johnstone informed me that these trees had been taken possession of by colonies of Weaver-birds (or

"Starlings" as they are called in those parts), and this bulk of brown nests was forming quite a new feature in the landscape. The eggs are three or four in number, variable in form, some roundish, others elongate, of a greenish-white colour, with bright reddish-brown spots and dots, more numerous towards the larger end. Length (A) 1·1 x 0·78 inch ; (B) 0·99 x 0·79 inch ; (C) 1·1 x 0·75 inch." (*Ramsay, P.Z.S.*, 1875, p. 594.)

Hab. Gulf of Carpentaria, Cape York, Rockingham Bay, Port Denison, South Coast New Guinea. (*Ramsay.*)

Family MELIPHAGIDÆ.

Genus MELIORNIS, *G. R. Gray.*

2 - 3   MELIORNIS NOVÆ-HOLLANDIÆ, *Latham.*

New Holland Honey-eater.

Gould, Handbk. Bds. Aust., Vol. i., sp. 296, p. 486.

This showy and attractive bird is the most common species of the genus in New South Wales and Victoria ; it is very abundant in the scrubby undergrowth and stunted Banksias in the neighbourhood of Botany and La Perouse in the former colony ; it is also found in our public parks and gardens both in Sydney and Melbourne, where it may be seen extracting the nectar from various flowers, with its brush-like tongue so well adapted for the purpose.

A nest of this species now before me, in the Australian Museum Collection, is rather roughly but compactly formed on the exterior with fine twigs, strips of bark and grasses, neatly lined on the inside, which is cup-shaped, with dried portions of the soft Flannel Fower, *Actinotus helianthus*, and downy tufts of the Banksia cones.   Exterior measurement four inches in diameter by two inches and a-half in depth, cavity two inches and a-quarter in diameter by one inch and a-half in depth.   The position of the nest varies with the locality in which it is built, sometimes being placed in a low tree about six or seven feet from the ground but more often in a bush close to the ground. Eggs two or three in number for a sitting, usually two.

Two eggs taken by Dr. Ramsay at Bondi, on the 25th of August 1875, are of a creamy-buff ground colour, spotted, and minutely freckled with markings of a rich reddish-brown, and a few of reddish-black, more particularly towards the larger end. Length (A) 0·8 x 0·58 inch ; (B) 0·81 x 0·58 inch.

A set taken by Dr. Hurst, near Botany, on the 21st of July 1888, give the following measurements :—length (A) 0·77 x 0·63 inch ; (B) 0·77 x 0·6 inch.

A nest of this species taken by myself at Melbourne, and several others found subsequently, each contained three eggs of *M. novæ-hollandiæ* and one of *Chalcites basilis*. July and the four following months constitute the breeding season of this species.

Hab. Wide Bay District, Richmond and Clarence Rivers New South Wales, Victoria and South Australia, Tasmania. (*Ramsay.*)

2 - 3   MELIORNIS LONGIROSTRIS, *Gould.*

Long-billed Honey-eater.

Gould, Handbk. Bds. Aust., Vol. i., sp. 297, p. 488.

This species is an inhabitant of Western Australia, and differs but slightly, if anything at all, from the eastern representative *M. novæ-hollandiæ.* Mr. George Masters procured a number of the birds, also nests and eggs, while at King George's Sound in 1868, but failed to find the distinction made by Mr. Gould between the birds from the eastern and western portions of the continent, specimens having been received from eastern Australia with the bill equally as long and robust as in that of *M. longirostris.*

A nest of this species now before me, taken by Mr. Masters on the 1st of October 1868 is very much neater in appearance on the outside than that of the preceding bird; it is cup-shaped, outwardly composed of fine strips of bark, fibrous roots and grass, lined inside with the soft downy tufts of the Banksia cones; the rim of the nest is thick and rounded. Exterior measurement four inches in diameter by two inches and a quarter in depth, cavity two inches and a-quarter in diameter by one inch and a-half in depth; rim about one inch in thickness. This nest, Mr. Masters informs me was built in a Banksia close to the ground. Eggs two or three in number for a sitting, of a pale buff, minutely but thickly freckled and spotted with chestnut and reddish-brown, in some instances forming a well defined zone, in others the markings are nearly obsolete and appear as if beneath the surface of the shell. Dimensions of a set in the Australian Museum Collection, taken October 1st 1868, length (A) 0·77 x 0·57 inch; (B) 0·8 x 0·58 inch.

The breeding season of this species commences in August and lasts till the end of November, but early in October the greater number of eggs were procured.

Hab. Wide Bay District, New South Wales, West and South-west Australia. (*Ramsay.*)

2   MELIORNIS SERICEA, *Gould.*

White-cheeked Honey-eater.

Gould, Handbk. Bds. Aust., Vol. i., sp. 298, p. 490. XII /9.

This bird is found breeding in the neighbourhood of Sydney; a nest of this species now before me, taken by Dr. Ramsay at Dobroyde on the 18th of September 1864, is cup-shaped, outwardly composed of fine twigs, strips of bark and coarse grasses, lined inside with the nests of spiders, and the soft downy tufts of Banksia cones; exterior diameter three inches and a-half, depth three inches, inside cavity two inches in diameter by one inch and three-eighths in depth; another nest in the Australian Museum Collection is almost entirely composed of strips of bark, with a lining of dried portions of the Flannel flower, *Actinotus helianthus.* The nest is usually placed in the fork of a Banksia or Hakea

M

partly resting with the rim of the nest attached to the branches holding it in position, but it is often found in orange trees in gardens, in which case the nest is always slung by the rim. Eggs two in number for a sitting, of a beautiful flesh colour before being blown ; when emptied of their contents, the ground colour approaches to a very pale yellowish-buff, finely but distinctly spotted with reddish-chestnut, in some instances forming a zone towards the larger end, in others scattered over the surface of the shell. Length (A) 0·81 x 0·57 inch ; (B) 0·81 x 0·56 inch. Taken at Dobroyde by Dr. Ramsay, 18th September 1864. Another set taken in the same locality in 1858 give the same measurements.

Two eggs taken by Dr. Hurst on the 28th of June 1884, at Sandringham, near the mouth of the George's River, New South Wales, are of a reddish-buff ground colour ; one specimen being thickly freckled with nearly obsolete markings of chestnut-brown ; the other is more boldly marked with the same colour, particularly towards the thicker end, where the markings become larger, and form an irregular shaped zone. Length (A) 0·8 x 0·58 inch ; (B) 0·8 x 0·59 inch.

The breeding season of this species commences in June and continues during the four following months.

Hab. Rockingham Bay, Wide Bay District, Richmond and Clarence Rivers Districts, New South Wales, Victoria and South Australia. (*Ramsay.*)

2   MELIORNIS MYSTACALIS, *Gould.*

Moustached Honey-eater.

Gould, Handbk. Bds. Aust., Vol. i., sp. 299, p. 491.

" The *Meliornis mystacalis* is a native of Western Australia, it beautifully represents the *M. sericea* of New South Wales. It is a very early breeder, young birds ready to leave the nest having been found on the 8th of August ; it has also been met with breeding as late as November ; it doubtless, therefore, produces more than one brood in the course of the season. The nest is

M—2

generally built near the top of a small, weak, thinly branched bush of about two or three feet in height, situated in a plantation of seedling mahogany or other Eucalypti ; it is formed of small dried sticks, grass, and narrow strips of soft bark, and is lined with Zamia wool, but in those parts of the country where that plant is not found, the soft buds of flowers, or the hairy flowering parts of grasses, form the lining materials, and in the neighbourhood of sheep-walks, wool collected from the scrub. The eggs are usually two in number. They are nine lines long by seven lines broad, and are usually of a dull reddish-buff, spotted very distinctly with chestnut and reddish-brown, interspersed with obscure dashes of purplish-grey." *(Gould, Handbk. Bds. Aust.*, Vol. i., p. 491.)

Hab. Wide Bay District, New South Wales, West and South-West Australia, *(Ramsay.)*

Genus LICHMERA, *Cabanis.*

3   LICHMERA AUSTRALASIANA, *Shaw.*

Tasmanian Honey-eater.

Gould, Handbk. Birds Aust., Vol. i., sp. 300, p. 493.   XII. /7.

This bird is distributed over the whole of Tasmania and portions of Queensland, New South Wales, Victoria, and South Australia. In Gippsland the nest of this species is a round cup-shaped structure, outwardly composed of fine strips of bark, lined inside with grasses and the downy covering of the young fronds of the Tree Fern, *Dicksonia antarctica;* it is generally placed in a low tree or thick bush not far from the ground. Eggs three in number for a sitting, of a light saturnine buff becoming darker towards the larger end, where they are marked with spots of a deeper tint of the same colour and chestnut-brown, with a few obsolete spots of dark lilac. In some instances the markings are indistinct, and not well defined ; in others they form a well defined zone. Dimensions of a set taken at South Gippsland, October 1879. Length (A) 0·76 x 0·58 inch ; (B) 0·74 x 0·58 inch ; (C) 0·75 x 0·56 inch.

A set taken near Hobart, Tasmania, in October 1885, measures as follows :—length (A) 0·75 x 0·56 inch ; (B) 0·75 x 0·58 inch ; (C) 0·74 x 0·57 inch.

The breeding season of this species commences in August, and continues until the end of December.

Hab. Wide Bay District, Richmond and Clarence Rivers Districts, New South Wales, Victoria and South Australia, Tasmania. (*Ramsay.*)

Genus GLYCIPHILA, *Swainson.*

GLYCIPHILA FULVIFRONS, *Lewin.*

Fulvous-fronted Honey-eater.

Gould, *Handbk. Bds. Aust.*, Vol. i., sp. 301, p. 495. *XIII 6.*

This bird is plentiful on the sterile and low scrubby Banksia covered tracts of land near the coast in the neighbourhood of Sydney, particularly at Botany and La Perouse. A nest of this species in the Australian Museum Collection, taken among others at Heathcote on the Illawarra Line by Dr. Hurst and myself on the 30th of October 1886, is a deep cup-shaped structure, outwardly composed of strips of bark and dried grasses, warmly lined with feathers, downy grass seeds, and velvety tufts from the Banksia cones ; it measures exteriorly three inches and a-half in diameter by three inches in depth, inside cavity two inches in diameter by two inches in depth. It was built close to the ground, beneath some ferns *(Pteris aquilina)*, to which the rim of the nest was attached. The bird as is usual with this species, allowed itself to be nearly trodden upon before leaving the nest. Eggs two in number for a sitting, usually elongated in form, and varying considerably in the disposition of their markings. The eggs taken from the above nest are pure white with a zone of reddish-chestnut spots towards the larger end. Length (A) 0·85 x 0·59 inch ; (B) 0·84 x 0·6 inch.

Two eggs in the Australian Museum Collection are elongated in form, white with a very few spots of chestnut-brown on the larger end. Length (A) 0·87 x 0·6 inch ; (B) 0·85 x 0·63 inch.

A pair in the Dobroyde Collection taken at Bondi, in September 1865, are in form rounded ovals, white with a few minute spots of chestnut-brown scattered over the surface. Length (A) 0·79 x 0·64 inch ; (B) 0·85 x 0·63 inch.

In New South Wales this species commences to breed in July and continues the five following months.

Hab. Wide Bay District, Richmond and Clarence Rivers Districts, New South Wales, Victoria and South Australia, Tasmania, West and South-West Australia. (*Ramsay.*)

2 GLYCIPHILA ALBIFRONS, *Gould.*

White-fronted Honey-eater.

Gould, Handbk. Bds. Aust. Vol. i., sp. 302, p. 497.

This bird is found in the Mallee country of Victoria and South Australia, and also in the interior of New South Wales, over which it is very sparingly distributed. A nest of this species now before me taken by Mr. K. H. Bennett, together with the eggs, in October 1886, at Ivanhoe, is a very flat structure, the base being composed of very thin dried stems of a climbing plant, and grasses, matted together with a little wool, over which is placed a layer of wool intermingled with a few wiry blades of dried grass. Diameter of base, four inches ; the layer of wool, two inches and three-quarters ; and the whole structure one inch and a-half in thickness. There is just sufficient depression in the centre of the nest to keep the eggs in position. Eggs two in number for a sitting, of a light saturnine red ground colour ; on the larger end they are thickly spotted, and in a few places blotched with irregular shaped markings of reddish-chestnut and chestnut-brown, but over the remainder of the surface the markings are much smaller, and very sparingly distributed ; on the larger end are obsolete spots of purplish-grey. Length (A) 0·78 x 0·57 inch ; (B) 0·82 x 0·58 inch.

Hab. New South Wales, Interior, Victoria and South Australia, West and South-West Australia. (*Ramsay.*)

GLYCIPHILA MODESTA, *Gray.*
(G. subfasciata, Ramsay.)
Plain-coloured Honey-eater.

Gray, P.Z.S., 1858, p. 174.   *XIII   10.*

"This species, although possessing nothing in its sombre plumage
to recommend it, is certainly very interesting on account of its
peculiarly shaped nest, being the only one of the Australian
Meliphagidæ that I have met with which constructs a dome-shaped
nest. It is a neat structure, composed of strips of bark, spiders'
webs, and grass, and lined with fine grasses &c. The opening
at the side is rather large ; but the nest itself is deep. The
eggs I did not obtain ; but one taken from the oviduct of a bird
is 0·75 inch in length and 0·5 inch in breadth, pure white, with a
few dots of black sprinkled over the larger end. The nests were
invariably placed among the drooping branches of a species of
Acacia, always overhanging some creek or running water."
(Ramsay, P.Z.S., 1868, p. 365.)   •

Hab. Cape York, Rockingham Bay, South Coast New Guinea.
(Ramsay.)

Genus STIGMATOPS, *Gould.*

2
STIGMATOPS OCULARIS, *Gould.*
Brown Honey-eater.

Gould, Handbk. Bds. Aust., Vol. i., sp. 304, p. 500.   *XIII 17.*

For the nest and several sets of eggs of this bird I am indebted
to Mr. George Barnard of Coomooboolaroo, Duaringa, Queensland.
The nest is a very neat cup-shaped structure, outwardly composed
of strips of bark and grasses, held together with the nests of spiders
and lined inside with finer grasses, the downy seeds of some
composite plant, and hair. The one now before me was attached
to the thin twigs of an orange tree in Mr. Barnard's garden,
within a few feet of the ground. It measures exteriorly two inches
in diameter by one inch and a-half in depth, internal diameter one

inch and a-half, depth one inch and a-quarter. Eggs two in number for a sitting, white, minutely freckled with reddish-brown and greyish-lilac, the latter colour appearing as if beneath the surface of the shell ; in some instances these markings form a zone, in others they are confined to a few nearly invisible spots on the larger apex. An average set measures as follows :—length (A) 0·67 x 0·52 inch ; (B) 0·67 x 0·48 inch.

Hab. Port Darwin and Port Essington, Rockingham Bay, Port Denison, Wide Bay District, Dawson River, Richmond and Clarence Rivers Districts, New South Wales, West and South-West Australia. (*Ramsay.*)

GENUS PTILOTIS, *Swainson.*

? PTILOTIS LEWINII, *Swainson.*
Lewin's Honey-eater.

Gould, Handbk. Bds. Aust., Vol. i., sp. 306, p. 503.   *XIII. /.*

" The nest of this species is like that of *P. chrysops*, cup-shaped, open at the top, and slung by the sides of the rim between the twigs of some leafy bough or vine ; it is composed of shreds of bark and grasses, webs of spiders, &c., and lined with similar material of a finer texture, or occasionally, when found in the neighbourhood of dwellings, with feathers, wool, or other soft substances. The eggs are two in number, pearly white with deep reddish dots." (*Ramsay, P.Z.S.*, 1875, p. 595.)

Two eggs taken by Dr. Ramsay on the 29th of December 1871, on the Mary River, Queensland, are nearly oval in form, white, with circular spots and dots of reddish-black confined to the larger end. Length (A) 1· x 0·7 inch ; (B) 0·99 x 0·68 inch. In specimens taken at Port Macquarie, the markings assume the form of a zone.

They breed during the months of October, November, and December.

Hab. Rockingham Bay, Port Denison, Wide Bay District, Dawson River, Richmond and Clarence River Districts, New South Wales, Victoria and South Australia. (*Ramsay.*)

2 -(3)   PTILOTIS VITTATA, *Cuvier.*

(P. sonora, Gould.)

Sonorous Honey-eater.

Gould, Handbk. Bds. Aust., Vol. i., sp. 307, p. 504.

This bird is distributed over a wide expanse of country, specimens having been received in nearly every collection formed in different parts of Australia ; it is particularly plentiful on the Lachlan and Darling Rivers, and in the interior of New South Wales, being the most common of all the species of Honey-eaters in that locality. According to Mr. K. H. Bennett the nests of this bird in the neighbourhood of Mossgiel are round, cup-shaped and somewhat scanty structures of dried grasses &c., matted and held together with spiders' webs, lined inside with fibrous roots, and attached by the rim to suitable twigs in some low bush. Eggs two, or occasionally three in number for a sitting.   •

A set in the Dobroyde Collection, taken by Mr. James Ramsay on the Bogan River in 1880, are of a light yellowish-buff, with a clouded band of reddish-buff towards the larger end. Length (A) 0·89 x 0·6 inch ; (B) 0·92 x 0·63 inch. In some specimens the markings are confined to a faint clouded patch on the larger end, others closely resemble small eggs of *Cuculus inornatus,* Latham.

A set taken by the late Mr. W. Liscombe, on the Darling in October 1883, are of a pale yellowish-buff, finely freckled with minute dots of reddish-brown particularly on the larger end. Length (A) 0·95 x 0·65 inch ; (B) 0·91 x 0·63 inch.

September and the three following months comprise the breeding season of this species.

Hab. Derby, N.W. Australia, Gulf of Carpentaria, Rockingham Bay, Port Denison, Wide Bay District, New South Wales, Interior, Victoria and South Australia, West and South-west Australia. (*Ramsay.*)

2 - 3 PTILOTIS FLAVICOLLIS, *Vieillot.*
(P. flavigula, Gould.)
Yellow-throated Honey-eater.
Gould, Handbk. Bds. Aust., Vol i. sp. 310, p. 508. XII /3.

I have never seen the nest of this species, but extract the following description from Mr. Gould's Handbook, Vol. i., p. 508 :—" The nest of this species, which is generally placed in a low bush, differs very considerably from those of all the other Honey-eaters, with which I am acquainted, particularly in the character of the material forming the lining ; it is the largest and warmest of the whole, and is usually formed of ribbons of stringy-bark, mixed with grass and the cocoons of spiders ; towards the cavity it is more neatly built, and is lined internally with opossum or kangaroo fur ; in some instances the hair-like material at the base of the large leaf stalks of the tree-fern is employed for the lining, and in others there is merely a flooring of wiry grasses and fine twigs."

Eggs two or three in number for a sitting, of a pale fleshy-buff, spotted and blotched on the larger end with round deep chestnut-red markings, together with irregular-shaped spots of very faint chestnut-brown, in other specimens the markings are of the latter colour very minutely and evenly dispersed over the surface of the shell, together with a few obsolete spots of dark lilac on the larger end.

Two specimens in the Dobroyde Collection taken near Hobart, in September, measure as follows :—length (A) 0·96 x 0·67 inch ; (B) 0·95 x 0·67 inch.

Hab. New South Wales, Victoria, South Australia, Tasmania. (*Ramsay.*)

2 PTILOTIS LEUCOTIS, *Latham.*
White-eared Honey-eater.
Gould, Handbk. Bds. Aust., Vol. i., sp. 311, p. 510. XIII 5

This bird is widely distributed over the eastern and southern portions of the continent of Australia, and is likewise found in the interior and in Western Australia, but in these places it is by no means a common bird. I found it particularly plentiful among the low saplings in the neighbourhood of Mount Buninyong in Victoria, and also in the scrubby undergrowth bordering portions of the shores of Western Port Bay.

The nest is a neat cup-shaped structure composed of bark and grasses, and warmly lined inside with opossum fur, or cow hair; it is slung by the rim to the small horizontal twigs of a bush, and placed within a few feet of the ground. Eggs two in number for a sitting, of a uniform delicate flesh colour, with a few minute freckles of a darker tint on the larger end, but in some instances I have obtained specimens with small but very distinct spots; others are nearly devoid of markings.

A set taken by Mr. James Ramsay at Merule, measure, length (A) 0·84 x 0·62 inch; (B) 0·84 x 0·61 inch. Sets from Mr. K. H. Bennett's and my own collection, give the same average measurements.

Two eggs taken by Dr. Hurst at Newington on the Parramatta River, New South Wales, differ somewhat from each other; one specimen (A) is perfectly white, with a few minute reddish-chestnut markings towards the larger end, the other is of a fleshy-buff ground colour, with irregular shaped markings of reddish-chestnut evenly dispersed all over, some of which appear as if beneath the surface of the shell. Length (A) 0·8 x 0·63 inch; (B) 0·81 x 0·63 inch.

September and the four following months constitute the breeding season of this species.

Hab. Derby, North-West Australia, Port Darwin and Port Essington, Gulf of Carpentaria, Wide Bay District, Richmond and Clarence Rivers Districts, New South Wales, Interior, Victoria and South Australia, West and South-West Australia. (*Ramsay.*)

2   PTILOTIS AURICOMIS, *Latham.*

Yellow-tufted Honey-eater.

Gould, Handbk. Bds. Aust., Vol. i., sp. 312, p. 511.   *XII. 14.*

"This species remains with us in the neighbourhood of Sydney,
throughout the whole year, breeding earlier than the generality
of Honey-eaters.   We have eggs in our collection taken early in
June, and as late as the end of October, during which month they
sometimes have a third brood.   August and September seem to
be their principal months for breeding.   Upon referring to my
note book, I find that I captured two young birds, well able to
fly, on the 18th of July 1863, but during some seasons birds breed
here much earlier than in others.   The nest is a neat but somewhat
bulky structure, open above, and composed of strips of the Stringy-
bark tree *(Eucalyptus obliqua).*   The total length of the nest is
about four inches by from two inches and a-half to three inches
wide, being two inches deep by one inch and a-half inside.   The
eggs which are usually two in number, are of a pale flesh-pink,
darker at the larger end, where they are spotted and blotched with
markings of a much deeper hue, inclining to salmon-colour ; in
some the markings form a ring upon the thick end, in others, one
irregular patch with a few dots upon the rest of the surface.
When freshly taken, they have a beautiful blush of pink, which
they generally lose a few days after being blown.   Their length
is from ten to eleven lines by seven to eight in breadth.   Some
varieties have a few obsolete dots of faint lilac ; others are without
markings, save one patch at the top of the larger end : like most
of our Australian bird's eggs, they vary much in shape and in tint
of colour.   The site selected for the nest is usually some low bushy
shrub, among the rich clusters of *Tecoma australis,* or carefully
hidden in the thick tufts of *Blechnum (B. cartilagineum),* which
often covers a space of many square yards.   In these clumps, where
it clings to the stems of the ferns, I have several times found
two or three pairs breeding at the same time within a few yards
of each other.   The ferns and *Tecomæ* seem to be their favourite
places for breeding, although the nests may often be found placed
suspended between forks in the small bushy oaks *(Casuarinæ).*

In the nest of this Honey-eater I have several times found the egg of *Cuculus inornatus.*" (*Ramsay, Ibis,* 1864, Vol. vi., p. 243.)

Two sets taken at Dobroyde measure as follows :—No. 1, length (A) 0·89 x 0·65 inch ; (B) 0·85 x 0·68 inch ; (C) 0·88 x 0·65 inch. No. 2, length (A) 0·87 x 0·67 inch ; (B) 0·86 x 0·65 inch ; (C) 0·89 x 0·68 inch.

In company with Dr. Hurst I obtained several nests of this species on the 6th of August 1887 at Newington, on the Parramatta River, each of which contained young ones.

Hab. Port Denison, Wide Bay District, Dawson River, Richmond and Clarence River Districts, New South Wales, Interior, Victoria and South Australia. (*Ramsay.*)

2-3   PTILOTIS ORNATA, *Gould.*

Graceful Ptilotis.

Gould, Handbk. Bds. Aust., Vol. i., sp. 314, p. 515.

This species is found rather plentifully in the Mallee Scrubs, which skirt the edges of the Murray River in Victoria and South Australia.   As usual with the genus *Ptilotis*, the nest is suspended by the rim to the horizontal twigs of a tree, the Mallee gum, (*Eucalyptus oleosa)* being particularly favoured in this respect. The nest is an open cup-shaped structure, composed of dried grasses, bark fibre &c., neatly woven together with spiders' webs, and lined inside with a little wool, and similar soft substances. The eggs are two or three in number for a sitting, of a pale reddish-buff, becoming a rich salmon colour towards the larger end, where they are minutely freckled and spotted with irregular shaped markings of reddish-brown, but very sparingly distributed over the rest of the surface.   There is very little material variation in a number of sets of the eggs of this species taken during October and November at Gunbower on the Murray River ; an average set measures as follows :—length (A) 0·75 inch x 0·54 inch ; (B) 0·77 x 0·56 inch. ; (C) 0·74 x 0·54 inch.

The months of September, October, November, and December constitute the breeding season of this species.

Hab. Victoria and South Australia, West and South-West Australia. (*Ramsay.*)

₹ PTILOTIS PLUMULA, *Gould.*
Plumed Ptilotis.

Gould, Handbk. Bds. Aust., Vol. i., sp. 315, p. 516.

"The small, elegant, cup-shaped nest of this species, is suspended from a slender horizontal branch, frequently so close to the ground as to be reached by the hand; it is formed of dried grasses lined with soft cotton-like buds of flowers. The breeding season continues from October to January; the eggs being two in number, ten lines long by seven lines broad, of a pale salmon colour, with a zone of deeper tint at the larger end, and the whole freckled with minute spots of a still darker hue." (*Gould, Handbk. Bds. Aust.,* Vol. i., p. 516.)

Hab. Port Darwin and Port Essington, Gulf of Carpentaria, Interior, Victoria and South Australia, West and South-West Australia. (*Ramsay.*)

₹ - ୭ PTILOTIS PENICILLATA, *Gould.*
White-plumed Honey-eater.

Gould, Handbk. Bds. Aust., Vol. i., sp. 318, p. 519. *XII /5.*

This is the most common species of Honey-eater in the neighbourhood of Melbourne, in Victoria, and Wellington and Dubbo, in New South Wales, and is equally plentiful throughout South Australia. The nest of this species is a neat cup-shaped structure, outwardly composed of grasses, spiders' nests, and the woolly portions of the dead flowers of the "Cape weed," lined inside with the same material and horsehair; in some instances a few. feathers being worked into the side; it is usually suspended by the rim to the fine drooping leafy twigs of a *Eucalyptus*, or to the horizontal fork of an *Acacia;* I have taken the nest of this species within hand's reach of the ground, and at other times at a height of fifty or sixty feet. This Honey-eater always betrays the position of its nest upon the approach of any intruder in its vicinity by its loud and frequent notes of displeasure. Eggs two or three in number for a sitting ; of thirteen sets now before me the ground colour varies from buffy-white to a light saturnine-yellow, the majority of them being minutely freckled and spotted uniformly all over with irregular shaped reddish-chestnut markings ; a few have on the larger end nearly round deep reddish-purple spots, with others of a deep lilac appearing as if beneath the surface of the shell.

A set in the Australian Museum Collection measures (A) 0·78 x 0·55 inch ; (B) 0·79 x 0·55 inch ; (C) 0·76 x 0·58 inch. This set is marked uniformly all over with reddish-chestnut markings.

A set taken in Albert Park, near Melbourne, September 14th 1878, measure as follows :—length (A) 0·83 x 0·57 inch ; (B) 0·8 x 0·57 inch; (C) 0·82 x 0·54 inch.

I have found this species breeding in Victoria as early as the middle of July, and as late as the end of February.

Hab. Wide Bay District, Dawson River, Richmond and Clarence Rivers Districts, New South Wales, Interior, Victoria and South Australia. (*Ramsay.*)

2 PTILOTIS FUSCA, *Gould.*

Fuscous Honey-eater.

Gould, Handbk. Bds. Aust., Vol. i., sp. 319, p. 520.

"The Fuscous Honey-eater breeds in September and the three following months, making a neat cup-shaped nest of stringy-bark, strengthened by the addition of a great quantity of cobweb ; it is lined with fine shreds of bark, hair, and sometimes the silky down from the seed-vessels of the wild cotton, (*Gomphocarpus fruticosus*). It is usually placed among the twigs at the end of some horizontal bough, or among the bushy tops of the young Eucalypti. The Turpentine trees (Syncarpia) also afford favourite sites for their nests, which are two inches and a-half across by two inches deep. The eggs are two in number, from eight and a-half to ten lines long, by six to seven lines in breadth ; the ground colour is of a deep yellowish-buff, with spots of a deeper and more reddish hue, and a few of faint lilac, in some sprinkled equally over the whole surface, in others crowded, or forming a zone at the larger end." (Ramsay, Proc. Phil. Soc., Sydney, 1865, p. 321, pl. i., fig. 4.)

Two sets taken at Dobroyde measure as follows:—No. 1, length (A) 0·76 x 0·54 inch ; (B) 0·76 x 0·55 inch ; No. 2, (C) 0·8 x 0·6 inch ; (D) 0·78 x 0·58 inch.

Hab. Rockingham Bay, Port Denison, Wide Bay District, Dawson River, Richmond and Clarence Rivers Districts, New South Wales, Interior, Victoria and South Australia. (Ramsay.)

2 - 1   PTILOTIS CHRYSOPS, Latham.

Yellow-faced Honey-eater.

Gould, Handbk. Bds. Aust., Vol. i., sp. 320, p. 521.   XII 16.

This Honey-eater is distributed over the whole of the southern portion of the Australian continent ; I found it breeding in great numbers on the shores of Western Port Bay in Victoria, and in the neighbourhood of Sydney it is one of the most common species of the genus Ptilotis, inhabiting also all the parks and gardens in the City. A nest of this species in the Australian Museum Collection taken by Dr. Ramsay at Dobroyde, in September 1865, is a round cup-shaped structure, outwardly composed of fine strips of stringy-bark and moss, lined inside with grasses, and thin wiry bark fibre. Exterior measurements two inches and a-half in diameter, by one inch and three-quarters in depth ; internal diameter two inches, by one inch and a-half in depth. The rim of the nest is worked over a thin forked horizontal branch of a Eucalyptus. The nests of this species I found in the Gippsland Ranges were very beautiful structures, being built of fibrous bark, and the whole exterior thickly covered with a very fine bright-green moss, and neatly lined inside with the soft downy covering of the young fronds of the Tree Fern *(Dicksonia antarctica)*. The nests of this bird are placed in a variety of situations, sometimes being built at the top of a tree, but more often within about six feet of the ground, any suitable tree being selected for the purpose according to the locality in which it is found. At Western Port I frequently found that the nests of this species were placed in prickly Acacia hedges around gardens.

The eggs of this species, are subject to much diversity in their tints of colour, and in the disposition of their markings. Var. A. two eggs in the Australian Museum Collection, taken at Dobroyde in October 1868, are of a fleshy-buff ground colour, with round, deep, reddish-black markings confined to the larger end of the egg. length (A) 0·77 x 0·55 inch ; (B) 0·8 x 0·58 inch.

Var. B. Another set taken in the same locality on July 5th 1860, are of a light saturnine-yellow, heavily blotched with markings of a deeper tint, intermingled with others of a purplish-grey. Length (A) 0·8 x 0·57 inch ; (B) 0·82 x 0·58 inch ; (C) 0·81 x 0·56 inch.

Var. C. A set taken at Hastings, Western Port, on the 30th October 1883, are of a reddish-white ground colour, minutely but thickly freckled all over with chestnut-brown and purplish-grey, particularly towards the larger end where the markings become confluent, and form a well defined zone. Length (A) 0·82 x 0·6 inch ; (B) 0·8 x 0·57 inch ; (C) 0·8 x 0·58 inch.

The two last are the most usual varieties found. A reddish or chestnut-brown tint seems more or less to pervade through most of the eggs of this species, but Dr. Ramsay informs me that he has occasionally taken a set with an almost white ground sprinkled with darkish red dots. The breeding season commences in July and continues during the four following months.

Hab. Rockingham Bay, Port Denison, Wide Bay District, Richmond and Clarence Rivers Districts, New South Wales, Victoria and South Australia. *(Ramsay.)*

PTILOTIS NOTATA, *Gould.*
Yellow-spotted Honey-eater.

Gould, Suppl. Bds. Aust., Pl. 41.					*XIII.* 4.

" The eggs of this species are very similar to those of the other members of the genus, being of a pinkish-white ground colour, with rich dark spots on the thicker end, some confluent, forming blotches larger than usual. They come nearest to those of *Ptilotis auricomis*, and measure (A) 0·9 x 0·65 inch ; (B) 0·91 x 0·64 inch. Taken by Mr. Boyer-Bower near Cairns, Northern Queensland. (*Ramsay, P.L.S., N.S.W.*, 2nd Series, Vol. i., p. 1150.)

Hab. From Port Denison northward to New Guinea. (*Ramsay.*)

Genus PLECTORHYNCHA, *Gould.*

3 - 4	## PLECTORHYNCHA LANCEOLATA, *Gould.*
Lanceolate Honey-eater.

Gould, Handbk. Bds. Aust, Vol. i., sp., 323, p. 525.		*XIII.* 2.

This bird is widely distributed throughout the interior of New South Wales, and is also found on the borders of the Mallee country of Victoria and South Australia. A beautiful nest of this species, taken in the Wimmera District, Victoria, by Mr. James Hill, together with the eggs in September 1881, is a deep cup-shaped structure, being nearly as wide at the bottom as at the top ; it is composed of fibrous roots and grasses, which are completely hidden with an almost snow-white covering of sheep's wool ; the bottom of the nest inside being neatly lined with very fine grasses and a small quantity of horsehair. It measures internally three inches in diameter, by four inches in depth, and was attached by the rim to the thin pendant branches of a *Casuarina*, hanging within a few feet of the ground. Mr. K. H. Bennett also found this bird breeding plentifully in the neighbourhood of Ivanhoe and Mossgiel, in the interior of New South Wales. Eggs three or four in number for a sitting, usually three, elongated in form.

N

Two eggs of this species in the Australian Museum Collection, are fleshy-white, minutely but thickly and uniformly freckled all over with reddish-brown markings, intermingled with others of a pale lilac. Length (A) 0·95 x 0·65 inch ; (B) 0·95 x 0·65 inch ; (C) 0·96 x 0·65 inch.

There is usually very little variation in the form or distribution of the markings on the eggs of this species, but in one exception a set taken by Mr. K. H. Bennett at Mossgiel in November, is much more rounded in form, and has large bran-like markings of faint chestnut and yellowish-brown intermingled with freckles and spots of pale lilac, the markings being mostly confined to the larger end. Length (A) 0·85 x 0·65 inch ; (B) 0·86 x 0·67 inch ; (C) 0·82 x 0·67 inch.

The months of September, October, and November constitute the breeding season of this species.

Hab. Rockingham Bay, Wide Bay District, Dawson River, New South Wales, Interior, Victoria and South Australia. (*Ramsay.*)

GENUS MELIPHAGA, *Lewin*.

2-3 MELIPHAGA PHRYGIA, *Latham*.

Warty-faced Honey-eater.

Gould, Handbk. Bds. Aust., Vol. i., sp. 324, p. 527. *XII. 8.*

" The nest of this species is a neat cup-shaped structure composed of stringy-bark, and lined with finer shreds of the same material. It is two inches and a-half across inside, by one inch and a-half deep, and placed between the upright forks of some tall sapling, or upon a horizontal bough.   They breed from October to December, or perhaps earlier in some localities, and lay two or three eggs, ten to eleven lines and a-half long, by eight lines and a-half to nine lines in breadth.   These, when freshly taken, are certainly among the most beautiful I have ever met with ; but unfortunately, as in most bird's eggs, the bloom goes off, and the bright tint soon fades. From my note book, I find that when first taken from the nest they are of a deep saturnine buff, spotted with irregular markings of a deeper hue in some, evenly distributed over their surface, in others, or crowded at the larger end; there are also a few indistinct dots of greyish lilac dispersed over the surface; but these lilac dots arè not visible in all specimens. I have one however, in which greyish-lilac spots predominate. This species of Honey-eater was one of the first known, and was described under various names, and placed in several genera by as many different authors; but as its habits and economy became more perfectly understood, and ornithologists began to classify their birds more from their habits, &c., this species was finally placed among the Honey-eaters, and a new genus formed for its reception." (*Ramsay, Proc. Phil. Soc., Sydney*, 1865, p. 319 pl. i., fig. 3.)

A set taken at Dobroyde on the 1st of October 1865, measures as follows:—length (A) 0·95 x 0·72 inch ; (B) 0·97 x 0·71 inch.

This is a very common species in the neighbourhood of Wellington, New South Wales.

Hab. Rockingham Bay, Wide Bay District, Richmond and Clarence Rivers Districts, New South Wales, Interior, Victoria and South Australia. (*Ramsay.*)

GENUS CERTHIONYX, *Lesson.*

2   CERTHIONYX LEUCOMELAS, *Cuvier.*

(*Melicophila picata*, Gould.)

Pied Honey-eater.

Gould, Handbk. Bds. Aust., Vol. i., sp. 325, p. 529.

"Mr. K. H. Bennett informs me that this species constructs a nest very similar to that of *Meliphaga phrygia*, but of much finer materials, and resembles that of a *Rhipidura;* it is placed on a horizontal branch, and is cup-shaped, composed of strips of fine bark and lined with fur and hair, it is about 1·5 inches high and three inches in diameter. The eggs taken in the Lachlan district were two in number, of a beautiful pale greenish-blue, with rich reddish dots, which cluster and form irregular patches towards the thicker end, but do not form a zone. Another specimen (2) has only a few faint reddish spots and a black dot here and there, very sparingly sprinkled over the surface, length (1) 0·82 x 0·6 inch ; (2) 0·82 x 0·62 inch. *From Mr. K. H. Bennett's Coll."* *(Ramsay, P.L.S., N.S.W.,* Vol. vii., p. 414.)

Hab. New South Wales, Interior, Victoria and South Australia, West and South-west Australia. *(Ramsay.)*

Genus ENTOMOPHILA, *Gould.*

3   ENTOMOPHILA ALBIGULARIS, *Gould.*

White-throated Honey-eater.

Gould, Handbk. Bds. Aust., Vol. i., sp. 327, p. 532.

Gilbert remarks on the nidification of this species as follows :—
" Its small pensile nest is suspended from the extremity of a weak projecting branch in such a manner that it hangs over the water, the bird always selecting a branch bearing a sufficient number of leaves to protect the entrance from the rays of the sun ; in form the nest is deep and cup-like, and is composed of narrow strips of the soft paper-like bark of the *Melaleuca,* matted together with small vegetable fibres, and slightly lined with soft grass. I found a nest in the latter part of November, and another in the early part of December which contained three eggs in each, while a third procured towards the end of January had only two ; the eggs are rather lengthened in form and not very unlike those of *Malurus cyaneus* in the colour and disposition of their markings ; their ground colour being white, thinly freckled all over with bright chestnut-red particularly at the larger end ; they are nine lines long and six broad." *(Gould, Handbk. Bds. Aust.,* Vol. i., p. 532.)

Hab. Derby, N.W. Australia, Port Darwin and Port Essington, Gulf of Carpentaria, Cape York, South Coast New Guinea. (*Ramsay.*)

2 -(3) ENTOMOPHILA RUFIGULARIS, *Gould.*

Red-throated Honey-eater.

Gould, Handbk. Bds. Aust., Vol. i., sp. 328, p. 533.

"This species is found commonly dispersed all over the Gulf of Carpentaria country. It has been found breeding in the neighbourhood of Normanton and Georgetown, during the months from September to March. The nest is a round, open, and neat cup-shaped structure, usually slung by the rim between forked twigs. The one before me, sent by Mr. Armit, was taken from a branch of an *Erythrina*. It is composed of fine grasses, matted outside with white 'cobwebs,' and lined with fine grasses alone. It is rather a deep nest, being three inches and a-half long by two inches in diameter. The eggs are usually two, but sometimes three in number, of a pearly-white, rather thickly spotted with bright reddish-brown. Length 0·64 x 0·49 inch in diameter at the thicker end. The young on leaving the nest have all the upper surface brown, and all the under surface white; the outer webs of the wing quills margined with olive-yellow." (*Ramsay, P.L.S., N.S. W.*, Vol. ii., p. 111.)

Hab. Derby, N.W. Australia, Port Darwin and Port Essington, Gulf of Carpentaria, Cape York. (*Ramsay.*)

Genus ACANTHOGENYS, *Gould.*

2 ACANTHOGENYS RUFIGULARIS, *Gould.*

Spiny-cheeked Honey-eater.

Gould, Handbk. Bds. Aust., Vol. i., sp. 329, p. 534. XII 10.

This bird is a lover of the scrubby portions of the interior of Australia, and the Mallee country of Victoria and South Australia.

A nest of this species in the Australian Museum Collection presented by Mr. K. H. Bennett of Mossgiel, is a round, cup-shaped, and somewhat scanty structure, composed of long dried grasses, and stalks of plants, the latter bent into position when green, and held together with spiders' webs ; a peculiarity in this and in all the nests that I have seen of this species, being that many of the grasses and stalks are not worked in horizontally around the structure, but perpendicularly, though not necessarily straight, from the rim to the bottom of the nest where they cross and recross each other.   Exterior diameter, three and three-quarters of an inch, by two inches and a-half in depth ; internal diameter three inches by two inches in depth.   The nest is attached by the rim to the thin branch of a tree, usually at no great height from the ground.   Eggs two in number for a sitting.

Two eggs taken by Mr. K. H. Bennett, at Mossgiel, in the month of October, are of a very pale yellowish-brown ground colour becoming darker towards the larger end, where they are deeply marked with dark umber-brown and slaty-brown spots, intermingled with others of a wood-brown and dark lilac tint, appearing as if beneath the surface of the shell.   Length (A) 1·02 x 0·75 inch ; (B) 1·01 x 0·71.

Specimens taken in the Wimmera District, Victoria, and on the Lachlan River, New South Wales, are all alike in colour and in the disposition of their markings.   The breeding season commences in September and continues during the two following months.

Hab. Dawson River, New South Wales, Interior, Victoria and South Australia, West and South-west Australia.   (*Ramsay.*)

Genus ACANTHOCHÆRA, *Vigors and Horsfield.*

2-3   ACANTHOCHÆRA INAURIS, *Gould.*

Tasmanian Wattled Honey-eater.

Gould, Handbk. Bds. Aust., Vol. i., sp. 330, p. 536.

This bird is generally dispersed over the whole of Tasmania, and is found breeding in the vicinity of Hobart.   The nest is slightly larger, but in other respects similar to that of the Australian species, *A. carunculata*, Latham. Eggs two or three in number for a sitting. A set taken near Hobart in September 1885, are of a very pale salmon, or reddish-buff ground colour, one specimen (A) being spotted all over with rich reddish-brown markings, while appearing as if beneath the surface of the shell are others of a deep bluish-grey, particularly towards the larger end; specimen B, has a few irregularly shaped markings of yellowish-red and light chestnut-brown, sparingly distributed over the surface of the shell, together with numerous sub-surface spots and blotches of deep bluish-grey. Length (A) 1·37 x 0·94 inch; (B) 1·35 x 0·93 inch.

Hab. Tasmania.

3   ACANTHOCHÆRA CARUNCULATA, *Latham.*

Wattled Honey-eater.

Gould, Handbk. Bds. Aust., Vol. i., sp. 331, p. 538.   *XII. ℐ.*

The Wattled Honey-eater is widely dispersed over all the eastern and southern portions of the continent of Australia. The nest is an open one, rather roughly but compactly formed outwardly of twigs, and lined inside with dried grasses; it is placed on the horizontal branch of a tree, often in a low gum sapling, but not unfrequently in a Eucalyptus thirty or forty feet from the ground. The eggs are usually three in number for a sitting. A set in the Australian Museum Collection, are elongated in form, of a light saturnine-red, becoming slightly darker in tint towards the larger end, where they are spotted and blotched with irregular shaped markings of chestnut and yellowish-brown, intermingled with sub-surface spots of deep bluish-grey. Length (A) 1·27 x 0·84 inch; (B) 1·3 x 0·83 inch.

A set taken in the Albert Park, near Melbourne, in October 1878, are of a rich flesh colour, thickly spotted all over with roundish markings of dark chestnut-brown, a few spots of deep violet appearing as if beneath the surface of the shell. Length (A) 1·25 x 0·88 inch ; (B) 1·3 x 0·86 inch.

This species commences to breed in August and continues during the three following months.

Hab. Wide Bay District, Richmond and Clarence Rivers Districts, New South Wales, Interior, Victoria and South Australia, West and South-West Australia. (*Ramsay.*)

Genus ANELLOBIA, *Cabanis.*

2 - 3   ANELLOBIA MELLIVORA, *Latham.*

Brush Wattle-bird.

Gould, Handbk. Bds. Aust., Vol. i., sp. 332, p. 541.   *XII . 6*

The Brush Wattle-bird is found throughout the eastern and southern portions of Australia, and the whole of Tasmania. I found this species breeding in great numbers during October 1882 near Hastings, at Western Port. The nest is an open one, outwardly composed of fine twigs, lined inside with roots, and placed in the fork of a tree, not far from the ground, low Eucalyptus saplings, being especially favoured in this respect. The eggs are two or three in number for a sitting.

A set in the Australian Museum Collection are of a rich salmon colour, spotted and blotched with irregularly shaped clouded markings of dark chestnut-brown, and a few of reddish-brown, together with sub-surface markings of deep bluish-grey, particularly towards the larger end, where they assume the form of a zone. Length (A) 1·04 x 0·75 inch ; (B) 1·02 x 0·76 inch.

There is very little variation shown in a number of sets taken at Dobroyde by Dr. Ramsay during 1858–60, and those procured by myself at Hastings in 1882.

The breeding season of this species extends during the months of September, October, November, and December.

Hab. Wide Bay District, Richmond and Clarence Rivers Districts, New South Wales, Victoria and South Australia, Tasmania. *(Ramsay.)*

ANELLOBIA LUNULATA, *Gould.*
Lunulated Wattle-bird.
Gould, Handbk. Bds. Aust., Vol. i., sp. 333, p. 543.

"A remarkable circumstance, says Gilbert, connected with the incubation of this bird is that it appears to lay but a single egg and to have no regular time of breeding, its nest being found in abundance from August to November. It is rather small in size, and is deposited in the fork of a perpendicular growing branch : the tree most generally chosen is that called by the colonists of Swan River the stink-wood, but it has been found in the parasitic clump of a Banksia, and also in a small scrubby bush two or three feet from the ground ; but it is more frequently constructed at a height of at least eight or twelve. It is formed of dried sticks, and lined with Zamia wool, soft grasses or flowers, and sometimes with sheep's wool. The egg is rather lengthened in form, being one inch and two lines long by nine and a-half lines broad ; its ground colour is a full reddish-buff thinly spotted and marked with deep chestnut-brown and chestnut-red, some of the spots and markings appearing as if beneath the surface of the shell, and being most thickly disposed near the larger end." *(Gould, Handbk. Bds. Aust.*, Vol. i., p. 543.)

Hab. West Australia.

GENUS TROPIDORHYNCHUS, *Vigors and Horsfield.*
TROPIDORHYNCHUS CORNICULATUS, *Latham.*
Friar Bird.
Gould, Handbk. Bds. Aust., Vol. i., sp. 334, p. 545.

This bird is plentifully dispersed over the eastern and south-eastern portions of the continent of Australia. A nest of this species in the Australian Museum Collection, is an open cup-shaped structure outwardly composed of long, narrow, strips of stringy-bark, wool, and grass, firmly matted together, and lined inside with fine dried grasses, it is attached by the rim to a thin forked branch of a Eucalyptus, which is entirely hidden by the materials of which the nest is composed being worked over it; it measures exteriorly six inches and three-quarters in diameter, by four inches in depth; internal diameter three inches and a-half, depth two inches and a-quarter.   Another nest in the same collection, measures one inch deeper, externally and internally. The nest is always built at the extremity of a thin branch, some-times within a few feet of the ground, but not unfrequently at an altitude of forty or fifty feet.

Eggs three in number for a sitting, usually elongated in form; among a number of sets now before me, no two have exactly the same tints of ground colour or markings; and fade very much after being exposed to the light for any length of time.   Two eggs in the Australian Museum Collection, are of a rich salmon colour, heavily blotched with chestnut-red and deep bluish-grey; the latter colour appearing as if beneath the surface of the shell. Length (A) 1·26 x 0·9 inch; (B) 1·22 x 0·9 inch.

A set in the Dobroyde Collection taken by Mr. James Ramsay at Cardington, in December 1867, are elongated in form, of a dull, light pinkish-salmon colour, spotted and blotched with faint chestnut, and intermingled with obsolete markings of slaty-lilac, particularly towards the larger end.   Length (A) 1·37 x 0·9 inch; (B) 1·44 x 0·92 inch; (C) 1·38 x 0·92 inch.

The breeding season commences in October and extends over the months of November and December.

Hab. Rockingham Bay, Port Denison, Wide Bay District, Dawson River, Richmond and Clarence Rivers Districts, New South Wales, Interior, Victoria and South Australia.   (*Ramsay.*)

Genus PHILEMON, *Vieillot.*

3   PHILEMON CITREOGULARIS, *Gould.*

Yellow-throated Friar Bir.d

Gould, Handbk. Bds. Aust., Vol. i., sp. 337, p. 549.

This bird is universally dispersed throughout the whole interior of Australia, along the banks of rivers and water courses, from which it is seldom found far away. The nest is a cup-shaped structure, outwardly composed of dried grasses, lined inside with wool or fur, and placed at the extremity of a drooping branch, near or overhanging water. The eggs are three in number for a sitting. A set taken by Mr. K. H. Bennett, on the Lachlan River, in October 1882, are of a reddish-salmon ground colour, which is nearly obscured by very faint, clouded and streaky markings of reddish-purple, and purplish-grey, the latter colour appearing as if beneath the surface of the shell, the markings becoming darker towards the larger end. Length (A) 1·15 x 0·8 inch; (B) 1·12 x 0·8 inch; (C) 1·03 x 0·75 inch.

This species breeds during the months of September, October, and November.

Hab. Cape York, Rockingham Bay, Port Denison, Wide Bay District, Richmond and Clarence Rivers Districts, New South Wales, Interior, Victoria and South Australia. (*Ramsay.*)

2-3   PHILEMON SORDIDUS, *Gould.*

Sordid Friar Bird.

Gould, Handbk. Bds. Aust., Vol. i., sp. 338, p. 550.   *XII.* 3

" The nest of this species is very similar to that of the members of the genus *Tropidorhynchus,* being a cup-shaped structure of bark and grass, slung by the rim to the forks of twigs at the end of some horizontal or drooping branch. It is about half the size of that of *Tropidorhynchus corniculatus,* and equal to that of *Philemon citreogularis,* of which this species is but a northern form. The eggs are two or three in number, of a rich salmon-red, spotted with a darker tint, some of the spots fleecy, confluent, and distributed alike all over the surface of the shell, rather closer near the thicker end, but not forming a zone there ; in A., a few are confluent on the thick end forming a blotch on the top of the egg. In B. the spots are more scattered and obsolete markings of pale lilac are dispersed here and there over the surface. Length (A) 1·04 x 0·7 inch ; (B) 1·05 x 0·75 inch. *From Mr. Barnard's Collection. (Ramsay, P.L.S., N.S.W.*, Vol. vii., p. 52.)

Hab. Derby, N.W. Australia, Port Darwin and Port Essington, Gulf of Carpentaria, Dawson River. *(Ramsay.)*

GENUS ACANTHORHYNCHUS, *Gould.*

2   ACANTHORHYNCHUS TENUIROSTRIS,*Latham.*
Spine-bill.

Gould, Handbk. Bds. Aust., Vol. i., sp. 339, p. 551.   *XII* . *2D.*

This pretty bird is widely distributed over the eastern and southern portions of the continent of Australia, it is a tame and familiar species being found alike in our public and private gardens where it may be seen extracting its food from the various flowers with its long slender spine-like bill.   The nest of this species is placed in a low tree, or thick bush ; one found at Heathcote on the Illawarra Line, on the 30th of October 1886, was built in the dead leafy twigs of a gum sapling ; it was cup-shaped, and rather roughly formed on the exterior with fine twigs and grasses, and lined inside with feathers, it contained two eggs in an advanced state of incubation.   Eggs two in number for a sitting, somewhat pyriform, of a pale buff, becoming deeper in tint and approaching a light saturnine-red at the larger end, where they are marked with dark reddish-brown spots, intermingled with others of a deep bluish-grey, in some instances forming a zone, in others scattered over the surface of the shell, but always becoming more numerous on the larger end. Two eggs in the Australian Museum Collection measure as follows :—length (A) 0·77 x 0·57 inch ; (B) 0·76 x 0·54 inch.

A set taken at Dobroyde in September 1860, give the following measurements :—length (A) 0·73 x 0·54 inch ; (B) 0·75 x 0·55 inch.

In New South Wales this species commences to breed in August and continues during the four following months.

Hab. Wide Bay District, Richmond and Clarence Rivers New South Wales, Victoria and South Australia. (*Ramsay.*)

2 ACANTHORHYNCHUS DUBIUS, *Gould.*

Gould, P.Z.S., part v., p. 25.

"Some ornithologists do not consider this a good species, Mr. Gould himself, who first pointed out its difference from the northern and eastern Australian continental forms, inclining to believe them identical ; but as the Tasmanian bird is smaller in all its admeasurements, and much richer and deeper in the tints of the under surface, I give the description of a set of eggs taken near Hobart in October 1885. Eggs two in number for a sitting of a pale buff, approaching a light saturnine-yellow on the larger end, where they are minutely spotted with irregularly shaped markings of deep chestnut-brown, and a few nearly obsolete spots of bluish-grey. Length (A) 0·73 x 0·53 inch ; (B) 0·75 x 0·54 inch." (*North, P.L.S., N.S.W.,* 2nd Series, Vol. ii., p. 408.)

Hab. Tasmania. (*Gould.*)

2 ACANTHORHYNCHUS SUPERCILIOSUS, *Gould.*
White-eyebrowed Spine-bill.
Gould, Handbk. Bds. Aust., Vol. i., sp. 340, p. 553.

"The nest, which is constructed among the large-leaved Banksias, is of a round compact form, and is composed of dried fine grasses,

222   MELIPHAGIDÆ.

tendrils of flowers, narrow threads of bark, and fine wiry fibrous
roots matted together with Zamia wool, forming a thick body,
which is warmly lined with feathers and Zamia wool mingled
together; the external diameter of the nest is three inches, and
that of the cavity about one inch and a-quarter. The eggs are
two in number, nine lines long by six and a-half broad; their
ground colour in some instances is a delicate buff, in others a very
delicate bluish-white, with a few specks of reddish-brown distributed
over the surface, these specks being most numerous at the larger
end, where they frequently assume the form of a zone. The
breeding season is in October." (*Gould, Handbk. Bds. Aust.*, Vol.
i., p. 553.)

Hab. West and South-West Australia. (*Ramsay*)

GENUS MYZOMELA, *Vigors and Horsfield.*

2-(?)   MYZOMELA SANGUINOLENTA, *Latham.*

Sanguineous Honey-eater.

Gould, Handbk. Bds. Aust., Vol. i., sp. 341, p. 555. *XIII*. 20.

"The Sanguineous Honey-eater breeds during the months of
October, November, and December, making a neat but somewhat
scanty nest of stringy-bark, seldom with any other lining. It is
suspended between a fork or twigs at the end of some bough in
a bush, or among the upright and topmost branches of the
Melaleuca. The nest is perhaps smaller than that of any other
Australian bird, being in some instances scarcely one inch and a-
half in diameter by one inch in depth. The eggs are two, seldom
three, in number, of a delicate white strongly marked with reddish-
and yellowish-brown spots, more numerous at the larger end.
They are from six to seven lines in length, and from five to six
lines in breadth." (*Ramsay, Ibis*, 1865, p. 304.)

The markings on these eggs frequently become confluent on the
larger end, and assume the form of a zone. Two sets taken at

Dobroyde in September 1861, measure as follows :—No. 1, length (A) 0·58 x 0·46 inch; (B) 0·61 x 0·47 inch. No. 2, length (A) 0·59 x 0·47 inch; (B) 0·58 x 0·46 inch.

Hab. Rockingham Bay, Port Denison, Wide Bay District, Dawson River, Richmond and Clarence Rivers Districts, New South Wales, Victoria and South Australia. (*Ramsay*)

3   MYZOMELA NIGRA, *Gould.*

Black Honey-eater.

Gould, Handbk. Bds. Aust , Vol. i., sp. 344, p. 558.

" The nest is a shallow cup-shaped structure of fine shreds of bark or similar material, usually placed between the horizontal fork of a branch. Mr. K. H. Bennett informs me that some years ago this species was found to be plentiful near Mossgiel, feeding in the Sandalwood trees *(Myoporum platycarpum)*. Eggs two for a sitting, they are of a dull white or cream-white with an indistinct zone, which in some consists of distinct dots of dull brown near the thicker end, in others clouded markings of light brown. Length (A) 0·6 x 0·47 inch ; (B) 0·63 x 0·48 inch." (*Ramsay, P.L.S., N.S.W.*, 2nd Series, Vol. i., p. 1151.)

Hab. Port Darwin and Port Essington, New South Wales, Interior, Victoria and South Australia, West and South-West Australia. (*Ramsay.*)

GENUS ENTOMYZA, *Swainson.*

2 -(3)   ENTOMYZA CYANOTIS, *Swainson.*

Blue-faced Honey-eater.

Gould, Handbk. Bds. Aust., Vol. i., sp. 346, p. 560.   XII. *1.*

This bird has an extensive range over the eastern portions of Australia from Rockingham Bay in the north to the vicinity of the Murray River in Victoria and South Australia in the south. I have never heard of this bird constructing a nest for itself, but it re-lines the deserted tenements of *Myzantha garrula*, *Acanthochæra caruncula*, or a depression in the top of the dome-shaped nest of *Pomatostomus temporalis*. The eggs are two in number for a sitting, but sometimes a third is added. Among a number of sets of the eggs of this species now before me, taken in Queensland, New South Wales, and Victoria, there is some variation in the tints and disposition of their markings, even when from the same locality, although they are nearly uniform in their measurements.

A pair in the Australian Museum Collection are of a rich salmon colour boldly blotched, but particularly towards the larger end, with deep reddish markings, where a few obsolete spots of bluish-grey appear as if beneath the surface of the shell. Length (A) 1·24 x 0·87 inch ; (B) 1·24 x 0·86 inch.

A set in the Dobroyde Collection, taken by Mr. John Ramsay at Cardington on the Bell River, on the 2nd of December 1867, are of a deep salmon colour, with a zone of small irregular shaped markings of reddish-brown, intermingled with others of bluish-grey appearing as if beneath the shell ; the remainder of the surface being but sparingly marked. Length (A) 1·25 x 0·85 inch ; (B) 1·3 x 0·86 inch.

A set taken at Gunbower on the Murray River, in September 1883, are of a rich flesh colour, with a zone of roundish clouded spots of dark reddish-brown, together with subsurface markings of different shades of lilac and bluish-grey. Length (A) 1·23 x 0·87 inch ; (B) 1·18 x 0·86 inch.

The breeding season of this species commences in September and lasts until the end of February.

Hab. Rockingham Bay, Port Denison, Wide Bay District, Dawson River, Richmond and Clarence Rivers Districts, New South Wales, Interior, Victoria and South Australia. (*Ramsay.*)

Genus MELITHREPTUS, *Vieillot.*

3   MELITHREPTUS VALIDIROSTRIS, *Gould.*

Strong-billed Honey-eater.

Gould, Handbk., Bds. Aust., Vol. i., sp. 347, p. 565.

The nest of this species is an open cup-shaped structure composed of dried grasses and the flowering portions of plants, matted together and slung by the rim usually to the fine twigs of a Eucalyptus. Eggs three in number for a sitting, of a fleshy-buff ground colour, becoming darker towards the larger end, where they are thickly spotted with purplish-brown and superimposed markings of deep greyish-lilac. Length (A) 0·89 x 0·64 inch ; (B) 0·87 x 0·65 inch ; (C) 0·87 x 0·66 inch.

A set in the Collection of Dr. James C. Cox measure as follows:— length (A) 0·84 x 0·62 inch ; (B) 0·83 x 0·61 inch ; (C) 0·81 x 0·63 inch.

Hab. Tasmania.

3 MELITHREPTUS BREVIROSTRIS, *Vigors and Horsfield.*

Short-billed Honey-eater.

Vig. and Horsf., Trans. Linn. Soc., Vol. xv., p. 315.

This bird ranges over the eastern, south-eastern, and interior portions of the continent of Australia. In the Dobroyde Collection are the nest and eggs of this species, together with the birds shot therefrom, they were obtained by Mr. J. Ramsay at Cardington, on the Bell River, in November 1867. Like those of the other members of the genus, the nest was suspended by the rim to the thin twigs, at the extremity of a branch of a Eucalyptus ; it is outwardly composed of grasses matted together with a little wool, and lined inside with opossum fur. Eggs three in number for a sitting, of a pale salmon ground colour, rather indistinctly marked with short wavy lines and spots of reddish-chestnut, but particularly towards the larger end, where together with subsurface markings

O

of a deeper tint, they form an ill-defined zone.   Length (A) 0·78 x 0·57 inch ; (B) 0·76 x 0·54 inch ; (C) 0·75 x 0·54 inch.

Hab. Rockingham Bay, Port Denison, Wide Bay District, Richmond and Clarence Rivers Districts, New South Wales, Interior, Victoria and South Australia. *(Ramsay.)*

2   MELITHREPTUS GULARIS, *Gould.*

Black-throated Honey-eater.

Gould, Handbk. Bds. Aust, Vol. i., sp. 348, p. 566.

In a list of birds met with in North-eastern Queensland, published in the Proceedings of the Zoological Society of London, Dr. Ramsay writes as follows regarding this bird :—

" This species appears to be plentiful, but not in the immediate vicinity of the coast.   It is not rare about Maryborough, and is also found on the Upper Herbert.   It has considerable powers of song, which may be heard often at daylight in the morning.   While camped on the banks of the Gregory, a pair of these birds frequented a Wattle-tree *(Acacia)* near to our 'tent' (a *sheet of bark !*), and delighted us every morning for many days by pouring out their varied and pleasing song, which often lasted for ten or fifteen minutes without ceasing.   I have since heard their song under more comfortable circumstances ; and my brother and I at once recognized our old friends.   The nest and eggs are similar, but slightly larger than those of *M. lunulatus;* eggs two in number, pale salmon-pink with deep reddish-salmon dots on the larger end ;   length 0·73 x 0·55 inch.   The nest is cup-shaped, slung by the rim between twigs at the end of a leafy bough, and composed of fine grasses and strips of bark webbed together with spiders' nests." *(Ramsay, P.Z.S.,* 1876, p. 118.)

Hab. Rockingham Bay, Port Denison, Wide Bay District, New South Wales, Victoria and South Australia. *(Ramsay.)*

3 - 3   MELITHREPTUS LUNULATUS, *Shaw.*

Lunulated Honey-eater.

Gould, Handbk. Bds. Aust., Vol. i., sp. 349, p. 568.

This bird is found breeding plentifully in New South Wales, Victoria, and South Australia. Amongst a number of nests presented by Dr. Ramsay to the Trustees of the Australian Museum, is one of this species, taken in October 1864; it is a deep cup shaped structure outwardly composed of shreds of stringy-bark *(Eucalyptus obliqua)*, closely matted and held together with cobweb, wool, &c., and lined inside with hair; it is slung by the rim to the leafy twigs of a Eucalyptus, exterior measurements two inches and a-quarter in diameter, by two inches and a-half in depth; internal diameter one inch and a-half across by one inch and a-half in depth. Eggs two or three in number for a sitting, of a yellowish-buff ground colour, with spots of a deeper and more reddish hue; some specimens being uniformly spotted all over, but more often assuming the form of a zone.

A set taken at Dobroyde in July 1861, measure as follows:— length (A) 0 73 x 0·54 inch; (B) 0·7 x 0·55 inch; (C) 0·75 x 0·52 inch.

Two eggs of this species in my collection, taken at Hastings in October 1883, measure as follows :—length (A) 0·8 x 0·6 inch; (B) 0·81 x 0·59 inch. I also procured an egg of *Cacomantus pallidus*, from the same nest.

This bird usually breeds during August and the three following months in Victoria, but there are eggs of this species in the Dobroyde Collection taken at Dobroyde, New South Wales, in June 1859, and July 1861.

Hab. Wide Bay District, Richmond and Clarence Rivers Districts, New South Wales, Victoria and South Australia. (*Ramsay*)

3   MELITHREPTUS CHLOROPSIS, *Gould.*

Swan River Honey-eater.

Gould, Handbk. Bds. Aust., Vol. i., sp. 350, p. 570.

"The nest of this species is usually suspended from the small branches near the top of the gum-trees, where the foliage is thickest, which renders it extremely difficult to detect. A nest found by Gilbert in October, was formed of sheep's wool and small twigs; another found by him in November was attached to a small myrtle-like tree, in a thick gum forest, not more than three feet from the ground ; both these nests contained three eggs, nine and a-half lines long by six and a-half lines broad, of a deep reddish-buff, thinly spotted all over, but particularly at the larger end with dark reddish-brown, some of the spots being indistinct, while others were very conspicuous." (*Gould, Handbk. Bds. Aust.*, Vol. i., p. 570.

Hab. Western Australia.

2   MELITHREPTUS ALBIGULARIS, *Gould.*

White-throated Honey-eater.

Gould, Handbk. Bds. Aust, Vol. i., sp. 351, p. 571.

"This species, which inhabits the northern and eastern parts of Australia, is very abundant on the Cobourg Peninsula. The nest, which is always suspended to a drooping branch, and which swings about with every gust of wind, is formed of dried narrow strips of the soft bark of the *Melaleuca*. The eggs which are generally two in number, are of a light salmon colour, blotched and freckled with reddish-brown, and are about nine lines long by six lines broad." (*Gould, Handbk. Bds. Aust.*, Vol. i., p. 571.)

Hab. Derby, N.W. Australia, Port Darwin and Port Essington, Gulf of Carpentaria, Cape York, Rockingham Bay, Port Denison, Wide Bay District, South Coast New Guinea. (*Ramsay.*)

Genus MYZANTHA, *Vigors and Horsfield.*

3 MYZANTHA GARRULA, *Latham.*

Garrulous Honey-eater.

Gould, Handbk. Bds. Aust., Vol. i., sp. 353, p. 574. *XII . 2.*

This bird is universally distributed over the open forest lands of the eastern and southern portions of the continent of Australia, and the whole of Tasmania. The nest is cup-shaped, outwardly composed of long thin twigs, and lined inside with fibrous roots, grasses, wool &c. ; one now before me measures exteriorly six inches in diameter, by four inches and three-quarters in depth ; internal diameter three inches and a-quarter by two inches and a quarter in depth. It is usually placed in the thin upright forks at the top of a small gum sapling, but not unfrequently in the branch of a Eucalyptus thirty or forty feet from the ground. The eggs are usually three in number for a sitting, and vary both in their size and markings, I give descriptions of the most usual varieties found.

Var. A. Ground colour buffy-white, thickly freckled and spotted all over, but particularly towards the larger end with reddish-brown and a few indistinct markings of pale lilac. Length (A) 1·02 x 0·75 inch ; (B) 1·04 x 0·75 inch ; (C) 1·04 x 0·75 inch. *Aust. Mus. Coll.*

Var. B. Ground colour light saturnine-red, boldly blotched all over with irregular shaped markings of deep reddish-chestnut and pale bluish-grey. Length (A) 1·02 x 0·75 inch ; (B) 1·04 x 0·75 inch. · Taken at Nanama near Yass, by Mr. James Ramsay, October 1860.

Var. C. Ground colour buffy-white, nearly obscured by bran-shaped markings of reddish-brown, which at the larger end form an irregular shaped coalesced patch. Length (A) 1·13 x 0·8 inch ; (B) 1·18 x 0·8 inch ; (C) 1·18 x 0·8 inch. Taken near Hastings, October 1883.

August and the five following months constitute the breeding season of this species.

Hab. Wide Bay District, Richmond and Clarence Rivers Districts, New South Wales, Victoria and South Australia, Tasmania. (*Ramsay*)

MYZANTHA OBSCURA, *Gould.*
Sombre Honey-eater.

Gould, Handbk. Bds. Aust., Vol. i., sp. 354, p. 576.

"This species inhabits Swan River and the south-western portion of Australia generally, where it beautifully represents the *Myzantha garrula* of New South Wales. The nest is built on an upright fork of the topmost branches of the smaller gum-trees, and is formed of small dried sticks lined with soft grasses and feathers. The eggs are eleven and a-half lines long by nine lines broad, of a rich orange-buff, obscurely spotted and blotched with a deeper tint, particularly at the larger end."· (*Gould, Handbk. Bds. Aust.*, Vol. i., p. 576.)

Hab. West Australia.

4-5　　　MYZANTHA FLAVIGULA, *Gould.*
Yellow-throated Miner.

Gould, Handbk. Bds. Aust., Vol. i., sp. 356, p. 578.　　XII　4.

"The nest is a neat round structure of fine twigs, occasionally ornamented with wool and the egg-bags of spiders &c., giving the outside a beautiful white appearance; the inside is lined with hair of different kinds and wool, the inside diameter is three inches, the depth two and a-quarter inches; it is placed among the branches of trees and shrubs frequently near the ground. The eggs are four or five for a sitting, of a rich salmon colour with dark salmon-red spots and dots all over the surface of the shell, but larger and closer on the thicker end.   Length (A) 1·02 x 0·75 inch ; (B) 1·02 x 0·76 inch ; (C) 1·02 x 0·75 inch ; (D) 1 inch x 0·75 inch.'' (*Ramsay, P.L.S., N.S.W.*, Vol. vii., p. 52.)

Hab. Gulf of Carpentaria, Dawson River, New South Wales, Interior, Victoria and South Australia.   (*Ramsay.*)

Genus MANORHINA, *Vieillot.*

2-3   MANORHINA MELANOPHRYS, *Latham.*

Bell Bird.

Gould, Handbk. Bds. Aust., Vol. i., sp. 357, p. 579.

This species is confined to the eastern and south-eastern portions of the continent of Australia, and is very local in its habits.   I have frequently observed it in a belt of scrub on swampy ground, about a quarter of a mile in diameter in South Gippsland where it used to breed, but during my many visits to the locality, extending over a period of nine years, I have never seen it elsewhere in that district, either in the ranges or any where in the vicinity. The nest is a frail structure, suspended by the rim to the thin twigs of a low bush or tree, it is an open cup-shaped structure, composed of roots, grasses, and lichens loosely matted together.   Eggs two or three in number for a sitting.   A set in the Australian Museum Collection are of a light saturnine-red ground colour, with a few bold blotches and spots of reddish-chestnut and purplish-brown, but more particularly towards the larger end, where a few sub-surface markings of purplish-grey also appear.   Length (A) 0·88 x 0·67 inch ; (B) 0·87 x 0·64 inch.

A set in my collection, taken at South Gippsland in August 1881, are similar in tint and markings.   Length (A) 0·87 x 0·67 inch ; (B) 0·86 x 0·66 inch.

This bird breeds during the months of August, September, and October.

Hab. New South Wales, Victoria and South Australia. (*Ramsay.*)

Genus NECTARINIA, *Illiger.*

2　　　CINNYRIS FRENATA, *Müll.*

(*Nectarinia australis*, Gould.)

Australian Sun-bird.

Gould, Handbk. Bds. Aust., Vol. i., sp. 359, p. 584.

"According to Mr. Rainbird, numbers of this beautiful little Sun-bird may be seen on bright mornings, among the leafy tops of the mangrove-belts near Port Denison. They are ever darting out to capture some insect on the wing, returning and disappearing again in the thick foliage, or perching upon some topmost twig to devour their captures, and show their shining purple breasts glittering in the sun. During the hottest part of the day the Sun-birds betake themselves to the thick scrub, which in many places runs down to the water's edge. They breed in the months of November and December. One pair chose a little break in the scrub within a few yards of the water, where facing the rising sun, they constructed their nest (which I now have) suspending it by the top from the dead twig of a small shrub, at the foot of a large " Bottle tree" (*Sterculia rupestris*). The nest is of an oval form, much resembling and suspended in the same way as that of *Acanthiza lineata*, with a small hood over the opening, which is near the top. It is composed of fibrous roots and shreds of cotton-tree (*Gomphocarpus fruticosus*) bark, firmly interwoven with the webs and cocoons of various spiders, and a few pieces of white sea-weed ornamenting the outside. It is lined with feathers and the silky native cotton, and is about five inches long by three inches and a-half in diameter." (*Ramsay, Ibis,* 1865, Vol. i., New Series, p. 85.)

Eggs two in number for a sitting, oval in form somewhat pointed at the smaller end, of a greenish-grey ground colour, which is almost obscured by freckles and dashes of light brown. Two specimens in the Australian Museum Collection measure as follows: length (A) 0·66 x 0·42 inch ; (B) 0·67 x 0·44 inch.

A pair in the Macleayan Museum Collection, taken on the Endeavour River, Queensland, give the following measurements:— length (A) 0·7 x 0·43 inch ; (B) 0·67 x 0·43 inch.

Hab. Cape York, Rockingham Bay, Port Denison, South Coast New Guinea. *(Ramsay.)*

Sub-Family ZOSTEROPINÆ.

Genus ZOSTEROPS, *Vigors and Horsfield.*

3-4. ZOSTEROPS CÆRULESCENS, *Latham.*

Grey-backed Zosterops.

Gould, Handbk. Bds. Aust, Vol. i., sp. 360, p. 587.

This bird is universally dispersed over the eastern and southern portions of the continent of Australia, and the whole of Tasmania, A nest of this species in the Australian Museum Collection, taken at Dobroyde in September 1860, is a round cup-shaped structure, outwardly composed of bark fibre, dried grasses, and mosses, and neatly lined inside with fine fibrous roots and grasses ; it measures exteriorly three inches in diameter, by one inch and three-quarters in depth ; internal diameter two inches, by one inch and three-eighths in depth. The nest of this species is attached by the rim to the thin twigs of a tree and is usually placed about six or seven feet from the ground, the top of a *Melaleuca* or *Leptospermum* being especially a favourite situation for it, although I have found it in a number of different trees, both native and acclimatized, when it resorts to our public gardens to breed. Eggs three or four in number for a sitting of a uniform pale blue. Dimensions of a set in the Australian Museum Collection, taken with the above described nest :—length (A) 0·65 x 0·49 inch ; (B) 0·67 x 0·47 inch ; (C) 0·64 x 0·5 inch ; (D) 0·65 x 0·47 inch.

The breeding season of this species commences in September and continues during the four following months. I have several times taken the eggs of *Chalcites basalis* from nests of the above species.

Hab. Rockingham Bay, Port Denison, Wide Bay District, Dawson River, Richmond and Clarence Rivers Districts, New South Wales, Interior, Victoria and South Australia. (*Ramsay.*)

ZOSTEROPS WESTERNENSIS, *Quoy et Gaimard.*

Quoy et Gaimard, Voy. de l'Astrolabe, p. 215, pl. 11, fig. 4.

This species differs from *Z. cærulescens* only in having the throat wax-yellow and the flanks not being so deeply tinged with rich buff or chestnut-brown. At Manly I have shot the two species in company with each other. The nest and eggs so closely resemble those of the above species that a second description is hardly required, some eggs are slightly more pointed than others, but are of the same pale blue or bluish-green tint ;. an average set measures, length (A) 0·66 x 0·47 inch ; (B) 0·66 x 0·47 inch ; (C) 0·66 x 0·48 inch.

Hab. Rockingham Bay, Port Denison, Wide Bay District, Dawson River, Richmond and Clarence Rivers Districts, New South Wales, Interior, Victoria and South Australia. (*Ramsay.*)

ZOSTEROPS GOULDI, *Bonaparte.*

Green-backed Zosterops.

Gould, Handbk. Bds. Aust., Vol i. sp. 381, p. 568.

Gilbert writes as follows of this species :—" The *Zosterops gouldi* is an inhabitant of the western coast of Australia, where it constitutes a beautiful representative of the *Z. cærulescens* of the southern and eastern coasts, The breeding season commences in August and ends in November ; those nests that came under my observation during the earlier part of the season, invariably contained two eggs ; but in October and November I usually found the number to be increased to three, and upon one occasion to four. The nest is small, compact, and formed of dried wiry grasses, bound together with the hairy tendrils of small plants and wool, the inside being lined with very minute fibrous roots ; its breadth is about two inches, and depth one inch ; the eggs are greenish-blue without spot or markings, eight lines long by six lines broad." *(Gould, Handbk. Bds. Aust.,* Vol. i., p. 588.)

Hab. West Australia.

4.   ZOSTEROPS FLAVOGULARIS, *Masters.*

Masters, P.L.S., N.S.W., Vol. i., p. 56.

" This very distinct and well marked species was found tolerably abundant at Cape York and the adjacent islands, by the members of the Chevert Expedition in 1875. A nest of this species now before me, taken by Mr. George Masters, at Warrior Island on the 27th of June 1875, is a deep cup-shaped structure, composed of the dried skeletons of leaves, held together with spiders' webs and neatly lined inside with fine wiry grasses, the whole exterior surface being covered with thin, broad strips of perfectly white semi-transparent paper-like bark of a *Melaleuca,* which gives it a very beautiful appearance. Exterior diameter three inches and one-eighth, depth two inches ; internal diameter one inch and three-quarters ; depth one inch and a-half. The nest was attached by the rim to the thin branches of a shrub about five feet from the ground. The eggs were two in number, but four is the full complement for a sitting, of a uniform pale bluish-green both specimens giving exactly the same measurements, viz., 0·72 inch in length, by 0·5 inch in breadth." *(North, P.L.S., N.S.W.,*Vol. ii., 2nd Series, p. 408.)

Mr. Sharpe, (Brit. Mus. Cat. Bds. Vol. ix., p. 164) considers this species identical with Zosterops à ventre blanc, Homb. et Jacq., Voy. Pôle Sud. Atlas, pl. 19, fig. 3, (1842).

Hab. Cape York, Islands of Torres Straits.

Family DICÆIDÆ.

Genus DICÆUM, *Cuvier.*

DICÆUM HIRUNDINACEUM, *Shaw.*

Swallow Dicæum.

Gould, Handbk. Bds. Aust., Vol. i., sp. 358, p. 581.

No bird seems to have a wider range over the continent of Australia than the present species, specimens having been received from all portions of it where collections have been formed. Especially is it to be found inhabiting the trees where the *Loranthus* and other parasitical plants abound ; berries of various kinds constituting its food. Mr. J. A. Thorpe found this bird breeding at Cape York in 1866-67, and obtained both nests and eggs. One of the nests now before me, is a beautiful pear-shaped structure with an entrance on one side close to the top, and is suspended to the thin leafy branch of a Eucalyptus. It is composed throughout of the soft downy seeds of plants, beautifully woven together, closely resembling felt, and has quite an elastic tendency ; total length three inches and a-half, breadth two inches and a-quarter; length of aperture which is pear-shaped, one inch and a-quarter, breadth one inch, meeting at a point at the top. Another nest from the same locality is slightly larger, and is ornamented on the outside with portions of the woolly buds of some flowering plant.

Dr. Hurst who obtained a nest and eggs of this bird in the grounds of Newington College, on the Parramatta River near Sydney, informs me that it took six weeks from the time the nest was first commenced, till it was finished and the full complement of eggs, three in number, laid therein. Eggs perfectly white ; the above set measure as follows :—length (A) 0·65 x 0·45 inch ; (B) 0·68 x 0·47 inch ; (C) 0·66 x 0·45 inch.

Measurements of a set in the Macleayan Museum, taken at Cairns, Northern Queensland, in 1886, length (A) 0·7 x 0·44 inch ; (B) 0·7 x 0·43 inch.

Hab. Derby, N.W. Australia, Port Darwin and Port Essington, Gulf of Carpentaria, Cape York, Rockingham Bay, Port Denison, Wide Bay District, Dawson River, Richmond and Clarence Rivers Districts, New South Wales, Victoria and South Australia, Tasmania. *(Ramsay.)*

Family CERTHIIDÆ.

Sub-Family CERTHIINÆ.

GENUS CLIMACTERIS, *Temminck.*

2. CLIMACTERIS SCANDENS, *Temminck.*

Brown Tree-creeper.

Gould, Handbk. Bds. Aust., Vol. i., sp. 366, p. 598. *XII* . 𝓎

This bird is to be found breeding freely throughout New South Wales, Victoria and South Australia; constructing a nest of grasses, fur, &c., usually in the hole of some decayed branch or spout of a Eucalyptus, and occasionally out of arms reach. The eggs two in number for a sitting are of a reddish-white ground colour, closely freckled all over with rich reddish markings towards the larger end, where in some instances they form a zone. Two eggs taken by Mr. James Ramsay at Tyndarie, on the 24th of August 1879, measure as follows:—(A) 0·93 x 0·74 inch; (B) 0·95 x 0·73 inch.

The breeding season commences in August and lasts till the end of December.

Hab. Rockingham Bay, Port Denison, Wide Bay District, Dawson River, Richmond and Clarence Rivers Districts, New South Wales, Interior, Victoria and South Australia. *(Ramsay.)*

3   ## CLIMACTERIS RUFA, *Gould.*

Rufous Tree-creeper.

Gould, Handbk. Bds. Aust., Vol. i., sp. 367, p. 600.

" This species makes a very warm nest of soft grasses, the down of flowers and feathers in the hollow part of a dead branch, generally so far down that it is almost impossible to reach it, and it is, therefore very difficult to find. I discovered one by seeing the old birds beating away a Wattle-bird that tried to perch near their hole ; the nest, in this instance, was fortunately within arm's length ; it contained three eggs of a pale salmon colour, thickly blotched all over with reddish-brown, eleven lines long by eight and a-half lines broad : this occurred during the first week in October." *(Gould, Handbk. Bds. Aust.*, Vol. i., p. 600.)

Hab. Derby, N.W. Australia, Western Australia. (*Ramsay.*)

2.   ## CLIMACTERIS ERYTHROPS, *Gould.*

Red-eyebrowed Tree-creeper.

Gould, Handbk. Birds Aust., Vol. i., sp. 368, p. 602. *IX.* //.

The nest of this species, like all other members of the genus, is built of grasses, lined with feathers and placed in one of the numerous spouts or hollow branches of the Eucalypts.

" I am indebted to Mr. K. H. Bennett, of Mossgiel for a fine set of the eggs of this species, the first I had seen ; they closely resemble some of the varieties of those of *Ptenœdus rufescens*, but have a climacterine look about them, and a smooth shell. The ground colour, apparently white, is obscured with evenly dispersed dots and freckles of a rich red, which, occasionally confluent, form elongated spots here and there ; some have a zone formed by confluent spots of red intermixed with slate or lilac-brown, and here the spots are largest, and the lilac marks appear beneath the shell. The following are the measurements :—(A) 0·83 x 0·65 inch ; (B) 0·85 x 0·63 inch ; (C) 0·82 x 0·63 inch." *(Ramsay, P.L.S., N.S.W.*, 2nd Series, Vol. i., p. 1149.)

The above eggs were taken by Mr. K. H. Bennett in November at Ivanhoe. Two eggs only are laid for a sitting, the other egg (C) was addled and was taken from a nest containing also a young bird. This bird breeds from September to December.

Hab. Wide Bay District, Richmond and Clarence Rivers Districts, New South Wales, Interior. *(Ramsay)*

CLIMACTERIS MELANURA, *Gould.*
Black-tailed Tree-creeper.
Gould, Handbk. Bds. Aust., Vol. i., sp. 370, p. 604. ⅨⅩ *12.*

The only eggs of this species I have yet seen, were taken by Mr. James Ramsay at Tyndarie, on the 19th of September 1880, from a nest built of grasses, &c., in the hollow branch of a Eucalyptus. The eggs two in number are of a light reddish ground colour, almost obscured by heavy longitudinal blotches of rich reddish-brown, and a few obsolete spots of lilac. Length (A) 0·9 x 0·73 inch ; (B) 0·89 x 0·73 inch.

Hab. Derby, N.W. Australia, Port Darwin and Port Essington, Gulf of Carpentaria, New South Wales. (*Ramsay.*)

CLIMACTERIS LEUCOPHŒA, *Latham.*
White-throated Tree-creeper.
Gould, Handbk. Bds. Aust., Vol. i., sp. 371, p. 605. ⅩⅢ *3.*

This, like all other members of the genus *Climacteris* constructs a nest of grasses, feathers, &c., in the hollow part of a dead branch. A set of two taken by Mr. John Ramsay at Macquarie Fields in 1861, are of a dull white, marked with round spots of reddish-brown towards the larger end, with very minute dots of the same colour scattered over the rest of the surface. Length (A) 0·85 x 0·67 inch ; (B) 0·89 x 0·66 inch.

This species breeds during September and the two following months.

Hab. Port Denison, Wide Bay District, Dawson River, Richmond and Clarence Rivers Districts, New South Wales, Interior, Victoria South Australia. (*Ramsay.*)

Sub-Family SITTINÆ.

Genus SITTELLA, *Swainson.*

ɔ‑ɣ   SITTELLA CHRYSOPTERA, *Latham.*

Orange-winged Sittella.

Gould, Handbk. Bds. Aust., Vol. i., sp. 373, p. 609.

The Orange-winged Sittella is universally dispersed over the eastern and south-eastern portions of the continent of Australia. It breeds freely in the neighbourhood of Sydney, and there are several of its singular and beautifully formed nests in the Australian Museum Collection. A nest now before me is placed at the junction of an upright forked branch of a Eucalyptus ; it is of an inverted cone shape, with a cup-like cavity at the top, the rim of the nest being sharp and thin, it is constructed of the downy tufts of Banksia cones interwoven together with cobwebs, but they are entirely hidden by the whole exterior surface being covered with short pieces of bark perpendicularly and neatly fastened on with cobwebs, which gives it a " shingled " appearance and renders it most difficult of detection : the inside is beautifully lined with soft mosses and mouse-eared lichens. It measures exteriorly, four inches and a-half in length, by two inches and a-quarter in diameter ; cavity, one inch and seven-eighths in diameter, by one inch and a-quarter in depth. The nest of this bird is usually built on the side of a dry upright forked branch of a Eucalyptus, Casuarina, or Acacia, varying according to the kind of tree in which it is placed, from fifteen to fifty or sixty feet from the ground. Eggs three or four in number for a sitting, usually the former.

A set in the Dobroyde Collection are of a greenish-white ground colour, heavily blotched all over with irregular shaped markings of slaty-black, but more particularly towards the larger end, where they become confluent, together with a few spots of a lighter tint appearing as if beneath the surface of the shell. Length (A) 0·68 x 0·52 inch; (B) 0·65 x 0·5 inch; (C) 0·68 x 0·52 inch. Taken at Dobroyde, October 1860.

Another set in the same collection, taken September 1861, are of a bluish-white ground colour, minutely freckled all over with slaty-grey, and slaty-black markings, with a few faint subsurface blotches of slaty-lilac; length (A) 0·68 x 0·53 inch; (B) 0·66 x 0·5 inch; (C) 0·68 x 0·52 inch.

September and the three following months constitute the breeding season of this species.

Hab. Port Denison, Wide Bay District, Richmond and Clarence Rivers Districts, New South Wales, Victoria and South Australia. *(Ramsay.)*

SITTELLA TENUIROSTRIS, *Gould.*

Gould, Handbk. Bds. Aust., Vol. i., p. 610.

"This is a somewhat doubtful species, and Dr. Hans Gadow, who has presumedly examined the type from Mr. Gould's collection, has made it still more doubtful by placing it as identical with *Sittella pileata,* Gould; but on reference to Mr. Gould's Handbook, Vol. i., p. 610, it will be seen that that author considered the bird a variety of *S. chrysoptera.* As I have specimens agreeing very well with Mr. Gould's description, from the interior provinces obtained by Mr. James Ramsay, I prefer to consider it more nearly allied to *S. chrysoptera* than to any other. The length of the bill is 0·7 inch. The nest is a very beautiful structure placed often between the upright forks of a dead branch; it is very deep, open above, the edges sharp not rounded, and composed of fine shreds of bark, lichens, and cobweb, the outside felted or 'shingled' with small scales of bark fastened on with cobwebs, and made to

P

so resemble the sides of the forked branch between which it is placed, as to be most difficult of detection ; the interior is usually lined with ' mouse-eared ' lichen, and the colour of the eggs closely resembles that of the lichen itself. The eggs are three, seldom four in number, of a delicate greenish-white, with dots and confluent irregular markings of slaty-lilac, and slate-black, the lilac freckles appearing beneath the shell, in some forming a zone of larger spots near the thicker end, in others the spots are nearly evenly dispersed over the whole surface. Length (A) 0·63 x 0·55 inch ; (B) 0·68 x 0·55 inch; (C) 0·66 x 0·53 inch ; (D) 0·62 x 0·52 inch." (*Ramsay, P.L.S., N.S.W.,* Vol. i., 2nd Series, p. 1149.)

Hab. New South Wales, Interior. (*Ramsay.*)

SITTELLA LEUCOCEPHALA, *Gould.*
White-headed Sittella.
Gould, Handbk. Bds. Aust., Vol. i., sp. 374, p. 610. •

This bird is found in New South Wales and Queensland. The nest is similar to that of other members of the genus. Two eggs taken by Mr. George Barnard of Coomooboolaroo, Duaringa, Queensland, during 1885, are of a delicate greyish-green ground colour, minutely freckled and spotted all over with slaty-grey and slaty-black markings, a few of a lighter tint appearing as if beneath the surface of the shell ; one specimen A, has a· large coalesced slaty-grey patch on the larger end. Length (A) 0·63 x 0·51 inch ; (B) 0·65 x 0·5 inch.

Hab. Wide Bay District, Dawson River, New South Wales, Interior. (*Ramsay.*)

3   ## SITTELLA PILEATA, *Gould.*
Black-capped Sittella.
Gould, Handbk. Bds. Aust., Vol. i., sp. 376, p. 612.

" For the nest and eggs of this species, together with the bird shot therefrom, I am indebted to Mr. James Hill, of Kewell,

Victoria, who procured them on the outskirts of the Mallee country in the Wimmera district, in September 1882. The nest was built in the upright fork of a *Casuarina* about fifteen feet from the ground, and is similar in every respect to that of *S. chrysoptera;* hence its description would be merely a repetition of that of the nest of the latter species. Eggs three in number for a sitting, the ground colour darker, and the blotches heavier, than in *S. chrysoptera*, being a deep bluish-white, with long slaty-black markings, while appearing underneath the surface of the shell are large superimposed blotches of dark lilac, which in some instances are confluent ; the markings on the under surface are much larger and more numerous than on the outer surface of the shell. Length (A) 0·66 x 0·51 inch ; (B) 0·66 x 0·53 inch ; (C) 0·67 x 0·54 inch." *(North, P.L.S., N.S.W.,* Vol. ii., 2nd Series, p. 409.)

Hab. New South Wales, Interior, Victoria and South Australia, West and South-West Australia. (*Ramsay.*)

Family CULCULIDÆ.

Genus CACOMANTIS, *Müller.*

CACOMANTIS PALLIDUS, *Latham.*

(Cuculus inornatus, Vigors and Horsfield.)

Pallid Cuckoo.

Gould, Handbk. Bds. Aust., Vol. i., sp. 378, p. 615.

"In the neighbourhood of Sydney this species usually deposits its egg in the nests of *Ptilotis auricomis,* and occasionally in those of *Ptilotis chrysops,* but rarely in those of *Melithreptus lunulatus;* in other districts, doubtless, in any nests suitable for the purpose. I have frequently observed that whenever the eggs of Cuckoos have been deposited in open nests, there is manifested a decided preference for those of birds which lay eggs similar to their own. The eggs of the Unadorned Cuckoo *(C. inornatus)**

* = *C. pallidus,* Latham.

closely resemble the large and almost spotless variety of that of the Yellow-tufted Honey-eater *(Ptilotis auricomis)*; they are, however somewhat more rounded, and of a much lighter tint, being of a pale flesh-colour, sprinkled with a few dots of a deeper hue, but often without any markings at all. In length they vary from eleven to twelve and a-half lines, being from eight and a-half to nine lines in breadth.

Eggs taken from the nests of *Ptilotis auricomis* and *P. chrysops* measure as follows :—length (A) 0·97 x 0·67 inch ; (B) 0·99 x 0·68 inch.

Hab. Derby, N.W. Australia, Gulf of Carpentaria, Cape York, Rockingham Bay, Port Denison, Wide Bay District, Dawson River, Richmond and Clarence Rivers Districts, New South Wales, Interior, Victoria and South Australia, Tasmania, West and South-West Australia. *(Ramsay.)*

CACOMANTIS FLABELLIFORMIS, *Latham.*

(Cuculus cineraceus, Vigors and Horsfield.)

Fan-tailed Cuckoo.

Gould, Handbk. Bds. Aust., Vol. i., sp. 379, p. 618. *XIII* 7.

"Among those species, the nests of which are favoured by visits from this 'parasite,' is *Acanthiza pusilla,* from a nest of which, in September 1863, we took no less than four eggs—two laid by the rightful owner of the nest, the other two by Cuckoos. One of these was a very fine specimen of *Chalcites plagosus,* the other an egg of the present species, *Cacomantis flabelliformis.* The entrance of this nest was greatly enlarged, being in width fully two inches ; and the hood, which usually conceals the entrance (which is near the top of the nest, and not generally wider than one inch across), was pushed back to such an extent that the eggs were rendered quite visible. I have now before me ten nests of *Acanthizæ* and four of *Maluri,* the former comprising *Acanthiza lineata, A. nana, A. pusilla,* and *A. reguloides;* the latter, *Malurus cyaneus,* and *M. lamberti.* Now, having compared the greatly enlarged entrances of those nests from which we have taken Cuckoo's eggs with the entrances of those which did not contain the egg of a Cuckoo, and which we took as soon as the bird had laid its full number of eggs for a sitting, I cannot but feel convinced more than ever that the eggs of these parasites are laid in the nests, and not deposited in any other manner. The average width of the entrances of the nests of *Acanthiza lineata* which have not been visited by a Cuckoo is one inch, while those which have contained Cuckoo's eggs vary from two to two inches and a-half. In addition to the nests of *Acanthiza pusilla,* we have known this Cuckoo *(C. flabelliformis)* deposit its eggs in the nest of *A. reguloides* and *Chthonicola minima.* How great is the difference between the Cuckoo's eggs and those of this last bird *(Chthonicola minima),* which are of a bright reddish-chocolate! The eggs of *Cacomantis flabelliformis* are from ten to ten lines and a-half in length by seven to seven lines and a-half in breadth. The ground colour is a delicate white, spotted and dotted with wood-brown, deep brownish-lilac, and faint lilac dots, which appear beneath the surface. Some specimens are faintly sprinkled all over, and the dots have a washed-out appearance; others are marked more strongly, and in these the markings formed are in a distinct zone at the larger end, which is sometimes broken by a batch of very deep coloured dots." *(Ramsay, P.Z.S.,* 1865, p. 463.)

Two eggs of this species taken from the nests of *Acanthiza pusilla* and *Chthonicola minima,* measures as follows :—length (A) 0·88 x 0·6 inch ; (B) 0·83 x 0·6 inch.

Hab. Gulf of Carpentaria, Cape York, Rockingham Bay, Port Denison, Wide Bay District, Dawson River, Richmond and Clarence Rivers Districts, New South Wales, Interior, Victoria and South Australia, Tasmania, West and South-west Australia *(Ramsay.)*

Genus LAMPROCOCCYX, *Cabanis et Heine.*

LAMPROCOCCYX PLAGOSUS, *Latham.*

Bronze Cuckoo.

Gould, Handbk. Bds. Aust. Vol. i., sp. 383, p. 623. *IX. /3.*

The Bronze. Cuckoo is universally found over the whole of Australia and Tasmania, depositing its single egg in any convenient nest, principally in those that are dome-shaped. I have taken most of the eggs of this species from· the nests of *Geobasileus chrysorrhœa.* The eggs are in form elongated ovals, being rounded and nearly equal in size at both ends, varying in colour from a uniform light ashy-grey to a rich dark olive-brown or bronze ; four average specimens, taken from the nests of *A. lineata, G. chrysorrhœa, G. reguloides,* and *Æ. temporalis,* are as follows :— length (A) 0·73 x 0·51 inch ; (B) 0·72 x 0·5 inch; (C) 0·72 x 0·52 inch ; (D) 0·73 x 0·52 inch.

Dr. Ramsay paid particular attention to the working out of this and the following species, by procuring the eggs of the Cuckoo's and placing them in nests convenient for observation, and when hatched obtaining the young birds in all the various stages of plumage, from the nestling upwards, thus enabling him to correctly identify the species to which the eggs belonged, and published the result of his labours in a series of interesting papers in the Proc. Zool. Soc. of London, (see P.Z.S., 1865, p. 460, and P.Z.S., 1869, p. 359), and also supplied Mr. Gould with the eggs and birds of both species at the same time.

A coloured plate of the eggs of *L. plagosus, L. basalis, Cacomantis pallidus,* and *C. flabelliformis,* also of those birds which are usually the foster-parents of the two former species in the neighbourhood of Sydney is given in the Proc. Zool. Soc. for 1869 see pl. xxvii., p. 358. Among the latter are the eggs of *Acanthiza lineata, A. pusilla, A. nana, Geobasileus reguloides,. Sericornis brevirostris,* and *Stipiturus malachurus.*

Hab. Port Darwin and Port Essington, Gulf of Carpentaria, Cape York, Rockingham Bay, Port Denison, Wide Bay District,

Dawson River, Richmond and Clarence Rivers Districts, New South Wales, Interior, Victoria and South Australia, Tasmania, West and South-west Australia. *(Ramsay.)*

LAMPROCOCCYX BASALIS, *Horsfield*.

Narrow-billed Bronze Cuckoo.

Gould, Handbk. Bds. Aust., Vol. i., sp. 385, p. 626. XIII /3.

The range of this bird does not extend so far as that of *L. plagosus*, not being found in either Northern or Western Australia. Like the preceding species, it is one of the first harbingers of Spring, and takes its departure again about the middle of Autumn. It deposits its single egg in the nest of any of the smaller birds, the first I found being in the nest of *Meliornis novæ-hollandiæ*, and I have also at various times taken it from the nests of the following species:—*Malurus cyaneus, Ephthianura albifrons, Zosterops caerulescens, Petrœca leggii, Estrelda temporalis, Geobasileus chrysorrhœa, Smicrornis brevirostris,* and *Acanthiza lineata;* it will be seen from the above, that this species evinces no decided preference for either those nests that are open or dome-shaped, but seems to bestow its favours pretty equally in the choice of a foster parent for its young.

From a nest of *Acanthiza nana*, Dr. Ramsay in 1856 took no less than six eggs, three of them being Bronze Cuckoo's, two of *Lamprococcyx plagosus*, and one of *L. basalis*. (See P.Z.S., 1865, p. 461).

The egg of this species is pinky-white minutely freckled all over the surface with light brownish-red or pinkish-red dots and spots, in some instances these markings are confluent forming coalesced patches on the egg but on no particular portion of it, sometimes being on one side only, at other times on the end. The dimensions of six eggs are as follows:—length (A) 0·68 x 0·48 inch ; (B) 0·76 x 0·5 inch ; (C) 0·72 x 0·5 inch ; (D) 0·71 x 0·5 inch ; (E) 0·66 x 0·47 inch ; (F) 0·68 x 0·48 inch. The colouring matter of the eggs of both this and the preceding species is easily rubbed off when moisture is applied to them.

Hab. Derby, N.W. Australia, Port Denison, Wide Bay District, Richmond and Clarence Rivers Districts, New South Wales, Interior, Victoria and South Australia, Tasmania. (*Ramsay.*)

Genus SCYTHROPS, *Latham.*

SCYTHROPS NOVÆ-HOLLANDIÆ, *Latham.*

Channel-bill.

Gould, Handbk. Bds. Aust., Vol. i., sp. 386, p. 628. *VII.* 3.

"This bird is universally distributed over the whole continent of Australia, and one or two stragglers have even been found in Tasmania. Dr. Hurst has kindly permitted me to describe an egg of this species from his collection, which, he informs me was taken from the oviduct of a bird shot at Kempsey, on the Macleay River, during the first week in November 1884, and which he exhibited at a meeting of this Society in the same month. Ground colour dull white, with faint washed-out pinkish spots and minute dots, also some of a light yellowish-brown tinge; appearing as if beneath the surface of the shell at the apex of the thick end are others of a light purplish-brown, becoming confluent, and forming a very indistinct patch, intermingled with some of a brownish shade. All the markings are very ill-defined, and the egg closely resembles a very large and washed-out specimen of the egg of *Grallina australis.* Length 1·5 x 1·05 inch." (*North, P.L.S., N.S.W.,* Vol. ii., 2nd Series, p. 410.)

Hab. Derby, N.W. Australia, Port Darwin and Port Essington, Gulf of Carpentaria, Cape York, Rockingham Bay, Port Denison, Wide Bay District, Richmond and Clarence Rivers Districts, New South Wales, Interior, Victoria and South Australia, Tasmania, South Coast New Guinea. (*Ramsay.*)

Genus EUDYNAMIS, *Vigors and Horsfield.*

EUDYNAMIS CYANOCEPHALA, *Latham.*

(*E. flindersi,* Gould.)

Australian Koel.

Gould, Handbk. Bds. Aust., Vol. i., sp. 387, p. 632.

" Mr. Geo. Masters obtained an egg of this species at Gayndah, Queensland, on the 25th of November 1870. Having shot at a female and broken her wing, while pursuing her on the ground the egg was dropped. It is a pointed oval in form, of a dull white minutely spotted with light brown, together with a few faint blotches here and there of purplish-brown, the smaller end being entirely devoid of markings. Whether this is the normal colour of the egg is yet to be proved, as the egg being dropped by the bird when wounded and the markings very faint, it is probable that it may not have been quite ready for laying. Long diameter 1·4 inch, short diameter 1·05 inch. A photograph of this egg sent by Dr. Geo. Bennett, F.Z.S., of Sydney, was exhibited at the June meeting of the Zoological Society of London, 1873, see P.Z.S., p. 519." (*P.L.S., N.S.W.,* Vol. ii., 2nd Series, p. 554.)

Hab. Derby, N.W. Australia, Port Darwin and Port Essington, Gulf of Carpentaria, Cape York, Rockingham Bay, Port Denison, Wide Bay District, Dawson River, Richmond and Clarence Rivers Districts, New South Wales, South Coast New Guinea. (*Ramsay.*)

Genus CENTROPUS, *Illiger.*

3-4. CENTROPUS PHASIANUS, *Latham.*

Pheasant-Coucal.

Gould, Handbk. Bds. Aust., Vol. i., sp. 388, p. 634.

This bird is found in the dense coastal brushes, and is very plentiful on the Herbert River, in Queensland, and the scrubby grass-lands of the Richmond and Clarence Rivers districts of

New South Wales. The nest is a bulky structure of dried leaves and grasses rounded and covered over above, with an opening on both sides, from which the head of the female on one side, and her tail on the other protrudes while sitting. It is usually placed in a tussock of long coarse grass. Eggs three or four in number for a sitting, an average specimen in the Dobroyde Collection taken by Mr. John Macgillivray, near Grafton on the Clarence River, in October 1864, is rounded in form, and of a dull dirty-white, having a thin coating of lime on it, one side showing scratches as if done by the bird while sitting. Length (A) 1·35 x 1·13 inch.

Hab. Cape York, Rockingham Bay, Port Denison, Wide Bay District, Richmond and Clarence River Districts, New South Wales, and North-West Australia. *(Ramsay.)*

Order SCANSORES.

Family PSITTACIDÆ.

GENUS CACATUA, *Vieillot.*

2   CACATUA GALERITA, *Latham.*

Great Sulphur-crested Cockatoo.

Gould, Handbk. Bds. Aust., Vol. ii., sp. 391, p. 2.

This bird is universally dispersed over the whole of Australia. It resorts to the hollow branches or boles of trees to nest and deposit its eggs, which are two in number, on the decaying wood usually found in such places, they are pure white, and vary in form from oval to pointed oval. Length (A) 1·65 x 1·21 inch; (B) 1·63 x 1·19 inch. A pair in the Dobroyde Collection measure : length (A) 1·62 x 1·18 inch ; (B) 1·61 x 1·2 inch.

August and the three following months constitutes the breeding season of this species.

Hab. Port Darwin and Port Essington, Gulf of Carpentaria, Cape York, Rockingham Bay, Port Denison, Wide Bay District, Richmond and Clarence Rivers Districts, New South Wales, Interior, Victoria and South Australia, Tasmania, West and South-West Australia, South Coast New Guinea. (*Ramsay.*)

3   CACATUA LEADBEATERI, *Vigors.*

Leadbeater's Cockatoo.

Gould, Handbk. Bds. Aust., Vol. ii., sp. 392, p. 5.   ~XIV~   2.

Mr. K. H. Bennett found this handsome bird breeding plentifully in the interior of New South Wales, between the Lachlan and the Darling Rivers. Like all other members of this genus it breeds in the hollow limbs of trees, usually of a lofty Eucalyptus. The eggs are three in number for a sitting, oval in shape, white ; a set taken on September the 5th 1884, measures as follows :—length (A) 1·38 x 1·1 inch ; (B) 1·39 x 1 inch ; (C) 1·41 x 1·12 inch.

This species breeds during the months of August, September, and October.

Hab. New South Wales, Interior, Victoria and South Australia, West and South-West Australia. (*Ramsay.*)

3   CACATUA ROSEICAPILLA, *Vieillot.*

Rose-breasted Cockatoo.

Gould, Handbk. Bds. Aust., Vol. ii., sp. 394, p. 8.   ~XIV~   3.

" Like all the members of this section the Rose Cockatoo nests in the hollow branches of large trees, laying its eggs on the débris of decaying wood usually found in such places, they are three in number, white, rather oblong in form and slightly granular ; length (A) 1·4 x 1·05 inch ; (B) 1·4 x 1·04 inch." (*Ramsay, P.L.S., N.S.W.*, Vol. vii., p. 53.)

Hab. Derby, North-West Australia, Port Darwin and Port Essington, Gulf of Carpentaria, New South Wales, Interior, Victoria and South Australia. (*Ramsay.*)

Genus LICMETIS, *Wagler.*

3   LICMETIS NASICA, *Temminck.*

(L. tenuirostris, Wagler.)

Long-billed Cockatoo.

Gould, Handbk. Bds. Aust., Vol. ii., sp. 395, p. 11.   XIV   ♀

The Long-billed Cockatoo usually exercises great care in placing its nest out of the way of human enemies, choosing one of the most inaccessible trees in the dead branch of which it deposits its eggs, which are white and three in number, oval, and rather pointed at the smaller end; shell inclined to be rough, an average specimen measures :—length 1·4 x 1·1 inch.

The breeding season commences in August and lasts during the two following months.

Hab. Gulf of Carpentaria, New South Wales, Interior, Victoria and South Australia. (*Ramsay.*)

Genus CALYPTORHYNCHUS, *Vigors and Horsfield.*

CALYPTORHYNCHUS NASO, *Gould.*

Western Black Cockatoo.

Gould, Handbk. Bds. Aust., Vol. ii., sp. 399, p. 17.

" This species breeds in the holes of trees, where it deposits its snow-white eggs on the soft dead wood. They are generally placed in trees so difficult of access that even the natives dislike to climb them. Those given to Gilbert by the son of the colonial chaplain were taken by a native from a hole in a very high white gum, in the last week of October; they are white, one inch and eight lines long by one inch and four lines broad." (*Gould, Handbk. Bds. Aust.*, Vol. p. 17.)

Hab. West Australia.

2 CALYPTORHYNCHUS FUNEREUS, *Shaw.*

Funereal Cockatoo.

Gould, Handbk. Bds. Aust., Vol. ii., sp. 401, p. 20.

" The eggs of this species are white and two in number about one inch and five-eights long by one inch and three-eighths broad, are deposited on the rotten wood in the hollow branch of a large gum." *(Gould, Handbk. Bds. Aust.*. Vol. ii., p. 20.)

Hab. Wide Bay District, Dawson River, Richmond and Clarence Rivers Districts, New South Wales, Victoria and South Australia, Tasmania. *(Ramsay.)*

2 CALYPTORHYNCHUS XANTHONOTUS, *Gould.*

Yellow-eared Black Cockatoo.

Gould, Handbk. Bds. Aust., Vol. ii., sp. 402, p. 22.

" This bird lays two white eggs in some large rotten gum-tree, generally where one of the large branches has rotted off at the fork ; inside this hole, which occasionally extends five or six feet down the bole of the tree, the bird scrapes and clears away some of the rotten wood until a sort of seat is formed ; for it is a very rude attempt at making a nest. The laying commences about the latter end of October or beginning of November. The eggs are one inch and eight lines long by one inch and four lines broad." *(Gould, Handbk. Bds. Aust.*, Vol. ii., p. 22.)

Dr. Ramsay, who has examined one of the types of this species considers it identical with *C. funereus.*

Hab. Tasmania.

2 CALYPTORHYNCHUS BAUDINII, *Vigors.*

Baudin's Cockatoo.

Gould, Handbk. Bds. Aust., Vol. ii., sp. 403, p. 25.

" This species breeds in the holes of the highest white gum-trees, often in the most dense and retired part of the forest. The eggs

are generally two in number, of a pure white; their average
length being one inch and three-quarters by one inch and three-
eighths in breadth. The breeding season extends over the months
of October, November, and December." (*Gould, Handbk. Bds.
Aust.*, Vol. ii., p. 25.)

Hab. Western Australia.

Genus MICROGLOSSUM, *Geoffroy*.

MICROGLOSSUM ATERRIMUM, *Gmelin*.

Great Palm Cockatoo.

Gould, Handbk. Bds. Aust., Vol. ii., sp. 404, p. 27.

"An egg of this species was taken from the débris at the
bottom of a hollow branch or bole of a tree about twenty-five feet
from the ground, the bird was seen to fly from the nest, and when
shot proved to be the female. The tree was situated in the open
forest country on the Astrolabe Range. The egg is white, pointed
at the thin end, rounded at the thicker end. Length 2 inches;
diameter near the thicker end 1·4 inch." (*Ramsay, P.L.S., N.S.
W.*, Vol. viii., p. 27.)

Hab. Cape York in Australia, and the North-west, South and
East Coasts of New Guinea.

Genus CALOPSITTA, *Lesson*.

4-6   CALOPSITTA NOVÆ-HOLLANDIÆ, *Wagler*.

Cockatoo-Parrakeet.

Gould. Handbk. Bds. Aust., Vol. ii., sp. 440, p. 84.   XIV 9.

This species, the only one of the genus known, breeds in the
hollow limbs or spouts of trees, depositing its eggs on the dry dust
or decaying wood contained therein, the eggs are four to six in
number, white, and in shape oval, pointed somewhat at the smaller

end. Three eggs of a set taken in October 1886 at Ivanhoe, measure as follows:—length (A) 1·1 x 0·8 inch ; (B) 1·03 x 0·78 inch ; (C) 1·09 x 0·8 inch.

This bird breeds during September and the three following months.

Hab. Derby, N.W. Australia, Port Darwin and Port Essington, Gulf of Carpentaria, Wide Bay District, New South Wales, Interior, Victoria and South Australia, West and South-West Australia. (*Ramsay.*)

Genus APROSMICTUS, *Gould.*

APROSMICTUS SCAPULATUS, *Bechstein.*

King Lory.

Gould, Handbk. Bds. Aust., Vol. ii , sp. 409, p. 35.

The hollow spouts of the lofty Eucalyptus trees in South Gippsland are frequently tenanted by these birds for the purpose of breeding, also in the same locality, others are taken possession of by the Gang-gang Cockatoo, *Callocephalon galeatum,* but always at such a height that rendered it impossible to take the eggs of either species ; on one occasion only was a nest of the former taken during my stay there, and this unfortunately contained four young birds. An egg of this species in Mr. George Masters' collection, is in form a swollen oval, pure white. Long diameter 1·27 inch, short diameter 1·04 inch.

Hab. Port Denison, Wide Bay District, Richmond and Clarence Rivers Districts, New South Wales, Victoria, and South Australia. (*Ramsay.*)

Genus PTISTES, *Gould.*

4. PTISTES ERYTHROPTERUS, *Gmelin.*

Red-winged Lory.

Gould, Handbk. Bds. Aust., Vol. ii., sp. 410, p. 37. XIV 5.

The Red-winged Lory is found both on the north-eastern and north-western coast of Australia, and inland as far as the Dawson

River, Queensland; a set of eggs taken from the hollow branch in a lofty Eucalyptus, by Mr. George Barnard in 1882 at the latter place, are four in number, white, and measure as follows:— length (A) 1·19 x 1 inch; (B) 1·25 x 0·98 inch; (C) 1·13 x 0·95 inch; (D) 1·15 x 0·92 inch.

This bird commences to breed in October, and continues the three following months.

Hab. Rockingham Bay, Port Denison, Wide Bay District, Dawson River. (*Ramsay.*)

Genus PLATYCERCUS, *Vigors.*

5   PLATYCERCUS BARNARDI, *Vigors and Horsfield.*

Barnard's Parrakeet.

Gould, Handbk. Bds. Aust., Vol. ii., sp. 412, p. 40.   XIV   7.

"This beautiful Parrakeet is distributed over the southern portions of the interior of Australia, and is found frequenting alike the neighbourhood of the Lachlan and Darling Rivers in New South Wales, as well as the dense Mallee districts of Victoria and South Australia. In the cultivated portions of the country the birds assemble together in small flocks, and commit great depredations on the crops, consequently a merciless warfare is waged against them by the farmers. For a set of the eggs of this species I am indebted to Mr. Joseph A. Hill, of 'Pine Rise,' Kewell, Victoria, who obtained them, after carefully watching a pair of birds for some time in the vicinity, on September 15th, 1887. They were deposited on the decaying wood, about two feet down the hollow limb of a Eucalyptus, at a height of thirty feet from the ground. The eggs are five in number for a sitting, pure white, oval in form, nearly equal in size at both ends, measuring as follows:—length (A) 1·11 x 0·9 inch; (B) 1·2 x 0·92 inch; (C) 1·16 x 0·91 inch; (D) 1·17 x 0·9 inch; (E) 1·18 x 0·92 inch." (*North, P.LS., N.S.W.,* Vol. ii., 2nd Series, p. 985.)

This species breeds during September and the three following months.

Hab. New South Wales, Interior, Victoria and South Australia. (*Ramsay.*)

PLATYCERCUS SEMITORQUATUS, *Quoy et Gaimard.*
Yellow-oollared Parrakeet.
Gould, Handbk. Bds. Aust., Vol. ii., sp. 413, p. 42.

"The *Platycercus semitorquatus* begins breeding in the latter part of September or beginning of October, and deposits its eggs in a hole in either a gum- or mahogany-tree, on the soft black dust collected at the bottom; they are from seven to nine in number and of a pure white. In most instances these eggs have a pinky blush before being blown." (*Gould, Handbk. Bds. Aust.*, Vol. ii., p. 42.)

An average specimen measures, 1·23 x 0·98 inch.

Hab. West Australia.

– 6   ## PLATYCERCUS PENNANTII, *Latham.*
Pennant's Parrakeet.
Gould, Handbk. Bds. Aust., Vol. ii., sp. 415, p. 44. *XIV*   6.

This bird is found plentifully throughout New South Wales and Victoria, and especially in the heavy timber clad ranges of South Gippsland. Splendid specimens of both sexes in fully adult plumage can be obtained in August, and in April the immature birds of the previous season may be flushed at every few steps in walking through the scrubs. It breeds in the holes of the lofty gum-trees depositing its eggs from four to six in number on the rotten wood, the eggs when first laid are white, but soon become stained with the decaying wood' or dust on which they are placed; in form they are rounded; length (A) 1·14 x 0·95 inch; (B) 1·17 x 0·97 inch; (C) 1·16 x 0·96 inch.

Q

This species breeds during the months of September, October, and November.

Hab. Rockingham Bay, Richmond and Clarence Rivers Districts, New South Wales, Victoria and South Australia. *(Ramsay.)*

PLATYCERCUS FLAVIVENTRIS, *Temminck.*
Yellow-bellied Parrakeet.
Gould, Handbk. Bds. Aust., Vol. ii., sp. 417, p. 48.

The Yellow-bellied Parrakeet is found throughout Tasmania, and the adjacent islands in Bass's Straits, where it deposits its eggs, six or seven in number, on the decaying wood in the hole of a gum tree. Eggs white, oval, length (A) 1·2 x 0·91 inch ; (B) 1·17 x 0·92 inch ; (C) 1·23 x 0·96 inch.

The breeding season lasts from September until January.

Hab. Victoria and South Australia, Tasmania. *(Ramsay)*

PLATYCERCUS PALLIDICEPS, *Vigors.*
Pale-headed Parrakeet.
Gould, Handbk. Bds. Aust, Vol. i., sp., 419, p. 51.

" The eggs of the Moreton Bay Rosella are from three to five in number, white, rounded or oblong-oval in shape ; length 1 inch x 0·88 inch to 1·06 x 0·9 inch, they are laid in the hollow boughs of trees during the months of August to December." *(Ramsay, P.L.S., N.S.W.,* Vol. vii., p. 53.)

Hab. Port Denison, Dawson River, Wide Bay District, New South Wales, Interior. *(Ramsay.)*

PLATYCERCUS EXIMIUS, *Shaw.*
Rose-hill Parrakeet.
Gould, Handbk. Bds. Aust., Vol. ii., sp. 422, p. 55.

This bird is found breeding plentifully in New South Wales and Victoria ; it lays its eggs which are five to seven in number in the hollow branch of a Eucalyptus. Eggs white, measuring as follows:—length (A) 1·08 x 0·9 inch ; (B) 1·06 x 0·9 inch ; (C) 1·05 x 0·87 inch ; (D) 1·03 x 0·85 inch.

This species breeds during September and the three following months.

Hab. Wide Bay District, Richmond and Clarence Rivers Districts, New South Wales, Interior, Victoria and South Australia, Tasmania. (*Ramsay.*)

ʟ-ʔ PLATYCERCUS ICTEROTIS, *Temminck.*

Yellow-cheeked Parrakeet.

Gould; Handbk. Bds. Aust , Vol. ii., sp. 424, p. 58.

"The eggs, which are white and six or seven in number, are eleven lines long and nine and a-half lines broad ; they are deposited in the holes of large trees without any nest." *(Gould, Handbk. Bds. Aust.,* Vol. ii., p. 58.)

Hab. South Australia, and West Australia. (*Ramsay.*)

Genus PURPUREICEPHALUS, *Bonaparte.*

ʔ-ʔ. PURPUREICEPHALUS PILEATUS, *Vigors.*

(Platycercus spurius, Kuhl.)

Red-capped Parrakeet.

Gould, Handbk. Bds. Aust., Vol. ii., sp. 425, p. 60.

"The Red-capped Parrakeet is an inhabitant of Western Australia, where it is rather numerously dispersed over the country about King George's Sound. The breeding season extends over the months of October, November, and December. The hollow dead branch of a gum- or mahogany-tree is the place usually chosen by the female for the reception of her eggs, which are milk-white and from seven to nine in number, about an inch and an eighth long by seven-eighths of an inch broad." (*Gould, Handbk. Bds. Aust.*, Vol. ii., p. 60.)

Hab. West and South-West Australia. (*Ramsay.*)

Genus PSEPHOTUS, *Gould.*

7   PSEPHOTUS HÆMATOGASTER, *Gould.*
Red-bellied Parrakeet.

Gould, Handbk. Bds. Aust., Vol. ii., sp. 426, p. 62.

" For a full set of the eggs of this bird I am indebted to Mr. J. Hill, who obtained them at 'Pine Rise,' Kewell, Victoria, from the hollow branch of a Eucalyptus, on September 15th, 1887. They are seven in number for a sitting, and when found were in a very advanced state of incubation. In form they are rounded ovals, a single specimen only (F) being somewhat sharply pointed at one end, pure white, and the shell very smooth but without any gloss. They measure as follows :—length (A) 0·94 x 0·8 inch; (B) 0·95 x 0·8 inch ; (C) 0·94 x 0·78 inch ; (D) 0·97 x 0·76 inch ; (E) 0·95 x 0·8 inch ; (F) 0·97 x 0·78 inch ; (G) 0·97 x 0·8 inch." (*North, P.L.S., N.S.W.*, Vol. ii., 2nd Series, p. 986.)

Hab. New South Wales, Interior, Victoria and South Australia, West and South-west Australia. (*Ramsay.*)

PSEPHOTUS HÆMATOGASTER, *Gould.*
Var. XANTHORRHOUS, *Gould.*
Yellow-bellied Parrakeet.

Gould, Handbk. Bds. Aust., Vol. ii., sp. 427, p. 63.

A set of the eggs of this variety taken from the hollow spout of a tree by Mr. James Ramsay at Tyndarie, on the 2nd of September, 1868, are similar to those of the preceding species *P. hæmatogaster*, being rounded in form, pure white, and the texture of the shell very fine, but lustreless. Length (A) 0·93 x 0·77 inch; (B) 0·87 x 0·79 inch ; (C) 0·94 x 0·8 inch. *(Dobr. Mus. Coll.)*

Hab. New South Wales, Interior, Victoria and South Australia, West and South-West Australia. *(Ramsay.)*

3-4   PSEPHOTUS PULCHERRIMUS, *Gould.*
Beautiful Parrakeet.
Gould, Handbk. Bds. Aust., Vol. ii., sp. 429, p. 67.

The Beautiful Parrakeet is found breeding in the neighbourhood of Maryborough, Queensland, in the hollow branches of trees ; occasionally it resorts to a deserted burrow of *Dacelo leachii,* or of *Halcyon macleayi* in the nests of the Termites. Eggs white and three or four in number ; a set taken in September 1883 measure as follows :—length (A) 0·88 x 0·73 inch ; (B) 0·86 x 0·71 inch ; (C) 0·9 x 0·74 inch.

Specimens from the Dawson River, taken by Mr. Geo. Barnard, measure as follows :—(A) 0·86 x 0·71 inch ; (B) 0·87 x 0·7 inch ; (C) 0·87 x 0·71 inch.

Hab. Port Denison, Wide Bay District, Dawson River, New South Wales. *(Ramsay.)*

4   PSEPHOTUS MULTICOLOR, *Temminck.*
Many-coloured Parrakeet.
Gould, Handbk. Bds. Aust., Vol. ii., sp. 430, p. 68.

This bird is plentifully dispersed throughout the open forest country of the interior of New South Wales and South Australia, which may be considered the stronghold of this species. It breeds like most of the other members of the genus, during the months of August, September, and October, in the hollow trunk or branch of a tree, usually of a Eucalyptus or Casuarina. A set of eggs taken by Mr. K. H. Bennett, on the 24th August 1884, are four in number, white, and measure as follows:—length (A) 0·88 x 0·75 inch ; (B) 0·9 x 0·78 inch ; (C) 0·84 x 0·73 inch ; (D) 0·88 x 0·75 inch.

Hab. New South Wales, Interior, Victoria and South Australia. (*Ramsay.*)

:4   PSEPHOTUS HÆMATONOTUS, *Gould.*
Red-rumped Parrakeet.
Gould, Handbk. Bds. Aust., Vol. i., sp. 431, p. 69.

The Red-rumped Parrakeet is dispersed throughout New South Wales, Victoria, and South Australia, it is chiefly found in the belts of trees skirting the margins of rivers and creeks. This bird lays its eggs, which are white, upon the dry dust in a hole or spout usually of a Casuarina or Eucalyptus. A set of eggs taken by Mr. John Ramsay on the 2nd of September 1868, at Cardington on the Bell River, New South Wales, are four in number, varying in form from round to oval, and measure as follows :—length (A) 0·82 x 0·82 inch, rounded ; (B) 0·95 x 0·8 inch ; (C) 0·92 x 0·78 inch ; (D) 0·95 x 0·78 inch.

This species breeds during the months of September, October, and November, and is still found in great numbers on the Bell and Macquarie Rivers.

Hab. Wide Bay District, Dawson River, Richmond and Clarence Rivers Districts, New South Wales, Interior, Victoria and South Australia. (*Ramsay.*)

Genus EUPHEMA, *Wagler.*

5   EUPHEMA VENUSTA, *Temminck.*
(E. chrysostoma, Kuhl.)
Blue-banded Grass-Parrakeet.
Gould, Handbk. Bds. Aust., Vol. ii., sp. 432, p. 71.

This beautiful species is found breeding in the hollow branches of the Eucalyptus and other trees ; an average specimen of the eggs taken from the bottom of a hollow stump in which the female was captured while sitting, is white, and the shell smooth, and measures :—length 0·85 x 0·67 inch. Eggs usually five for a sitting.

The breeding season commences in September and lasts the three following months.

Hab. New South Wales, Victoria and South Australia, Tasmania. (*Ramsay.*)

EUPHEMA PULCHELLA, *Shaw.*
Chestnut-shouldered Grass-Parrakeet.
Gould, Handbk. Bds. Aust., Vol. ii., sp. 436, p. 77.

This lovely species resorts to a hole in the branch of a tree or a fallen log to deposit its eggs, of either a Eucalyptus or Casuarina. The eggs are white, and four in number, a set taken by Mr. Percy Ramsay, at Macquarie Fields in August 1859, vary in form from round to oval and measure as follows :—length (A) 0·8 x 0·71 inch; (B) 0·8 x 0·7 inch, rounded ; (C) 0·85 x 0·7 inch ; (D) 0·88 x 0·72 inch, oval.

The breeding season commences in August and continues the three following months.

Hab. Wide Bay District, Richmond and Clarence Rivers Districts, Interior, Victoria and South Australia. (*Ramsay.*)

EUPHEMA ELEGANS, *Gould.*
Elegant Grass-Parrakeet.
Gould, Handbk. Bds. Aust., Vol. ii., sp. 433, p. 73.

"The breeding season of this species is in September and October ; the eggs being from four to seven in number, of a pure white, eleven lines long by eight and a-half lines broad." (*Gould, Handbk. Bds. Aust.*, Vol. ii., p. 73.)

Hab. Wide Bay District, Richmond and Clarence Rivers Districts, New South Wales, Interior, Victoria and South Australia, West and South-West Australia. (*Ramsay*)

EUPHEMA BOURKII, *Gould.*
Bourke's Grass-Parrakeet.
Gould, Handbk. Bds. Aust., Vol. ii., sp. 438, p. 80.

Bourke's Grass Parrakeet is very sparingly distributed throughout the timbered back country of the interior of New South Wales. It usually breeds in the hollow limbs of the Eucalyptus or Casuarina, where it deposits its eggs which are white, oval in form, and four in number on the decaying wood; an average specimen from a set taken by Mr. K. H. Bennett on the 20th of August 1884, measures 0·9 inch in length by 0·7 inch in breadth.

The breeding season commences in August and continues during the three following months.

Hab. New South Wales, Interior, Victoria and South Australia. (*Ramsay.*)

Genus MELOPSITTACUS, *Gould.*
MELOPSITTACUS UNDULATUS, *Shaw.*
Warbling Grass-Parrakeet.
Gould, Handbk. Bds. Aust., Vol. ii., sp. 439, p. 81.

The Warbling Grass Parrakeet arrives in New South Wales and Victoria early in September to breed, and departs again in February, in some seasons it is seen in countless numbers covering the ground, where it obtains its food among the various grasses. It deposits its eggs, which are white and usually six in number, in the hollow branch of a tree. A set taken by Mr. W. I. Liscombe in September 1884, measure as follows:—length (A) 0·7 x 0·6 inch; (B) 0·68 x 0·59 inch; (C) 0·67 x 0·6 inch; (D) 0·68 x 0·6 inch.

A set in the Macleayan Museum give the following measurements: length (A) 0·71 x 0·6 inch; (B) 0·69 x 0·6 inch; (C) 0·7 x 0·61 inch; (D) 0·7 x 0·59 inch; (E) 0·68 x 0·61 inch.

The breeding season commences in September and lasts till the end of December.

Hab. Gulf of Carpentaria, New South Wales, Interior, Victoria and South Australia, West and South-West Australia. (*Ramsay.*)

Genus PEZOPORUS, *Illiger*.

PEZOPORUS FORMOSUS, *Latham*.

Ground-Parrakeet.

Gould, Handbk. Bds. Aust., Vol. ii., sp. 441, p. 86.

" Dr. Ramsay informs me this bird used to breed freely in the neighbourhood of Appin in the long tussocky grass, during the months of September, October, and November, and that the young birds afforded excellent sport about the end of January. A nest before me is composed of rushes and wiry grass, bitten into suitable lengths, and bent round and interwoven here and there into a platform of about half an inch thick ; a piece of *Lycopodium* also being worked into it. The diameter of the nest is 4·5 inches. Eggs white, three in number for a sitting, and smooth shelled, length (A) 1·03 x 0·85 inch ; (B) 1·01 x 0·85 inch ; (C) 1·06 x 0·85 inch." *Dobr. Mus. Coll.*" (*North, P.L.S., N.S.W.*, Vol. ii., 2nd Series, p. 410.)

With the exception perhaps of *Geopsittacus occidentalis*, it is the only Australian Parrot known that builds a nest.

Hab. Wide Bay District, Richmond and Clarence Rivers Districts, New South Wales, Victoria and South Australia, Tasmania, West and South-West Australia. (*Ramsay.*)

Genus LATHAMUS, *Lesson.*

LATHAMUS DISCOLOR, *Shaw.*

Swift Lorikeet.

Gould, Handbk. Bds. Aust., Vol ii. sp. 443, p. 90.

The Swift Lorikeet usually lays two white eggs, in the hollow of a dead branch of a Eucalypt. Specimens taken near Hobart, Tasmania, measure as follows:—length (A) 1·05 x 0·87 inch; (B) 1·05 x 0·86 inch.

This species breeds during the months of October, November, and December.

Hab. Wide Bay District, Richmond and Clarence Rivers Districts, New South Wales, Interior, Victoria and South Australia, Tasmania. (*Ramsay.*)

Genus TRICHOGLOSSUS, *Vigors and Horsfield.*

TRICHOGLOSSUS NOVÆ-HOLLANDIÆ, *Gmelin.*

(*T. multicolor,* Gmelin.)

Blue-bellied Lorikeet.

Gould, Handbk. Bds. Aust., Vol. ii., sp. 444, p. 93. XIV 8.

This species lays its eggs two in number, on the decayed wood in a hole of a dead branch of a Eucalyptus. Two average specimens received from Mr. Geo. Barnard of the Dawson River, Queensland, are dull white, one specimen (A) being a true oval, the other (B), inclined to be somewhat pyriform in shape. Length (A) 1·1 x 0·90 inch ; (B) 1·11 x 0·91 inch.

The usual breeding season of this species is during October and the three following months, but Mr. J. A. Boyd found a nest on the Herbert River, Queensland, containing young ones in the month of May 1888.

Hab. Gulf of Carpentaria, Cape York, Rockingham Bay, Port Denison, Wide Bay District, Dawson River, Richmond and Clarence Rivers Districts, New South Wales, Interior, Victoria and South Australia, Tasmania. (*Ramsay.*)

TRICHOGLOSSUS CHLOROLEPIDOTUS, *Kuhl.*

Scaly-breasted Lorikeet.

Gould, Handbk. Bds. Aust, Vol. ii., sp. 446, p. 96. XIV 12.

" The Scaly-breasted Lorikeet is plentifully dispersed over the greater part of Queensland and the northern portion of New South Wales, but is seldom found farther south than the Murray River, the natural boundary of the latter colony. Mr. George Barnard of Coomooboolaroo, who has contributed largely towards a knowledge of the nidification of many of the birds of Central Queensland, informs me that he found this species breeding in the hollow spouts of the lofty Eucalypts in the neighbourhood of the Dawson River, and that all the nests, seven in number, taken by his sons, unlike those of any other member of the family, each contained but a single egg, which in several instances was in a very advanced state of incubation. Two eggs taken during the month of November are pure white, in form oval, slightly tapering at one end, the texture of the shell being fine and smooth, but without any lustre. Length (A) 0·95 x 0·79 inch ; (B) 0·97 x 0·8 inch." *(North. P.L.S., N.S.W.,* Vol. ii., 2nd Series, p. 986.)

Hab. Rockingham Bay, Port Denison, Wide Bay District, Dawson River, Richmond and Clarence Rivers Districts, New South Wales, Interior, Victoria and South Australia. (*Ramsay.*)

Genus GLOSSOPSITTA, *Bonaparte.*

GLOSSOPSITTA AUSTRALIS, *Latham.*

(G. concinnus, Shaw.)

Musk-Lorikeet.

Gould, Handbk. Bds. Aust., Vol. ii., sp. 448, p. 100.

The Musk Lorikeet resorts to the hollow branch of a Eucalyptus or Casuarina for the purposes of breeding. Four eggs taken by Mr. K. H. Bennett during November 1885, are white, three of them are oval in form, the remaining one (D) round; length (A) 0·98 x 0·8 inch ; (B) 0·95 x 0·8 inch ; (C) 0·98 x 0·82 inch ; (D) 0·95 x 0·85 inch.

This species breeds during the months of October, November, and December.

Hab. Rockingham Bay, Port Denison, Wide Bay District, Dawson River, Richmond and Clarence Rivers Districts, New South Wales, Interior, Victoria and South Australia, Tasmania. (*Ramsay.*)

GLOSSOPSITTA PUSILLUS, *Shaw.*

Little Lorikeet.

Gould, Handbk. Bds. Aust., Vol. ii., sp. 450, p. 103.

The Little Lorikeet, like all other members of the genus *Trichoglossus,* generally lays its eggs in a hole in the branch of a lofty Eucalyptus. A set of four taken by Mr. Geo. Barnard in 1883, vary in form from round to oval, colour white ; length (A) 0·76 x 0·69 inch; (B) 0·77 x 0·68 inch, rounded ; (C) 0·82 x 0·68 inch ; (D) 0·81 x 0·68 inch, oval.

The breeding season commences in October and lasts till the end of January.

Hab. Rockingham Bay, Port Denison, Wide Bay District, Dawson River, Richmond and Clarence Rivers Districts, New South Wales, Interior, Victoria and South Australia, Tasmania. (*Ramsay.*)

Order RASORES.

Family COLUMBIDÆ.

GENUS LAMPROTRERON, *Bonaparte.*

LAMPROTRERON SUPERBUS, *Temminck.*

Superb Fruit-Pigeon.

Gould, Handbk. Birds Aust., Vol. ii., sp. 453, p. 108.

The egg of this fruit-dove is remarkably small in comparison with eggs of other pigeons of a smaller size. I have received specimens taken by Mr. Boyer-Bower in the brushes near Cairns, Queensland. Eggs white, oval, rather elongated and pointed, 1·2 x 0·83 inch; two only are laid for a sitting." (*Ramsay, P.L.S., N.S.W.,* 2nd Series, Vol. i., p. 1151.)

Hab. Port Darwin and Port Essington, Cape York, Rockingham Bay, Port Denison, Tasmania, South Coast New Guinea. (*Ramsay.*)

GENUS MYRISTICIVORA, *Reichenbach.*

MYRISTICIVORA SPILORRHOA, *G. R. Gray.*

White Nutmeg-Pigeon.

Gould, Handbk. Bds. Aust., Vol. ii., sp. 457, p. 114.

" During the months from October until the end of April, when they leave, this species is very numerous all over the Rockingham Bay district. Early in the morning, as soon as it is light enough they leave their roosting-places in large flocks, and betake themselves to their feeding-grounds, dispersing over the scrubs and among the various species of *Acmena* and *Jambosa* which line the margins of the Herbert River. Towards evening they assemble and leaving the feeding-grounds return to roost on the mangrove islands in Hinchinbrook Channel, and around the coast and mouths of the rivers, flying a distance of often forty miles night and morning. The tops of the mangroves on which they roost are literally white with birds; and, notwithstanding the disturbance and havoc committed among them by shooting parties, they continue to arrive until dark. They breed on these islands, building little or no nest, a few sticks placed so as to prevent the eggs from rolling away being considered sufficient." (*Ramsay, P.Z.S.*, 1875, p. 115.)

Eggs oval in form, white. Two eggs in the Australian Museum Collection presented by Mr. J. Macgillivray, and taken by him on the 22nd of October 1860, on Hope Island, off the north coast of Australia, measure as follows :—Length (A) 1·64 x 1·2 inch; (B) 1·65 x 1·2 inch.

Dimensions of two eggs in the Macleayan Museum Collection, taken at Cape Sydmouth, North-Eastern Queensland, in 1871; length (A) 1·7 x 1·17 inch; (B) 1·67 x 1·16 inch. A specimen in the Dobroyde Collection, measures 1·67 inch in length by 1·2 inch in breadth.

Hab. Port Darwin and Port Essington, Cape York, Rockingham Bay, Port Denison. (*Ramsay.*)

Genus MEGALOPREPIA, *Reichenbach.*

2 MEGALOPREPIA ASSIMILIS, *Gould.*
Allied Fruit-Pigeon.
Gould, Handbk. Bds. Aust., Vol. ii., sp. 455, p. 111.

The Allied Fruit-Pigeon is universally dispersed over the Cape York peninsula, and as far south as the neighbourhood of Rockingham Bay. A nest of this species found at Cape York by Mr. George Masters, on the 17th of September 1875, from which the bird was flushed and procured, was simply a few dried sticks placed cross-wise on the horizontal branch of a tree about eight feet from the ground; it contained two eggs, perfectly white, rather elongated in form and pointed at the smaller ends, in a very advanced state of incubation. An average specimen measures

1·4 inch in length by 0·95 inch in breath. *From the Macleayan Mus. Coll.* (*North, P.L.S., N.S.W.,* 2nd Series, Vol. ii., p. 410.) *Hab.* Cape York, Rockingham Bay. (*Ramsay.*)

Genus LOPHOLAIMUS, *G. R. Gray.*

LOPHOLAIMUS ANTARCTICUS, *Shaw.*

Top-knot Pigeon.

Gould, Handbk. Bds. Aust., Vol. ii., sp. 458, p. 116.

The habitat of this bird is the rich brushes of the eastern coast of the continent of Australia, such as are to be found clothing the sides of the Richmond and Clarence Rivers in New South Wales, and Rockingham Bay in Queensland. An egg of this species taken by Mr. Theodore McLennan from the oviduct of a bird shot on the Bellinger River during the month of December, is quite white and rough to the touch, without any gloss, which may be accounted for by its being immature ; in form a long oval slightly pointed at the thinner end. Long diameter 1·85 ; short diameter 1·25 inches.

Hab. Cape York, Rockingham Bay, Port Denison, Wide Bay District, Richmond and Clarence Rivers Districts, New South Wales, Tasmania. (*Ramsay*)

Genus CHALCOPHAPS, *Gould.*

2 CHALCOPHAPS CHRYSOCHLORA, *Wagler.*

Little Green Pigeon.

Gould, Handbk. Bds. Aust., Vol. ii., sp. 459, p. 118.

Mr. George Barnard, of Coomooboolaroo, Duaringa, Queensland, informs me that a nest of this species taken by one of his black boys on the Dawson River, on the 6th of February 1887, was a very frail structure of twigs placed in a low tree. The nest contained two eggs, one of which he sent me for description, is in form a nearly perfect oval, creamy-white, the texture of the shell being fine and smooth to the touch. Long axis 1·05 inch, short axis 0·8 inch.

Hab. Port Darwin and Port Essington, Gulf of Carpentaria, Cape York, Rockingham Bay, Port Denison, Wide Bay District, Dawson River, Richmond and Clarence Rivers Districts, New South Wales. (*Ramsay.*)

Genus LEUCOSARCIA, *Gould.*

2   LEUCOSARCIA PICATA, *Latham.*

Wonga-wonga.

Gould, Handbk. Bds. Aust., Vol. ii., sp. 461, p. 120.

This bird is found in the thickly wooded country that skirts the eastern and south-eastern portions of the continent of Australia, likewise on the mountain ranges farther inland ; it is particularly plentiful in the neighbourhood of Eden, New South Wales, and many birds are trapped and sent up to Sydney every season, and during the winter months they may be seen alive in the markets, or killed and exposed for sale in the poulterers shops. The nest is rather a frail structure of sticks and twigs placed crosswise on the horizontal branch of a tree about ten feet from the ground. Eggs two in number for a sitting, pure white. Two specimens in the Australian Museum Collection, measure as follows :—length (A)1·5 x 1·1 inch; (B) 1·48 x 1·1 inch.

Two eggs in Dr. James C. Cox's Collection, taken at Mulgoa, on the 12th of January 1875, measure as follows :—length (A) 1·41 x 1·05 inch ; (B) 1·43 x 1·15 inch.

Two in the Dobroyde Collection, taken at Toowoomba, Queensland, in October 1873, give the following measurements :—length (A) 1·45 x 1·08 inch ; (B) 1·48 x 1·09 inch.

The breeding season of this species commences in July, and continues until the end of January.

Hab. Rockingham Bay, Port Denison, Wide Bay District, Dawson River, Richmond and Clarence Rivers Districts, New South Wales, Victoria and South Australia. (*Ramsay.*)

Genus PHAPS, *Selby.*

2   PHAPS CHALCOPTERA, *Latham.*

Common Bronze-wing.

Gould, Handbk., Bds. Aust., Vol. ii., sp. 462, p. 122.

This fine Pigeon is universally distributed over the whole of Australia. The nest is a frail structure of twigs placed upon a horizontal branch of a tree, varying in height, according to the locality ; stunted *Eucalypti, Angophoræ* and the *Acaciæ*, being especially favoured trees in this respect. Eggs two in number for a sitting, pure white.

Two eggs in the Australian Museum Collection measure as follows :—length (A) 1·48 x 0·97 inch ; (B) 1·37 x 0·98 inch.

Two other sets taken by Mr. K. H. Bennett at Ivanhoe, give the following measurements :—length (A) 1·38 x 1 inch ; (B) 1·33 x 1 inch ; (C) 1·4 x 0·98 inch ; (D) 1·37 x 0·95 inch.

Mr. K. H. Bennett states that this bird has no fixed breeding time, as he has found nests containing eggs or young, in every month of the year, they are however more frequently found breeding during the months of October and November.

Hab. Port Darwin and Port Essington, Gulf of Carpentaria, Rockingham Bay, Port Denison, Wide Bay District, Dawson River, Richmond and Clarence Rivers Districts, New South Wales, Interior, Victoria and South Australia, Tasmania, West and South-West Australia. (*Ramsay*)

R

2
PHAPS ELEGANS, *Temminck.*
Brush Bronze-wing.
Gould, Handbk. Bds. Aust., Vol. ii., sp. 463, p. 125.

This bird was at one time found breeding rather freely in the scrubs in the neighbourhood of Sydney, the nest is similar to that of the preceding species *P. chalcoptera.* Eggs two in number for a sitting, pure white. A set taken at Bondi on the 4th of August 1863, measure as follows :—length (A) 1·33 x 0·95 inch ; (B) 1·4 x 0·95 inch.

Dimensions of a set taken near Hobart, Tasmania, on the 15th of April 1883 :—length (A) 1·28 x 0·97 inch ; (B) 1·33 x 0·99 inch.

Like the preceding species this bird has no fixed breeding season but nests containing eggs or young ones, are usually found during August and the three following months.

Hab. Port Darwin and Port Essington, Wide Bay District, Richmond and Clarence Rivers Districts, New South Wales, Victoria and South Australia, Tasmania, West and South-West Australia. (*Ramsay.*)

2
PHAPS HISTRIONICA, *Gould.*
Harlequin Bronze-wing.
Gould, Handbk. Bds. Aust., Vol. ii., sp. 464, p. 127.

"In some seasons this beautiful pigeon is to be found in countless numbers on the vast plains of the interior of Australia ; its range also extends to Port Darwin, and Derby in North-western Australia, specimens having been procured both by Mr. E. J. Cairn and the late Mr. T. H. Boyer-Bower during the latter part of 1886. In the evening these birds arrive in large flocks at the dams and water tanks to drink, but at the slightest indication of danger they take flight. This species is terrestrial in its habits, and for the purpose of breeding generally resorts to the shelter of a cotton bush, forming little or no nest it often lays its eggs upon the bare ground. A pair of these birds have bred in the avairy of the Hon. William Macleay, of Elizabeth Bay during 1887 and 1888. Mr. George Masters, the Curator of the Macleayan Museum, informs me that the eggs were two in number for a sitting. They are in form oval, of a faint creamy-white, the surface of the shell being both slightly granular and glossy. Length (A) 1·3 x 0·93 inch ; (B) 1·22 x 0·92 inch. These eggs are similar to specimens in the Dobroyde Collection, taken by Mr. J. B. White, on the Barcoo River during July, 1868. July and August are the usual breeding season of his species." *(North, P.L.S, N.S.W.,* Vol. iii., 2nd Series, p. 148.)

Hab. Derby, N.W. Australia, Port Darwin and Port Essington, Dawson River, New South Wales, Interior, South Australia. *(Ramsay.)*

Genus GEOPHAPS, *Gould.*

2   GEOPHAPS SCRIPTA, *Temminck.*

Partridge Bronze-wing.

Gould, Handbk. Bds. Aust., Vol. ii., sp. 465, p. 130.

The "Squatter" or Partridge Bronze Wing is widely dispersed over the interior of Australia, and is never found far away from the vicinity of water, it is terrestrial in its habits and will often allow itself to be almost trodden upon, before taking flight. As an article of food it is considered a great dainty, its flesh fully equalling in delicacy that of the Wonga-Wonga. Its eggs, two in number, are deposited upon the bare ground, sometimes under the shelter of any scanty herbage ; in form they are oval, being somewhat slightly pointed at one end, of a faint creamy-white, and the shell smooth and slightly glossy. Specimens taken in New South Wales by Mr. W. Liscombe, and by Mr. George Barnard on the Dawson River, Queensland, measure as follows :—length (A) 1·22 x 0·9 inch ; (B) 1·21 x 0·89 inch.

The season of incubation in New South Wales is generally during the months of September and October, but Mr. Barnard informs me that in Queensland he has taken the eggs in nearly every month of the year.

Hab. Rockingham Bay, Port Denison, Wide Bay District, Dawson River, Richmond and Clarence Rivers Districts, New South Wales, Interior, South Australia. (*Ramsay.*)

GEOPHAPS SMITHII, *Jardine and Selby.*

Smith's Partridge Bronze-wing.

Gould, Handbk. Bds. Aust., Vol. ii., sp. 466, p. 133.

"This bird incubates from August to October, making no nest but merely smoothing down a small part of a clump of grass and forming a slight hollow, in which it deposits two eggs, which are greenish-white, one inch and a-quarter long by seven-eights of an inch in breadth." (*Gould, Handbk. Bds. Aust.*, Vol. ii., p. 133.)

Hab. Derby, N.W. Australia, Port Darwin and Port Essington, Gulf of Carpentaria. (*Ramsay.*)

Genus LOPHOPHAPS, *Reichenbach.*

LOPHOPHAPS LEUCOGASTER, *Gould.*

Plumed Bronze-wing.

Gould, Suppl. Bds. Aust., pl. 69.

"Sub-Inspector Armit, late of the Native Police, Queensland, obtained specimens in the neighbourhood of Normanton, Gulf District, and forwarded a set of the eggs. The nest is placed on the ground, and, like that of *Geophaps scripta*, consists merely of a few blades of grass lining a slight hollow on the lee side of a tussock or tuft of grass. Eggs four in number for a sitting; of a pale cream colour. Length 1·05 x 0·8 inch. Mr. T. H. Boyer-Bower collected a fine series of this rare bird, in the vicinity of Derby, North Western Australia." (*Ramsay, P.L.S., N.S.W.,* 2nd Series, Vol. i., p. 1095.)

Hab. Derby, N.W. Australia, Gulf of Carpentaria, Interior, South Australia. (*Ramsay.*)

Genus OCYPHAPS, *Gould.*

2

OCYPHAPS LOPHOTES, *Temminck.*

Crested Bronze-wing.

Gould, Handbk. Bds. Aust., Vol. ii., sp. 469, p. 139. XIV *10.*

According to Mr. K. H. Bennett the principal habitat of this species is on the borders of rivers and watercourses, particularly where thickets of *Polygonum* abound, the scrubby bushes of which are frequently chosen as sites for its nests. Like *Phaps chalcoptera* it does not appear to have any fixed breeding time, for it is not uncommon to find nests containing eggs or young in almost every month of the year. The nest is a very frail structure of a few thin sticks loosely put together and placed amongst the thick twigs of some low tree or bush. Eggs two in number for a sitting varying in form from long to short pointed ovals. A set taken by Mr. Bennett during November, measure as follows:—length (A) 1·2 x 0·87 inch; (B) 1·25 x 0·88 inch.

Hab. Derby, N.W. Australia, Port Darwin and Port Essington, Gulf of Carpentaria, Dawson River, New South Wales, Interior, Victoria and South Australia. (*Ramsay.*)

Genus ERYTHRAUCHÆNA, *Bonaparte.*

2 ERYTHRAUCHÆNA HUMERALIS, *Temminck.*

Barred-shouldered Dove.

Gould, Handbk. Bds. Aust., Vol. ii., sp. 471, p. 142.

This bird ranges over the whole of Queensland, North-western Australia, and the greater portion of New South Wales. It constructs a frail nest of sticks, and places it upon the lower branches of a thickly foliaged tree. Eggs two in number for a sitting, oval in form, of a delicate fleshy-white. Dimensions of a set taken by Mr. George Barnard of Coomooboolaroo, Queensland, during 1883 :—length (A) 1·12 x 0·85 inch; (B) 1·9 x 0·83 inch.

Specimens in the Macleayan Museum Collection, taken in New South Wales, measures as follows :—length (A) 1·07 x 0·86 inch; (B) 1·06 x 0·85 inch.

Hab. Derby, N.W. Australia, Port Darwin and Port Essington, Gulf of Carpentaria, Rockingham Bay, Port Denison, Wide Bay District, Dawson River, Richmond and Clarence Rivers Districts, New South Wales, South Australia. (*Ramsay.*)

Genus GEOPELIA, *Swainson.*

GEOPELIA TRANQUILLA, *Gould.*

Peaceful Dove.

Gould, Handbk. Bds. Aust., Vol. ii., sp. 472, p. 144.

"The nest, like that of all the members of this genus, is a frail scanty structure of a few sticks and twigs placed usually near the end of a bushy bough, or top of a broken off thick limb. They are slightly larger than those of *G. cuneata,* oval in form and of a pure white colour. Length 0·8 x 0·6 ; 0·78 x 0·58. *Dobr. Mus.*" (*Ramsay, P.L.S., N.S.W.,* Vol. vii., p. 54.)

These birds were very plentiful on the Bell River, New South Wales, during August 1887.

Hab. Port Darwin and Port Essington, Rockingham Bay, Port Denison, Wide Bay District, Richmond and Clarence Rivers Districts, New South Wales, Victoria and South Australia. (*Ramsay.*)

GENUS STICTOPELIA, *Reichenbach.*

2. STICTOPELIA CUNEATA, *Latham.*

Little Turtle-dove.

Gould, Handbk. Bds. Aust., Vol. ii., sp. 474, p. 146.

"The nest is similar to that of the last species, Mr. John S. Ramsay, found this bird breeding in numbers at Cardington Station, on the Bell River, the nests were placed on the flattened top of the vine stakes in the vineyard, the birds were remarkably tame and would allow themselves to be almost taken with the hand. The eggs were invariably two in number, oval, pure white, length 0·7 x 0·55 inch ; they breed during September and the two months following. *Dobr. Mus.*" (*Ramsay, P.L.S., N.S.W.*, Vol. vii., p. 54.)

Hab. Derby, N.W. Australia, Port Darwin and Port Essington, Gulf of Carpentaria, Dawson River, New South Wales, Interior, Victoria and South Australia, West and South-West Australia. (*Ramsay.*)

Family MEGAPODIDÆ.

GENUS TALEGALLUS, *Lesson.*

TALEGALLUS LATHAMI, *Gray.*

Wattled Talegallus.

Gould, Handbk. Bds. Aust., Vol. ii., sp. 476, p. 150.

Dr. Ramsay writing in 1876 on the birds of north-eastern Queensland, remarks as follows :—

"However plentiful this species may have been formerly in the Rockingham Bay district, it is now very scarce, only one having been obtained during my visit. They are still plentiful in the New South Wales scrubs. I found that two or more females visited the same mound to lay their eggs in ; and when this is the case the mound is often twice as large as an ordinary mound. It seems probable that several individuals assist in scratching the mound together, when a space often fifty yards in diameter (on level ground) is cleared of almost every fallen leaf and twig. The mounds are often six feet in height, and twelve to fourteen yards wide at the base ; sometimes they are more conical. The central portion consists of decayed leaves mixed with fine débris, the next of coarser and less rotted materials ; and the outside is a mass of recently gathered leaves, sticks, and twigs not showing signs of decay. In opening the nest these are easily removed, and must be carefully pushed backwards over the sides, beginning at the top. Having cleared these, and obtained plenty of room remove the semidecayed strata, and below it where the fermentation has begun, in a mass of light fine leaf-mould will be found the eggs placed with the *thin end downwards*, often in a circle, with three or four in the centre, about six inches apart. At one side, where the eggs have been first laid they will probably be found more or less incubated, but in the centre where the eggs are placed last, quite fresh ; and if only one pair of birds have laid in the mound, about twelve to eighteen eggs will be the complement, and will be found arranged as described above. On the other hand, if several females resort to the same nest the regularity will be greatly interfered with and two or three eggs in different stages of development will be found close to one another, some quite fresh, others within a few days of being hatched. There are usually ten eggs in the first layer, five or six in the second, three or four only in the centre. I found that the females return every second day to lay, but never succeeded in ascertaining which of the parent birds opens the nest. The aborigines informed me that the male bird always performs this office ; and I usually found my black boys very correct in their statements of this kind. After robbing a nest it is necessary to replace the different layers as they were found, if the lowermost is too much mixed up with the others, or the the top tumbled into the excavations made in the bottom one, the birds will invariably forsake the mound ; so that I found it always necessary to carefully replace the different layers as I found them. It is not so with the *Megapodius tumulus*, which species does not seem to care how much the mound is tumbled about, so that there is sufficient débris left to burrow in ; and indeed should there not be they quietly set to work and scratch it together again. The mounds of the *Tallegallus* are seldom found on a great incline when a level spot can be obtained. They frequently bring the débris from a considerable distance, and in one instance on the Richmond River I noticed a place where about a cartload had been scratched through a shallow part of a creek three or four inches deep in water, and up the other side of the bank to the mound, which was over forty yards distant. The débris is *always thrown behind them.* The greatest number of eggs taken from one mound at one time was thirty-six. This was a very old mound and resorted to by several individuals. The eggs vary much in size, and in shape from almost round to a long oval, or pointed at the thin end ; their usual form is an oval slightly smaller at one end. The shell is very thin, minutely granulated, and snow-white in colour." (*Ramsay, P.Z.S.*, 1876, p. 116.)

Dimensions of thirteen eggs taken from an egg-mound on Taranya Creek, a branch of the Richmond River, in November 1866, by Dr. Ramsay are as follows:—length (A) 3·55 x 2·33 inches; (B) 3·6 x 2·42 inches; (C) 3·55 x 2·4 inches; (D) 3·27 x 2·43 inches; (E) 3·59 x 2·4 inches ; (F) 3·65 x 2·25 inches ; (G) 3·58 x 2·27 inches ; (H) 3·58 x 2·39 inches ; (I) 3·9 x 2·4 inches ; (J) 3·47 x 2·47 inches ; (K) 3·65 x 2·45 inches ; (L) 3·53 x 2·55 inches ; (M) 3·67 x 2·5 inches.

Hab. Cape York, Rockingham Bay, Port Denison, Wide Bay District, Dawson River, Richmond and Clarence River Districts, New South Wales. (*Ramsay.*)

Genus LEIPOA, *Gould.*

LEIPOA OCELLATA, *Gould.*

Ocellated Leipoa.

Gould, Handbk. Bds. Aust., Vol. ii., sp. 477, p. 155.

This bird is an inhabitant of the scrubs and plains of the inland portions of the continent of Australia, and is also met with in the

Mallee country of Victoria and South Australia. It scrapes up
huge mounds of sand and decayed vegetable matter, leaves, grass,
&c., and deposits its eggs, usually six or eight in number at the
bottom, leaving the young birds when hatched by the heat of the
mound to scramble out and shift for themselves the best way they
can. Eggs when fresh, are of a delicate pinky-white, but after
remaining in the mound a few days they become a dirty reddish-
brown. The shell is very thin, and when removing the eggs from
the mound unless a great amount of care is used in uncovering
them, they are easily broken.

Five eggs taken by Mr. James Ramsay, from a mound in the
Merule Scrubs, in October 1869, measure as follows :—(A) 3·4 x
2·24 inches ; (B) 3·47 x 2·3 inches ; (C) 3·55 x 2·35 inches ; (D)
3·36 x 2·4 inches ; (E) 3·67 x 2·27 inches.

Four eggs taken by Mr. K. H. Bennett in the Lachlan District
in October 1883, give the following measurements :—length (A)
3·62 x 2·35 inches ; (B) 3·43 x 2·25 inches; (C) 3·45·x 2·24 inches;
(D) 3·5 x 2·36 inches.

For full and exhaustive accounts of the mound raising habits of
this bird, see Gould's Handbook to the Birds of Australia, Vol. ii.,
p. 155 ; and "Habits of the Mallee Hen, *Leipoa ocellata*, by K.
H. Bennett," in the Proc. Linn. Soc. of New South Wales, Vol.
viii., p. 183.

Hab. New South Wales, Interior, Victoria and South Australia,
West and South-West Australia. (*Ramsay.*)

GENUS MEGAPODIUS, *Quoy et Gaimard.*
MEGAPODIUS TUMULUS, *Gould.*
Australian Megapode.
Gould, Handbk. Bds. Aust, Vol. ii., sp., 478, p. 167.

"This Mound-raiser is very plentiful north after passing Port
Denison ; I found it also in tolerable numbers as far south as the
Pioneer River. They are strictly confined to the dense scrubs,

and seldom if ever seen elsewhere. Their noisy cackling at night frequently disturbed us when encamped near one of their favorite resorts; and during the day their hoarse note at once betrays their presence. On the Herbert River they are not much sought after as an article of food, either by the natives or whites, for as their eggs are esteemed a delicacy, the birds themselves are not much molested. I examined several nests in March; and although it was not the regular breeding season, yet fresh eggs were obtained, and newly hatched young were found singly here and there throughout the denser parts of the brushes. Some of the mounds were very ruthlessly destroyed by the whites, and scattered over the ground, this however, did not cause the birds to forsake the place, and out of one large mound which had been very roughly handled, two new ones were formed about ten yards apart, on the base of the old one, which was so matted and interlaced with roots from the neighbouring trees that it appeared to me a marvel how the birds could burrow into it the great length they did; and having once laid their eggs there, however the young birds found their way out through the maze of roots is still a mystery. Once out however and their wings dry they are able to take care of themselves, but remain about the mounds for a day or so, as if waiting for some of their companions, but in less than a week from the day they are hatched, they may frequently be seen at least quarter of a mile away, and well able to fly about. I met one little fellow, only 5·5 inches in total length, fully a mile away from the nearest mound, he flew up and settled in a tree, about twenty feet from the ground, the wings and feet were remarkably developed for so small a bird, which could be scarcely more than four weeks old. Upon more than one occasion I have seen the birds busy at their mound, or feeding near it, but was never so fortunate as to meet with them in the act of burrowing. The largest mound I met with was about fifty feet in length, ten in height, and fourteen feet in width at the base, eight or ten on the summit. It seemed to be more like several mounds combined; and certainly more than two pairs of birds frequented it. While stationed, gun in hand, watching, for Cassowaries *(Casuaris australis*), I noticed on one occasion five birds arrive at this mound in company, they came very close to me, making a chuckling noise jerked out from their throat, and not unlike that of a domestic fowl when driven from its nest, but not so loud. Usually only a pair are met with together. Their flight is heavy, and they do not readily take wing, unless pursued by a dog, when they rise with a considerable flapping to the most convenient branch, where they are easily approached and shot. Their flesh is dark, rank, and tough. The young, about five inches in length, are of a dull brown, ashy-brown on the sides of the face, neck, and mantle, and on the abdomen of a lighter ashy-brown, rufous-brown on the flanks, and brown washed with rufous on the breast; the back, rump, and tail of a rich rufous-brown, primaries dark, brown, interscapular region and upper wing-coverts dark brown, tipped with light rufous, the secondaries and scapulars freckled, and margined on the outer web with light rufous, the outer series of secondary-coverts and outer scapulars barred and freckled with the same colour, iris dark brown, feet yellow. Total length 5·5 inches, bill ·45,. wing 4·5, tarsus 1 inch, tail a tuft of down about 1 inch in length." *(Ramsay, P.Z.S.*, 1876, p. 118.)

Eggs in form elongated ovals, being nearly equal at both ends, the normal colour when newly laid, being of a pale coffee-brown but after remaining in the mound a few days they become darker, and by the time they are partly incubated, the outer surface or epidermis of the shell easily chips off in places, revealing a snow-white surface underneath; the colouring matter can also be rubbed off when moisture is applied.

Dimensions of six eggs taken at Cairns, are as follows :—length (A) 3·58 x 2·07 inches ; (B) 3·55 x 2·05 inches ; (C) 3·52 x 2·01 inches ; (D) 3·6 x 2·01 inches ; (E) 3·58 x 2·05 inches ; (F) 3·45 x 2·0 inches. Specimens taken at Cape York in 1864 measure (A) 3·41 inches x 1·98 inch; (B) 3·45 x 2·01 inches.

Hab. Port Darwin and Port Essington, Gulf of Carpentaria, Cape York, Rockingham Bay, Port Denison. (*Ramsay.*)

Family TURNICIDÆ.

GENUS TURNIX, *Bonnaterre*.

TURNIX MELANOGASTER, *Gould*.

Black-breasted Turnix.

Gould, Handbk. Bds. Aust., Vol. ii., sp. 479, p. 178. *XVI* *//*.

This bird is found breeding in the eastern portions of Queensland. The eggs, three or four in number for a sitting are placed in a slight depression beside a tuft of grass. A specimen in the Australian Museum Collection taken by Mr. Geo. Masters, is of a pale buffy-white ground colour, minutely and thickly freckled all over with light reddish-brown; towards the larger apex are bold blotches and spots of chestnut-brown, purplish-grey, and inky-black, and a few indistinct superimposed markings of slaty-lilac. Long axis 1·12 inch; short axis 0·9 inch.

Two specimens from a sitting of three, taken by Mr. George Barnard, of Coomooboolaroo, Queensland, measure as follows :— length (A) 1·18 x 0·9 inch ; (B) 1·16 x 0·91 inch.

Hab. Wide Bay District, Dawson River, New South Wales. (*Ramsay.*)

TURNIX VARIUS, *Latham*.

Varied Turnix.

Gould, Handbk. Bds. Aust., Vol. ii., sp. 480, p. 179.

This bird is found all over the eastern and southern portions of Australia, and the whole of Tasmania, and is the commonest species we have of the genus. Its eggs are usually laid in a slight cavity lined with dried grasses, close to a tuft of grass. Eggs four in number for a sitting, in form swollen ovals, slightly pointed at the smaller end, of a buffy-white ground colour thickly freckled and minutely spotted all over with reddish-chestnut, wood-brown and slaty-grey, in some instances the markings are larger and very much darker towards the larger apex, but never assume the form of a zone. Two eggs in the Australian Museum Collection, taken by Mr. George Masters at King George's Sound, measure as follows :—length (A) 1·1 x 0·8 inch ; (B) 1·12 x χ·86 inch.

A set taken at Dobroyde on the 9th of October 1864, by Dr. Ramsay, give the following measurements :—length (A) 1·11 x 0·9 inch ; (B) 1·12 x 0·9 inch ; (C) 1·14 x 0·91 inch ; (D) 1·1 x 0·9 inch.

This bird sits very close and will allow itself to be almost trodden upon before leaving its eggs or young.

September and the four following months constitute the breeding season of this species.

Hab. Rockingham Bay, Port Denison, Wide Bay District, Dawson River, Richmond and Clarence Rivers Districts, New South Wales, Interior, Victoria and South Australia, Tasmania. (*Ramsay.*)

TURNIX VELOX, *Gould.*

Swift-flying Turnix.

Gould, Handbk. Bds. Aust., Vol. ii., sp. 483, p. 184.

This bird is distributed over the greater portion of the Australian continent, and is particularly abundant during the breeding season, September and the three following months, on the grassy plains of the Lachlan and Darling Districts of New South Wales. Like that of all other members of the genus, the nest is formed of grasses placed in a hollow of the ground behind some convenient tuft of grass. Eggs four in number for a sitting, in form swollen ovals, and extremely variable in the disposition of their markings ; some being minutely freckled, and closely resembling miniature eggs of *T. varius,* others being boldly blotched and spotted like those of *T. melanogaster.*

A set in the Australian Museum Collection are of a buffy-white ground colour, thickly freckled and blotched with dark reddish-brown, light brown, and slaty-grey markings, the blotches becoming larger on the thicker end. Length (A) 0·9 x 0·7 inch ; (B) 0·87 x 0·7 inch ; (C) 0·85 x 0·67 inch.

A set in K. H. Bennett's Collection, taken on the 18th of October 1886, at Mossgiel, are of a bluish-white ground colour, minutely freckled all over, intermingled with markings of light chestnut, chestnut-brown, and slaty-grey, the latter colour in one or two places forming small clouded blotches. Length (A) 0·93 x 0·75 inch ; (B) 0·93 x 0·76 inch ; (C) 0·95 x 0·76 inch ; (D) 0·93 x 0·75 inch.

A set taken by Mr. W. Liscombe in October 1883, has the entire surface of the shell obscured by very faint confluent markings of reddish-chestnut and slaty-grey. Length (A) 0·91 x 0·75 inch ; (B) 0·89 x 0·73 inch ; (C) 0·9 x 0·72 inch ; (D) 0·92 x 0·73 inch.

Hab. Derby, N.W. Australia, Wide Bay District, Richmond and Clarence Rivers Districts, New South Wales, Interior, Victoria and South Australia, West and South-West Australia. (*Ramsay.*)

TURNIX PYRRHOTHORAX, *Gould.*

Red-chested Turnix.

Gould, Handbk. Bds. Aust., Vol. ii., sp. 484, p. 186.

The nidification of this bird is similar to that of the preceding species. A nest found by Dr. Ramsay during 1864 at "Manar," New South Wales, was merely a shallow grass-lined depression in the earth, sheltered by a wind-bent tuft of grass. Eggs four in number for a sitting, in form swollen ovals, of a dull white ground colour, which is almost obscured by indistinct markings of chestnut and slaty-brown ; in some specimens the markings are larger and more pronounced, even in the same set. Length (A) 1 x 0·79 inch ; (B) 0·98 x 0·76 inch.

Hab. Cape York, Wide Bay District, Richmond and Clarence Rivers Districts, New South Wales, Interior, Victoria and South Australia. (*Ramsay.*)

GENUS PEDIONOMUS, *Gould.*

PEDIONOMUS TORQUATUS, *Gould.*

Collared Plain-Wanderer.

Gould, Handbk. Bds. Aust., Vol. ii., sp. 485, p. 187. XVI 12

This bird inhabits the large grassy plains of New South Wales and South Australia, and has likewise been found breeding as far south as the neighbourhood of Melbourne and Adelaide, but in both of the latter localities it is considered rare. The nest is constructed of dried grasses, and is placed in a slight depression in the ground underneath the shelter of some convenient shrub or tuft of grass. Eggs four in number for a sitting; in shape pyriform, of a stone-white ground colour, thickly freckled and blotched, and a few smudges here and there of different shades of umber-brown and slaty-grey, a few nearly obsolete blotches of the latter colour appearing as if beneath the surface of the shell. An average specimen in the Dobroyde Collection, taken at Springfield, near Goulburn in December 1875, measures 1·35 x 0·94 inch. Another from a set of four, taken near Adelaide, measures 1·2 inch in length by 0·89 inch in breath.

Specimens in my own collection taken near Pyramid Hill in Victoria, measure as follows:—length (A) 1·33 x 0·94 inch; (B) 1·34 x 0·95 inch.

October and November are the principal breeding months of this species.

Hab. New South Wales, Interior, Victoria and South Australia (*Ramsay.*)

Family PERDICIDÆ.

Genus COTURNIX, *Mœhring*.

COTURNIX PECTORALIS, *Gould.*
Pectoral Quail.
Gould, Handbk. Bds. Aust., Vol. ii., sp. 486, p. 190. ‾XVL‾ *10.*

This is a very common bird all over the eastern and south-eastern portions of Australia, as well as the whole of Tasmania. It breeds alike in the grassy flats or in the paddocks under cultivation. Eggs ranging from seven to fourteen in number for a sitting ; in form swollen ovals, and varying considerably in their markings even in the same nest.

Four eggs in the Australian Museum Collection are of a yellowish white ground colour, thickly blotched and minutely spotted all over with very irregular shaped markings of dark umber-brown. Length (A) 1·12 x 0·87 inch ; (B) 1·12 x 0·87 inch ; (C) 1·14 x 0·87 inch ; (D) 1·17 x 0·87 inch.

A set of seven taken at Macquarie Fields by Dr. Ramsay in September 1859, have a yellowish-white ground colour, with markings varying from minute freckles of umber-brown to large marbled blotches of a darker tint. Length (A) 1·2 x 0·94 inch ; (B) 1·24 x 0·95 inch; (C) 1·21 x 0·94 inch ; (D) 1·23 x 0·92 inch ; (E) 1·1 x 0·87 inch ; (F) 1·27 x 0·95 inch ; (G) 1·17 x 0·87 inch.

September and the four following months comprise the breeding season of this species.

Hab. Rockingham Bay, Port Denison, Wide Bay District, Richmond and Clarence Rivers Districts, New South Wales, Interior, Victoria and South Australia, Tasmania. (*Ramsay.*)

Genus SYNOICUS, *Gould.*

SYNOICUS AUSTRALIS, *Latham.*
Swamp Quail.
Gould, Handbk. Bds. Aust., Vol. ii., sp. 487, p. 193.

This bird is distributed over the greater portion of Australia, and the whole of Tasmania. It constructs its nest which is

s

composed of dried grasses, &c., on the ground in the long rank grass bordering water courses and rivers. Eggs of a pale bluish-white finely freckled all over with very minute light brown markings, which appear foreign to the shell, and when moistened are easily rubbed off. The dimensions of four eggs in the Australian Museum Collection are as follows :—length (A) 1·26 x 0·97 inch ; (B) 1·24 x 0·94 inch ; (C) 1·21 x 0·93 inch ; (D) 1·2 x 0·92 inch.

A set of five taken at Dobroyde on the 20th March 1867, give the following measurements :—length (A) 1·18 x 0·94 inch ; (B) 1·13 x 0·9 inch ; (C) 1·17 x 0·9 inch ; (D) 1·18 x 0·92 inch ; (E) 1·17 x 0·91 inch.

The eggs of this species are usually swollen ovals in form, and rather pointed at one end.

This bird commences to breed in September, and continues until the end of March, during which time two or more broods are reared.

Hab. Derby, N.W. Australia, Gulf of Carpentaria, Cape York, Rockingham Bay, Port Denison, Wide Bay District, Richmond and Clarence Rivers Districts, New South Wales, Interior, Victoria and South Australia, Tasmania, West and South-West Australia. (*Ramsay.*)

SYNOICUS DIEMENENSIS, *Gould.*

Tasmanian Swamp-Quail.

Gould, Handbk. Bds. Aust., Vol ii. sp. 488, p. 194.

This bird is confined to Tasmania, and is found breeding among rank herbage in the vicinity of swamps and rivers. Eggs twelve to fourteen in number for a sitting, in form swollen ovals, being rather sharply pointed at the smaller end, of a greenish-white ground colour, thickly freckled all over with very minute spots of brown. Dimensions of three average specimens taken in November 1885 in Tasmania, are as follows :—length (A) 1·22 x 0·93 inch ; (B) 1·2 x 0·93 inch ; (C) 1·23 x 0·94 inch.

September and the four following months constitute the ordinary breeding season of this species.

Hab. Tasmania.

SYNOICUS CERVINUS, *Gould.*
Northern Swamp-Quail.

Gould, Handbk. Bds. Aust, Vol. i., sp. 490, p. 195.

This bird is confined to the northern portions of Australia, the Islands of Torres Straits, and the South Coast of New Guinea. It lays its eggs on the ground in a slightly constructed nest of dried grasses. In the Dobroyde Collection are five eggs of this species taken in 1873, they are of a creamy-white ground colour, minutely dotted with brown ; like all the eggs of the genus *Synoicus*, the shell is very thick and strong. Length (A) 1·2 x 0·91 inch ; (B) 1·19 x 0·93 inch ; (C) 1·19 x 0·94 inch ; (D) 1·2 x 0·93 inch ; (E) 1·18 x 0·91 inch.

Hab. Port Darwin and Port Essington, Gulf of Carpentaria, Cape York, South Coast New Guinea. *(Ramsay.)*

Genus EXCALFATORIA, *Bonaparte.*
EXCALFATORIA AUSTRALIS, *Gould.*
Least Swamp-Quail.

Gould, Handbk. Bds. Aust., Vol. ii., sp. 491, p. 197.

"The Little Swamp-Quail is found tolerably abundant in the marshy parts about Botany Bay and South-head, in which situations it breeds freely, rearing often three broods in the year. It usually lays five eggs, in shape resembling those of *Synoicus australis,* Lath., but much smaller in size, being 1·1 inch in length by 0·8 in breadth, and when fresh of a pale light green colour, dotted all over with blackish-umber ; in some the ground colour is a dirty olive-yellow ; others again are almost brown, with black dots. This species is known by our Sydney sportsmen under the name of the "King Quail," and is by most people considered a rare bird ; but if its natural haunts be visited it will be found plentiful enough, although hard to "rise." It shows preference for the long tall grass in low damp situations, particularly bordering swamps and lagoons. The nest is like that of the rest of the family, a few pieces of grass, upon which the eggs are laid, but on the whole greatly depending on the nature of the ground. The breeding season lasts from August to January, but in confinement they will lay at almost any time of the year. The young upon leaving the shell are of a dusky hue, almost black." (*Ramsay, Ibis*, 1868, Vol. iv., New Series, p. 279.)

Hab. Rockingham Bay, Port Denison, Wide Bay District, Richmond and Clarence Rivers Districts, New South Wales, Victoria and South Australia. (*Ramsay.*)

Order GRALLATORES.
Family STRUTHIONIDÆ.
GENUS DROMAIUS, *Vieillot.*
DROMAIUS NOVÆ-HOLLANDIÆ, *Latham.*
Emu.

Gould, Handbk. Bds. Aust., Vol. ii., sp. 492, p. 200. XV /.

This fine bird is found all over Australia, with the exception of the North-western portions. It used to be at one time common near the coast, but as the country became more thickly populated, it was driven towards the interior, where its numbers are fast decreasing, owing to the ruthless manner in which it is hunted and shot down, and the wholesale destruction of its eggs by men employed to search for, and break them, on account of the damage done by the birds to the wire fences, and the quantity of grass consumed by them. On a station in the Riverina

District, during the breeding season of 1881 no less than 1,500 eggs were destroyed ; and on another station in the Cobar District, more than that number were broken during 1887. No less than 10,000 Emus were destroyed in the Wilcannia District, New South Wales, during the first nine months of 1888.* The breeding season commences as early as June and continues till the end of September.

"Mr. K. H. Bennett in MSS. notes states that the nests vary in different situations : on the plains the eggs are deposited on the bare ground, and if amongst the low polygonum, so common on large areas, the eggs are surrounded by a ring of short broken pieces of the stems of this plant, apparently for the purpose of keeping them in position and preventing them from rolling away. At other times they are placed on a thin layer of grass and without any protecting ring, whilst in the timbered country, and particularly where "Leopard" trees abound, a low flat mound some three inches high is formed by scraping together the scales of bark, thrown off by the above mentioned trees, on which the eggs are placed."

Eggs from seven to ten in number for a sitting, usually oval in form, and in some instances slightly pointed at each end, in others at the smaller apex only. The colour varies in all shades of dark green, and occasionally one egg in a sitting is found of a very pale green. The shell is pitted and granulated all over, closely resembling shagreen in appearance.

Dimension of two eggs in the Australian Museum Collection. Length (A) 5·4 x 3·65 inches ; (B) 5·45 x 3·6 inches.

Three average specimens from a set of seven in the Dobroyde Collection, taken by Mr. James Ramsay at Tyndarie, in July, measure as follows :—length (A) 5·4 x 3·6 inches ; (B) 5·5 x 3·5 inches ; (C) 5·5 x 3·55 inches.

Hab. Cape York, Rockingham Bay, Port Denison, Wide Bay District, Dawson River, Richmond and Clarence Rivers Districts, New South Wales, Interior, Victoria and South Australia. (*Ramsay*)

---

* *Sydney Morning Herald*, October 15th, 1888.

DROMAIUS IRRORATUS, *Bartlett.*

Spotted Emu.

Gould, Handbk. Bds. Aust , Vol. ii., sp. 493, p. 204.

This rather doubtful species is found in the north-western portion of the continent of Australia. A single egg of this bird in the Dobroyde Collection, taken at Alligator Creek, about one hundred miles south-west of Port Darwin, is finely pitted all over, and entirely without granulations, the surface of the shell being glossy and smooth, of a very dark green colour, almost approaching to black. Long diameter 5·3 inches ; short diameter 3·5 inches.

Hab. Western Australia.

GENUS CASUARIS, *Linnæus.*

CASUARIS AUSTRALIS, *Wall.* *

Australian Cassowary.

Gould, Handbk. Bds. Au t., Vol. ii., sp. 494, p. 206.

Dr. Ramsay writing in 1876 on the birds of North-eastern Queensland, remarks as follows :—

"The Australian Cassowary is a denizen of the dense dark scrubs scattered over the district of Rockingham Bay, and extending as far north as the Endeavour River." It was tolerably plentiful only a few years ago even in the neighbourhood of Cardwell ; but since the advent of sugar-planters &c., on the Herbert River and adjacent creeks, these fine birds have been most ruthlessly shot down and destroyed for the sake of their skins several of which I saw used for hearth-rugs and door-mats. Formerly they were procured easily enough, but latterly so wary have they become, and their numbers so decreased, that it is only with the greatest amount of patience even a stray shot can be obtained. I know of no bird so wary and timid ; and although their fresh tracks may be plentiful enough, and easily found in the soft mud on the sides of the creeks, or under their favourite feeding-trees, yet the birds themselves are now seldom seen. During the day they remain in the most dense parts of the scrubs wandering about the sides of the watercourses and creeks, diving in through the bushes and vines at the slightest noise. Towards evening and early morning they usually visit their favourite feeding-trees, such as the native figs, Leichhardt-tree *(S. leichhardti)*, and various species of *Acmena, Jambosa, Davidsonia* &c. ; they appear to be particularly fond of the astringent fruit of a species of *Maranta*, which produces bunches of large seed-pods filled with juicy pulp, resembling in appearance the inside of a ripe passion fruit *(Passiflora edulis)*. They breed during the months of August and September. The first nest procured was found by some of Inspector Johnstone's black troopers, from whom Mr. Miller, a settler on the Herbert River, purchased some of the eggs. One which he kindly presented to me is of the *light-green* variety mentioned hereafter. The nest consists of a depression among the fallen leaves and débris with which the ground in the scrubs is covered, with the addition of a few more dry leaves. The place selected is always in the most dense part, and well concealed by entangled masses of vegetation. The eggs were five in number in the two instances recorded ; and in both cases one of the eggs in each set differed from the others, being of a light-green colour, and having a much smoother shell. The others all have a rough shell, covered rather sparingly with irregular raised patches of dark but bright green on a lighter green and smooth ground. In the pale (No. 1) variety, the raisings on the shell are close together and not so well developed ; in both varieties they are more thinly spread over the central portion than at the ends. On the whole they closely resemble the eggs of *Casuaris bennetti*, in which similar variations are noticeable ; but they are larger, and of a greater diameter, being greatest in the middle. I am indebted to Inspector Robert Johnstone for the fine series of the eggs of this species, which at present grace my collection. The following are measurements of some of the specimens of the eggs : No. 1, Light-green, smooth shell, length 5·33 x 3·73 inches ; No. 2, Dark-green, rough shell, length 5·3 x 3·88 inches." *(Ramsay, P.Z.S.*, 1876, p. 119.)

A specimen in my own collection, taken at Cairns, Northern
Queensland, in November, is of a beautiful pale green, and
measures 5·45 inches in length by 3·68 inches in breadth.

Hab., Cape York, ? Rockingham Bay.   (*Ramsay.*)

Family OTIDIDÆ.

·Genus CHORIOTIS, *Bonaparte.*

CHORIOTIS AUSTRALIS, *Gray.*

Australian Bustard.

Gould, Handbk. Birds Aust., Vol. ii., sp. 495, p. 208.

" The Australian Bustard breeds during September, October,
and November, and lays but two eggs, on the ground without any
nest—a small bare spot being selected among the trees on the
hillside ; a few small sticks and blades of grass are sometimes
found gathered round the eggs.   The eggs vary both in shape and
size, some are thickest at an equal distance from each end ; others
are more elongated, and widest an inch from the thicker end.   In
length they are from 3 to 3·3 inches, and from 2·1 to 2 3 inches
in breadth.   The ground colour varies from light olive-green to
olive-brown, having longitudinal smears, spots, and dashes of
olive-brown, equally dispersed over the surface.   In a valuable
collection, for which I am indebted to my brother, Mr. J. Ramsay
of Nanama, there are seven Bustard's eggs ; one particularly fine
one measures 3·3 x 2·1 inches ; it is of a light olive-green sparingly
marked with reddish olive-brown.   The smallest Bustard's egg in
our collection measures 2·3 x 1·6 inches, and is of an olive-brown,
thickly spotted and dashed with dark olive-brown.   I have seen
small eggs of the same colour with very few faint and longitudinal
markings extending nearly the whole length of the egg : these I
take to be the eggs of the younger birds.   So far as I am aware
the Australian Bustard has but one brood in the season." (*Ramsay,
Ibis,* 1867, Vol. iii., New Series, p. 417, pl. ix., fig. 1.)

The nest of this bird, if worthy of the name of a nest, is often found with a single egg only. In the Dobroyde Collection there are twenty-four eggs, of these eight sets were found in pairs. Mr. K. H. Bennett usually found nests with one egg. Mr. Edward Lord Ramsay during 1887, found two nests at "Kerriegundah," near Louth, New South Wales, each of which contained but a single egg. During 1868, however, Mr. James Ramsay found no less than four nests containing two eggs each, at Nanama, New South Wales. Four averaged sized specimens measure as follows: length (A) 3·2 x 2·2 inches; (B) 3·11 x 2·07 inches; (C) 3·25 x 2·22 inches; (D) 3·18 x 2·25 inches. Eggs of this bird are occasionally found of a pale sky-blue tint.

Hab. Derby, N.W. Australia, Gulf of Carpentaria, Cape York, Rockingham Bay, Port Denison, Wide Bay District, Dawson River, Richmond and Clarence Rivers Districts, New South Wales, Interior, Victoria and South Australia, West and South-west Australia. (*Ramsay.*)

Family CHARADRIADÆ.

GENUS ŒDICNEMUS, *Temminck.*

ŒDICNEMUS GRALLARIUS, *Latham.*

Southern-Stone Plover.

Gould, Handbk. Bds. Aust., Vol. ii., sp. 496, p. 210. XVIII 3.

This bird is found over the greater portion of the continent of Australia. Its peculiarly mournful and dismal note, uttered at night time, has a most depressing effect on any belated traveller. The eggs, two in number, are deposited on the bare ground in open forest-lands. The eggs vary much in their markings, but the most usual variety are of a light stone colour, thickly blotched all over with irregular shaped markings of umber-brown; others are so closely marked as to nearly obscure the ground colour;

specimens in the Dobroyde Collection, taken by Dr. Ramsay at Macquarie Fields in 1860 are of a stony-white colour with very obscure markings of blackish and pale brown. This bird often resorts to the same spot to breed year after year, even though its eggs be repeatedly taken.

Two eggs in the Australian Museum Collection measure as follows :—length 2·18 x 1·67 inches ; (B) 2·18 x 1·65 inches. Two eggs in the Dobroyde Collection, taken at Nanama near Yass, by Mr. John Ramsay in October 1866, measure, length (A) 2·35 x 1·6 inches ; (B) 2·28 x 1·55 inches.

Specimens taken at Yendon, in December 1878, give the following measurements :—length (A) 2·22 x 1·65 inches ; (B) 2·3 x 1·63 inches.

This bird breeds in September and continues during the four following months.

Dr. Ramsay describing the eggs of this species in the Proc. of the Zool. Soc. for 1877, p. 335, writes as follows :—"A curious variety of the egg of this species is sometimes found ; it is of a rich creamy-buff, clouded with a duller tint, or irregularly and indistinctly blotched with dull brownish-buff. On showing some of this variety to the late Mr. John Macgillivray, author of the "Voyage of the Rattlesnake" &c., he assured me they were so remarkably similar to the one found by the late Commander J. M. R. Ince, at Port Essington, and described by Mr. Gould in his "Handbook to the Birds of Australia, Vol. ii., p. 213, as that of *Esacus magnirostris*, that no doubt as to their identity remained in his mind. On every occasion that I have obtained the buff coloured egg, the accompanying one was of the usual heavily blotched variety, with but a few markings at the thin end."

Hab. Derby, North-West Australia, Cape York, Rockingham Bay, Port Denison, Wide Bay District, Dawson River, Richmond and Clarence Rivers Districts, New South Wales, Interior, Victoria and South Australia, West and South-West Australia. (*Ramsay.*)

Genus HÆMATOPUS, *Linnæus*.

HÆMATOPUS LONGIROSTRIS, *Vieillot*.

White-breasted Oyster-catcher.

Gould, Handbk. Bds. Aust., Vol. ii., sp. 498, p. 215.

This bird is distributed all over the coast line of Australia. It deposits its eggs two or three in number on the small islands and rocky headlands, during August and the four following months. Specimens in the Australian Museum Collection, taken by Mr. J. A. Thorpe at Fraser Island, Wide Bay, Queensland, are of a pale stone colour, thickly spotted and blotched all over with irregular shaped brownish-black markings, and nearly obsolete spots and dashes of the same colour, appearing as if beneath the surface of the shell. Length (A) 2·25 inches x 1·57 inch ; (B) 2·27 inches x 1·55 inch.

A set in the Dobroyde Collection taken by Mr. John Ramsay at Cape Upstart on the 28th of August 1882, are a light stone colour, with blotches, smears, and angular lines of different shades of dark umber and black, together with subsurface markings of the same colour, in one specimen (A), forming an ill-defined zone at the larger apex. Length (A) 2·25 x 1·54 inches ; (B) 2·27 x 1·55 inches ; (C) 2·24 x 1·55 inches.

Specimens in my collection taken on King Island, Bass's Straits, and the coast of South Australia, give the same average measurements.

Like many sea-birds' eggs they show great variation, both in the colour and disposition of their markings.

Hab. Port Darwin and Port Essington, Gulf of Carpentaria, Cape York, Rockingham Bay, Port Denison, Wide Bay District, Richmond and Clarence Rivers Districts, New South Wales, Victoria and South Australia, Tasmania, West and South-west Australia, South Coast New Guinea. *(Ramsay.)*

HÆMATOPUS UNICOLOR, *Wagler.*

(H. fulginosus, Gould.)

Sooty Oyster-catcher.

Gould, Handbk. Bds. Aust., Vol. ii., sp. 499, p. 217. _XX_   2.

Like the preceding species this bird is found all over the
Australian coast, it is however particularly plentiful on the rocky
islets of Bass's Straits, whither it resorts to breed during September
and the three following months. The Seal Rocks about three miles
distant from the "Nobbys," Phillip Island in Western Port Bay,
Victoria, is a favorite breeding ground of this species, but the
difficulty of access to it on account of the risk of a boat being
dashed to pieces on the rocks by the heavy surf continually breaking
upon it, renders the place a pretty secure retreat for countless
numbers of gulls, terns, oyster-catchers, &c. Eggs usually two in
number for a sitting varying from a pale stone colour to light
brown, in some instances being uniformly covered with round
spots and dots of dark brown; in others, heavily blotched and
covered with figures resembling Egyptian hieroglyphics in every
shade of dark umber and black.

Two eggs taken at Phillip Island from off a grassy ledge on
the 25th of October 1883, which are average sized specimens,
measure as follows:—length (A) 2·63 x 1·65 inches; (B) 2·58
x 1·64 inch.

Hab. Port Darwin and Port Essington, Gulf of Carpentaria,
Cape York, Rockingham Bay, Port Denison, Wide Bay District,
Richmond and Clarence Rivers Districts, New South Wales,
Victoria and South Australia, Tasmania, West and South-West
Australia. (*Ramsay.*)

GENUS LOBIVANELLUS, *Strickland.*

LOBIVANELLUS LOBATUS, *Latham.*

Wattled Plover.

Gould, Handbk. Bds. Aust., Vol. ii., sp. 500, p. 218.

"The Spur-winged Plover breeds during September and the two
following months, in some localities a month earlier or later. The

eggs, which are four in number, are placed with the smaller end inwards, and laid upon the ground by the side of some tuft of grass or rushes in a slight hollow made for their reception, with occasionally a few blades of grass placed under and around them, but as often as not without any sign of a nest. The ground colour of the eggs varies from yellowish- and olive-brown to bright deep olive-green, evenly spotted with deep blackish-brown and yellowish-brown, which latter appear beneath the surface of the shell, the majority of the markings being towards the larger end. They vary from 1·9 to 2 inches in length, and from 1·3 to 1·4 in breadth. My brother has given me a most beautiful set of these eggs in which the ground colour is of a bright deep olive-green evenly spotted with deep blackish-brown. The Spur-winged Plover shows great anxiety for its eggs and young, fluttering off as you approach and using all the enticing actions in its power to draw you away from the spot ; should a horse, cow, or any other quadruped approach, it uses quite different means to save its treasures, and by flying up in the animal's face and flapping it with its wings it quickly produces the desired effect." (*Ramsay, Ibis*, 1867, Vol. iii., New Series, p. 419, pl. ix., fig. 2.)

Hab. Rockingham Bay, Port Denison, Wide Bay District, Richmond and Clarence Rivers Districts, New South Wales, Interior, Victoria and South Australia. (*Ramsay.*)

LOBIVANELLUS MILES, *Bodd.*

(*L. personatus*, Gould.)

Masked Plover.

Gould, Handbk. Bds. Aust., Vol. ii., sp. 501, p. 220.

"This is a very common bird in the Cobourg Peninsula, inhabiting swamps, the borders of lakes and open spots among the mangroves, and like its near ally, is mostly seen associated in small families. It is rather a noisy species, frequently uttering a note, which is not unlike its native name, both while on the wing and on the ground. The task of incubation is performed during the months of August and September, the eggs which are two or three in number, being laid in a hollow on the bare ground at the edge of a flat adjoining a salt-marsh ; they are of a dull olive-yellow, dashed all over with spots and markings of blackish-brown and dark olive-brown, particularly at the larger end ; they are one inch and five-eights long by one inch and three sixteenths broad, somewhat pointed at the smaller end." (*Gould, Handbk. Bds. Aust.*, Vol. ii., p. 220.)

Hab. Derby, N.W. Australia, Port Darwin and Port Essington, Gulf of Carpentaria, Cape York. (*Ramsay*.)

Genus SARCIOPHORUS, *Strickland.*

SARCIOPHORUS PECTORALIS, *Cuvier.*

Black-breasted Plover. •

Gould, Handbk. Bds. Aust., Vol. ii., sp. 502, p. 222. XVII 6

"The habits and actions of this pretty species closely resemble those of the Spur-winged Plover ; it breeds during August and three following months, laying its eggs on the bare ground in places similar to those chosen by *L. lobatus*, but is more local, and frequents drier tracts of country. I have frequently met with flocks in the ploughed fields, where they would be found sitting down and basking in the sun, or in a long string in the shade of a fence. In their flight they differ greatly from their ally, and are seldom heard except when flushed or separated. At night they separate and spread about over the fields in scarch of food. The eggs of this species are four in number, 1·7 inch in length by 1·2 in breadth. Some specimens vary to the extent of a tenth either way. The ground-colour is a light olive-brown, tinged with yellowish- or greenish-olive, spotted with brown and grey, which latter appears beneath the surface of the shell. In some the spots incline to reddish-brown. and are equally dispersed over the whole surface ; in others the markings are crowded on the larger end." (*Ramsay, Ibis*, 1867, Vol. iii., New Series, p. 420, pl. ix., fig. 3.)

A set of four taken by Mr. James Ramsay at Tyndarie, measure as follows :—length (A) 1·67 x 1·22 inch ; (B) 1·72 x 1·23 inch ; (C) 1·75 x 1·25 inch ; (D) 1·58 x 1·18 inch.

Hab. Rockingham Bay, Port Denison, Wide Bay District, Richmond and Clarence Rivers Districts, New South Wales, Interior, Victoria and South Australia, West and South-west Australia. (*Ramsay.*)

Genus EUDROMIAS, *Boie.*

EUDROMIAS AUSTRALIS, *Gould.*

Australian Dotterel.

Gould, Handbk. Bds. Aust., Vol. ii., sp. 505, p. 227. *XVI* ♀

"The habitat of this species is the interior portion of the Province of South Australia, and the interior of New South Wales but as far as is yet known, it is nowhere plentiful ; sometimes it is met with in the Melbourne markets during the game season, and is considered a rare bird by the dealers. Mr. E. G. Vickery has been fortunate enough to obtain the nest and eggs during a surveying trip in the Darling River District near Wilcannia. The eggs were placed on the ground among a few loose stones near the summit of a small hillock or "rise" in the level country, and placed on a little mound about two inches high, probably an old ant-hill ; they were three in number, a pair measure as follows :— length (A) 1·45 x 1·05 inch ; (B) 1·45 x 1·03 inch. In form they are rather less pointed than the usual pyriform shape of Plover's eggs ; the ground colour is of a deep rich cream or buff, sparingly sprinkled all over with irregular spots, and some elongated crooked markings of chocolate black with a few minute dots and dashes of a lighter tint, the markings look black in certain lights, but of a chocolate tint in others. Specimens in Mr. Bennett's Collection were taken during the month of October, on the Lachlan River near Mossgiel." (*Ramsay, P.L.S., N.S.W.*, Vol. vii., p. 410.)

A fine set of these eggs, obtained by Mr. W. A. Mackay in 1878, on the One Tree Plain near Deniliquin, measure (A) 1·47 x 1·05 inch; (B) 1·45 x 1·03 inch.

Hab. New South Wales, Interior, Victoria and South Australia. (*Ramsay.*)

Genus ÆGIALITIS, *Boie*.

ÆGIALITIS MONACHUS, *Geoffroy*.

Hooded Dotterel.

Gould, Handbk. Bds. Aust., Vol. ii., sp. 508, p. 231. *XVI* /.

For the eggs of this bird, I am indebted to Mr. W. Gordon, who procured them during his residence on King Island, Bass's Straits, while superintending the erection of the lighthouse at that place during 1878-9. The eggs, two in number for a sitting were simply deposited in a slight hollow on the beach, and resembled very closely their surroundings ; they are of a light stone colour, sprinkled all over with irregular shaped brownish-black spots, which in some places become confluent, together with superimposed markings of a lighter colour appearing as if beneath the surface of the shell. Length (A) 1·37 x 1·06 inch ; (B) 1·38 x 1·04 inch.

This species breeds during the months of September, October, and November.

Hab. Wide Bay District, Richmond and Clarence Rivers Districts, New South Wales, Victoria and South Australia, Tasmania, West and South-West Australia. (*Ramsay*)

ÆGIALITIS NIGRIFRONS, *Cuvier*.

Black-fronted Dotterel.

Gould, Handbk. Bds. Aust., Vol. ii., sp. 509, p. 232. *XVI* 4

"This species is found dispersed over the whole of the eastern and southern portions of Australia, even venturing far inland. I have met with it high up on the Bogan and Bell Rivers, and on the Murrumbidgee River, near Yass; it gives preference to the margins of inland lakes and lagoons rather than the sea-coast. Mr. J. S. Ramsay, a most persevering and successful oologist, found it breeding during the months of October, November, and December on the margins of the Bell River, at Cardington. There was always a difficulty in discovering the eggs, from their similarity to the adjacent ground on which they were laid, it being necessary to watch the birds to their eggs. Mr. James Ramsay of Nanama, near Yass, has also sent me authentic eggs of this species, taken in that district, while others which I have received from Melbourne' and South Australia, all exhibit a similar style and colour in their markings. The eggs are two or three in number for a sitting, and are usually placed with the thinner ends together, in a slight depression in the sand or pebbles near water; the ground colour is of a rich creamy-white when fresh, nearly obscured by numerous irregular angular markings and hair-lines of blackish-brown, dark brown, and bluish slate-colour, the last appearing as if beneath the surface of the shell. In some specimens these markings are close together, giving a clouded appearance to the eggs, in others, about equally dispersed over the whole surface. In some light varieties they are less numerous at the thinner end; and these specimens are slightly smaller in size. Length 1·05 x 0·8 inch; the darker and most usual variety 1·1 inch in length by 0·85 inch in breadth." (*Ramsay, P.Z.S.*, 1877, p. 336.)

I have frequently taken the eggs of this species in the Albert Park, near Melbourne, and have received them from nearly every portion of Australia.

A set taken by Mr. James Ramsay, at Nanama, measures as follows :—length (A) 1·08 x 0·83 inch ; (B) 1·12 x 0·85 inch.

Two eggs taken by the late Mr. T. H. Boyer-Bower, at Derby, North-western Australia, in 1886, give the following measurements Length (A) 1·06 x 0·81 inch ; (B) 1·03 x 0·78 inch.

Hab. Derby, North-West Australia, Rockingham Bay, Port Denison, Wide Bay District, Dawson River, Richmond and Clarence

T

Rivers Districts, New South Wales, Interior, Victoria and South Australia, West and South-West Australia. (*Ramsay*)

ÆGIALITIS RUFICAPILLUS, *Temminck*.

Red-capped Dotterel.

Gould, Handbk. Bds. Aust., Vol. ii., sp. 510, p. 235. *XVI 6.*

This species differs from the preceding one in preferring the bays and inlets of the coast, and adjacent salt-water marshes, although its eggs have been procured in the interior of New South Wales. During a period of ten years, I have very frequently, taken the eggs of this species in the vicinity of Melbourne. The Albert Park and the stretches of the then sandy wastes of Middle Park, near St. Kilda, were favourite breeding places of this bird. The nest in the former place consisted merely of a slight depression in the ground, lined with a few short pieces of dried grass, and small fresh-water shells; in the latter place the eggs were simply deposited on the sand, with a few small pebbles placed around them to keep them from rolling away. Eggs two or three in number for a sitting, usually the former, varying considerably in their shape and markings.

A set in the Australian Museum Collection, procured in 1878, are ovate in form, of a light stone colour, blotched all over with small irregular shaped brownish-black markings. Length (A) 1·22 x 0·9 inch; (B) 1·22 x 0·87 inch. Taken at Albert Park, October, 1877.

Two other specimens are pyriform in shape, of a light cream colour, heavily blotched and sparingly lined with blackish markings, a few dots appearing as if beneath the surface of the shell. Length (A) 1·25 x 0·87 inch; (B) 1·26 x 0·88 inch; others again have a faint greenish-tinge in the ground colour and the markings confined to the larger end, in some instances assuming the form of a zone.

T—2

The eggs of this bird may be found as early as the month of August, and as late as the end of March ; October and the two following months however, are the principal months to obtain them.

Hab. Port Darwin and Port Essington, Gulf of Carpentaria, Cape York, Rockingham Bay, Port Denison, Wide Bay District, Richmond and Clarence Rivers Districts, New South Wales, Interior, Victoria and South Australia, Tasmania, West and South-West Australia, South Coast New Guinea. (*Ramsay.*)

Genus ERYTHROGONYS, *Gould.*

ERYTHROGONYS CINCTUS, *Gould.*

Red-kneed Dotterel.

Gould, Handbk. Bds. Aust., Vol. ii., sp. 513, p. 240. XVI 5.

"This species of upland plover breeds during October and November, sometimes in December. It is a bird never as far as I know, met with on the coast, but I have received specimens from the Clarence, shot near Grafton. Its stronghold seems to be the interior of New South Wales and South Australia. The eggs are four for a sitting, placed in a slight depression on the ground; Mr. Bennett informs me he found them in several instances on the mud at the waters' edge of large inland lagoons and lakes in the Lachlan district, and smeared over with mud as if the birds had been shifting them from place to place, or perhaps they were purposely smeared over to prevent them being detected. On the whole they resemble those of *Ægialitis nigrifrons,* varying from light to dark stone colour, thickly covered all over with irregular angular and curved lines, and irregular shaped markings of black, which cross and recross each other in various directions, the lines vary in thickness from that of a fine hair to that of coarse thread, on the thicker end here and there they loop and form tangles. Measurements of three from one nest :—(A) 1·18 x 0·85 inch ; (B) 1·15 x 0·85 inch ; (C) 1·22 x 0·87 inch. *From Mr. Bennett's Coll.*" (*Ramsay, P.L.S., N.S.W.,* Vol. vii., p. 412.)

Hab. Derby, N.W. Australia, Gulf of Carpentaria, Richmond and Clarence Rivers Districts; New South Wales, Interior, Victoria and South Australia. (*Ramsay.*)

Sub-Family GLAREOLINÆ.

GENUS GLAREOLA, *Brisson.*

GLAREOLA GRALLARIA, *Temminck.*

Australian Pratincole.

Gould, Handbk. Bds. Aust., Vol. ii., sp. 515, p. 243. XVI. *9*

"The home of the Australian Pratincole is the interior of New South Wales, and the northern portion of the province of South Australia, it is also found occasionally during the wet seasons in the neighbourhood of Cape York and Port Darwin. In New South Wales I have received specimens from the Lachlan and Darling Rivers, and Mr. James Ramsay has noticed it at Tyndarie, in the Bourke district. Mr. E. G. Vickery has kindly permitted me to describe an egg from his collection, taken near Wilcannia on the Darling River, in September 1880. He informs me that the parent bird was seen to fly from the eggs, and before they were taken to return again and sit on the nest. The eggs were three in number, the ground colour is of a creamy-white, dull light stone-brown or light buff, well covered with irregularly shaped blotches, dots, and spots and freckles of dull umber and sienna-brown with a few dots and dashes almost black, and obsolete spots here and there of slaty-grey ; length 1·3 inch by 1 inch, in shape they are slightly oval, slightly swollen at the thicker end and not pointed. An egg of this species in the collection of Mr. K. H. Bennett measures 1·24 x 0·95 inch ; none differ materially from Mr. Vickery's specimens. Mr. Bennett informs me that they select a bare spot on the ground where the earth or sand assimilates to the colour and markings of the eggs. They breed during October." (*Ramsay, P.L.S., N.S.W.*, Vol. vii., p. 410.)

Specimens taken by the late Mr. W. Liscombe at Wilcannia, in October 1883. measure as follows:—length (A) 1·25 x 0·93 inch; (B) 1·27 x 0·95 inch; (C) 1·28 x 0·95 inch.

Hab. Derby, N.W. Australia, Port Darwin and Port Essington, Cape York, Wide Bay District, Richmond and Clarence Rivers Districts, New South Wales, Interior. (*Ramsay.*)

Sub-Family HIMANTOPODINÆ.

GENUS RECURVIROSTRA, *Linnæus.*

RECURVIROSTRA RUBRICOLLIS, *Temminck.*

Red-necked Avocet.

Gould, Handbk. Bds. Aust., Vol. ii., sp. 519, p. 249. XVII /.

" Mr. K. H. Bennett informs me that this species lays four eggs for a sitting, and breeds during the months of September to December, laying its eggs on the bare ground without making any nest, and sometimes close to the water's edge. The present specimens were found among the herbage usually growing about the sheep tanks in the interior of the country, and were taken in the Lachlan district; the ground colour varies from light stone colour to creamy-yellow, some of the former tint have a faint olive-green shade, some are heavily blotched towards the thicker end, others sparingly covered with spots, dots, and freckles of dark umber-brown and black, with a few obsolete spots of slate-grey. A set measures as follows:—length (A) 2 x 1·4 inches; (B)
1·35 x 0·95 inch; (C) 1·4 x 0·95 inch; (D) 1·3 x 0·95 inch. *From Mr. Bennett's Collection.*" (*Ramsay, P.L.S., N.S.W.*, Vol. vii., p. 411.)

During 1887 Mr. Bennett found these birds breeding in large colonies on the margin of a lake in the interior of New South Wales.

Hab. Derby, N.W. Australia, Rockingham Bay, Port Denison, Wide Bay District, Richmond and Clarence Rivers Districts, New South Wales, Interior, Victoria and South Australia, Tasmania, West and South-West Australia. *(Ramsay.)*

GENUS HIMANTOPUS, *Brisson.*

HIMANTOPUS LEUCOCEPHALUS, *Gould.*

White-headed Stilt.

Gould, Handbk. Bds. Aust., Vol. ii., sp. 517, p. 246. XVII 3

"The Stilted Plover must be considered rather a scarce than a rare bird in New South Wales, its visits being few and far between. When it does come, however, which is usually in some very dry or remarkably wet season, it appears in great numbers and in all stages of plumage. In 1865 large flocks arrived in company with the Straw-necked and White Ibises, *(Geronticus spinicollis* and *Threskiornis strictipennis),* and took up their abode in the lagoons and swamps in the neighbourhood of Grafton, on the Clarence River, where, on my visit to that district in September 1866, all three species were still enjoying themselves. A few days previous to my arrival in Grafton, a black in the employ of Mr. J. Macgillivray, and a very intelligent collector, discovered a nest of this species containing four eggs, which have been secured for our collection. The nest was a slight structure, consisting merely of a few short pieces of rushes and grass, placed in and around a depression at the foot of a clump of rushes growing near the water's edge of a lagoon in the neighbourhood of South Grafton. The eggs vary slightly in form, two being pyriform, the other two rather long. The ground colour is of a yellowish-olive or light yellowish-brown, lighter when freshly taken—in some sparingly, in others thickly blotched and spotted with umber and black, the black spots running together and forming large patches on the thick ends. Length from $1\frac{9}{10}$ inch to $1\frac{7}{10}$ inch ; breadth $1\frac{1}{4}$ inch to $1\frac{1}{3}$ inch." *(Ramsay, P.Z.S.,* 1867, p. 600.)

Specimens taken by M. H. Bennett during October 1886, at Ivanhoe, New South Wales, measure as follows :—length (A) 1·72 x 1·3 inch ; (B) 1·75 x 1·34 inch ; (C) 1·73 x 1·31 inch ; (D) 1·78 x 1·35 inch.

Hab. Derby, N.W. Australia, Port Darwin and Port Essington, Gulf of Carpentaria, Rockingham Bay, Port Denison, Wide Bay District, Richmond and Clarence Rivers Districts, New South Wales, Interior, Victoria and South Australia, West and South-West Australia, South Coast of New Guinea. (*Ramsay.*)

Genus CLADORHYNCHUS, *G. R. Gray.*

CLADORHYNCHUS PECTORALIS, *Dubus,*

Banded Stilt.

Gould, Handbk. Bds. Aust., Vol. ii., sp. 518, p. 248.

" The eggs are four in number for a sitting placed in a scanty nest of a few dry reeds and water grasses ; the ground colour varies from an olive-brown to creamy-brown irregularly spotted and blotched with black, in shape oval but slightly pointed. Length (A) 1·9 x 1·4 inch ; (B) 1·88 x 1·3 inch." (*Ramsay, P.L.S., N.S.W.,* Vol. vii., p. 57.)

Hab. New South Wales, Victoria and South Australia, West and South-West Australia. (*Ramsay.*)

Family SCOLOPACIDÆ.

Genus RHYNCHÆA, *Cuvier.*

RHYNCHÆA AUSTRALIS, *Gould.*

Australian Rhynchæa.

Gould, Handbk., Bds. Aust., Vol. ii., sp. 534, p. 274. *XVI* 3.

The eggs of this bird have several times been erroneously described as those of *Gallinago australis,* but recently on receipt of well authenticated sets from Mr. K. H. Bennett and Mr. George Masters, Dr. Ramsay corrected his mistake in describing them as the eggs of the latter bird. (See P.L.S., N.S.W., Vol. i., New Series, 1886, p. 1060.)

A single egg of this species in the Australian Museum Collection is of a light stone colour, the entire surface of which is nearly obscured by very irregular shaped black patches and lines· Length 1·38 x 1·04 inch. Taken by Mr. F. Morrow near Orange, on the 27th of November 1878.

The handsome set of eggs taken by Mr. K. H. Bennett near the margin of a swamp at Ivanhoe in October, from a slight depression in the ground, neatly lined with broad Eucalyptus leaves, are four in number, rather elongated in form, of a creamy-white ground colour, the blotches and lines being smaller and closer together than in the former specimen, in some places forming confluent brownish-black patches ; where the ground colour is visible appear a few superimposed spots of a lighter colour, appearing as if beneath the surface of the shell. Length (A) 1·4 x 1 inch ; (B) 1·47 x 0·99 inch; (C) 1·4 x 1 inch ; (D) 1·35 x 1 inch.

Another set of four measures as follows :—length (A) 1·31 x 1 inch; (B) 1·39 x 0·98 inch; (C).1·37 x 1 inch; (D) 1·38 x 0·98 inch.

Three eggs in the Macleayan Museum Collection give the following dimensions :—length (A) 1·42 x 1 inch; (B) 1·4 x 1·02 inch; (C) 1·37 x 1 inch.

Hab. Rockingham Bay, Wide Bay District, Richmond and Clarence River Districts, New South Wales, Interior, Victoria and South Australia, West and South-West Australia. *(Ramsay)*

Family TANTALIDÆ.

Genus GERONTICUS.

GERONTICUS SPINICOLLIS, *Jameson.*

Straw-necked Ibis.

Gould, Handbk. Bds. Aust , Vol. ii., sp., 538, p. 282.

The late Mr. W. Liscombe of Moonee Ponds, Victoria, found this bird breeding in great numbers during wet seasons in the immense inundated areas of *Polygonum* bushes that skirt the edges of the Lachlan River in the neighourhood of Hillston, New South Wales, breaking down the bushes and forming a nest of sticks, twigs, and reeds, just above the surface of the water. The eggs are very faint greenish-white on the outer surface becoming much darker in tint on the inside. The shell is lustreless and minutely pitted all over. In a number of the eggs of this species now before me, there is a great variation in their form, some being true ovals, others elongated, swollen, and pointed ovals.

Specimens in the Australian Museum Collection give the following measurements :—length (A) 2·58 inches x 1·78 inch ; (B) 2·59 inches x 1·76 inch. Dimensions of specimens in the Dobroyde Collection are as follows :—length (A) 2·6 inches x 1·7 inch ; (B) 2·52 inches x 1·75 inch ; (C) 2·52 inches x 1·75 inch ; (D) 2·71 inches x 1·68 inch ; (E) 2·57 inches x 1·77 inch ; (F) 2·49 inches x 1·7 inch.

This species breeds during the months of October, November, and December.

Hab. Port Darwin and Port Essington, Gulf of Carpentaria, Rockingham Bay, Port Denison, Wide Bay District, Richmond and Clarence Rivers Districts, New South Wales, Interior, Victoria and South Australia, South Coast New Guinea. (*Ramsay.*)

Genus PLATIBIS, *Bonaparte.*

PLATIBIS FLAVIPES, *Gould.*

Yellow-legged Spoonbill.

Gould, Handbk. Bds. Aust., Vol. ii., sp. 542, p. 288.

From a description of the breeding places of this bird given by Mr. K. H. Bennett in the P.L.S. of N.S.W., Vol. vii., p. 324, I have made the following extract :—

"About December 1879, I found this species breeding in great numbers in company with *Ardea pacifica*, in a belt of low trees, some fifty yards long, in a swamp thirty miles north of the Lachlan River. The nests were large structures of sticks, loosely interlaced together with a considerable depression in the centre, lined with the soft fibre of decayed bark, and placed on the crooked and gnarled branches of low trees which formed a capital foundation for nests, an advantage the birds had evidently recognized, for every available place was occupied by a nest either of *P. flavipes* or *A. pacifica*. The eggs in every instance were four in number, white, rather limey, long and pointed in shape and minutely pitted all over the surface of the shell. Length (A) 2·7 x 1·85 inches ; (B) 2·73 x 1·85 inches ; (C) 3·05 x 1·8 inches ; (D) 2·78 x 1·9 inches.

Hab. Rockingham Bay, Wide Bay District, Richmond and Clarence Rivers Districts, New South Wales, Interior, Victoria and South Australia. (*Ramsay.*)

Family GRUIDÆ.

GENUS GRUS, *Linnæus.*

GRUS AUSTRALASIANUS, *Gould.*

Australian Crane.

Gould, Handbk. Bds. Aust., Vol. ii., sp. 543, p. 290.

This bird is distributed over the greater portions of Australia. It deposits its eggs two in number in a slight depression of the ground, usually on the plains ; they are of a rich cream colour blotched and spotted all over with light chestnut and purplish-brown markings, the latter colour appearing as if beneath the surface of the shell.

Dimensions of two specimens in the Australian Museum Collection are as follows :—(A) 3·87 x 2·5 inches ; (B) 3·83 x 2·45 inches. Two eggs in the Dobroyde Collection are elongated in form and gradually tapering to a nearly sharp point at the smaller apex. Length (A) 3·93 x 2·22 inches ; (B) 3·92 x 2·32 inches.

A pair of eggs in my collection taken by Mr. W. Liscombe, near Tumut, are of a dull white, uniformly and sparingly spotted all over with blood-red markings, a few nearly obsolete spots of purplish-brown appearing as if beneath the surface of the shell. Length (A) 3·6 x 2·27 inches ; (B) 3·65 x 2·3 inches.

The eggs of this species are subject to much variation in the colour and disposition of their markings, and the shell is minutely pitted over the whole surface. September and the two following months constitute the breeding season of this species.

Hab. Derby, N.W. Australia, Gulf of Carpentaria, Rockingham Bay, Port Denison, Wide Bay District, Dawson River, Richmond and Clarence Rivers Districts, New South Wales, Interior, Victoria and South Australia. (*Ramsay.*)

Family CICONIIDÆ.

GENUS XENORHYNCHUS, *Bonaparte.*

XENORHYNCHUS AUSTRALIS, *Latham.*

Australian Jabiru.

Gould, Handbk. Bds. Aust., Vol. ii., sp. 544, p. 293.

" The Jabiru of the Australian Continent, at one time thought to be specifically distinct from that of India but now recognised to be one and the same species, is widely dispersed over the northern portions of Australia. It is found frequenting the estuaries of rivers as well as the inland marshes and lagoons, from the Clarence River on the east coast to Cambridge Gulf on the north-west, specimens having been procured at the latter place by the late Mr. T. H. Boyer-Bower, and it will undoubtedly be found much farther south when our knowledge of the range of the Western Australian avi-fauna is fully worked out. The great stronghold, however, of this species is the Indian Empire, over the principal portion of which it has been found breeding, and accounts of which have been given by various writers. Allan Hume in his valuable work on the "Nests and Eggs of Indian Birds," deals exhaustively with the subject ; but it is only within the last few years that it has been found breeding on the Australian Continent. The nest is a huge flat structure composed of sticks, lined with twigs and grasses, and is usually placed in the high boughs of a lofty tree in the near vicinity of water. Mr. Geo· Barnard found a nest near Rockhampton, Queensland, but the Jabirus did not remain long in indisputed possession of it, owing to the repeated attacks of a pair of Wedge-tailed eagles, *(Aquila audax)* which ultimately caused them to desert it. I am indebted to Mr. John Leadbeater of the National Museum, Melbourne, for the opportunity of describing these rare Australian eggs, which, he informs me, were taken in August 1887, in the Clarence River district, New South Wales. They are oval in form, being nearly equal in size at both ends, of a dull yellowish-white or whity-brown colour, the surface of the shell being smooth but minutely pitted all over, similar to those of the Yellow-legged Spoonbill *(Platibis flavipes)*, and measure as follows :—Length (A) 2·93 x 2·1 inches ; (B) 2·92 x 2·12 inches. These eggs in shape, colour, and size, agree with those described by Mr. A. Hume ;* the average of forty-five eggs measured by him being 2·91 x 2·12 inches. So also do those of the Australian specimens obtained in the neighbourhood of Ingham at the mouth of the Herbert River, Queensland, in March 1885, and described in a joint paper contributed to the Royal Society of Queensland by Messrs. W. T. White and Henry Tryon on the 6th of August, 1886.† Dr. Ramsay informs me that a pair of these birds were found breeding on the border of Lake Macquarie in 1860· During the last few years young birds have been occasionally forwarded to the Australian Museum and Botanic Gardens, Sydney, showing that the birds breed freely in New South Wales." *(North, P.L.S., N.S.W., 2nd Series, Vol. ii., p. 987.)*

---

* See "Nests and Eggs of Indian Birds" by Allan Hume, p. 608.
† See Proceedings of the Royal Society, Queensland, p. 139.

Hab. Derby, N.W. Australia, Port Darwin and Port Essington, Gulf of Carpentaria, Cape York, Rockingham Bay, Port Denison, Wide Bay District, Richmond and Clarence Rivers Districts, New South Wales, South Coast New Guinea. (*Ramsay.*)

Family ARDEIDÆ.

GENUS ARDEA, *Linnæus.*

ARDEA SUMATRANA, *Raffles.*

(*Ardea rectirostris,* Gould.)

Great-billed Heron.

Gould, Handbk. Bds. Aust., Vol. ii., sp. 546, p. 296.

" The nest of this species found by Gilbert on Cobourg Peninsula was built in an upright fork of a large and lofty *Melaleuca* at about eighty feet from the ground, and was formed of an outer layer of very strong sticks, with a few small twigs as a lining, and contained two eggs of a light ash-grey." (*Gould, Handbk. Bds. Aust.,* Vol. ii., p. 296.)

Hab. Port Darwin and Port Essington, Rockingham Bay, Port Denison, Wide Bay District, Richmond and Clarence Rivers Districts, New South Wales. (*Ramsay.*)

ARDEA PACIFICA, *Latham.*

Pacific Heron.

Gould, Handbk. Bds. Aust., Vol. ii., sp. 547, p. 297. ⟪XVIII⟫ 2.

This fine Heron was observed by Mr. K. H. Bennett breeding in company with *Platibis flavipes,* on trees in swamps, in the Lachlan district, the nest is composed of sticks laid crossways over some horizontal fork or flat portion of a thick bough, it is a scanty structure, through which the eggs, four for a sitting can be seen. They are of a beautiful pale blue, average specimens measure as follows:—length (A) 2·12 x 1·55 inches ; (B) 2·2 x 1·52 inches ; (C) 1·83 x 1·37 inch ; (D) 1·83 x 1·35 inch ; pairs taken from different nests in the same tree.

Specimens in the Macleayan Museum Collection, give the following dimensions :—length (A) 2·1 x 1·46 inches ; (B) 2·07 x 1·55 inches ; (C) 2·07 x 1·5 inches ; (D) 2·08 x 1·52 inches.

An interesting account of the breeding places of *Ardea pacifica* and *Platalea flavipes*, is given by Mr. K. H. Bennett in the P.L.S. N.S.W., Vol. vii., p. 324.

Hab. Derby, N.W. Australia, Port Darwin and Port Essington, Gulf of Carpentaria, Rockingham Bay, Port Denison, Wide Bay District, Dawson River, Richmond and Clarence Rivers Districts, New South Wales, Interior, Victoria and South Australia, Tasmania, West and South-West Australia. *(Ramsay)*

•

ARDEA NOVÆ-HOLLANDIÆ, *Latham.*

White-fronted Heron.

Gould, Handbk. Bds. Aust., Vol. ii., sp. 548, p. 299.

The White-fronted Heron constructs a nest of sticks and leaves, in the topmost branch of a tree (usually a Eucalyptus) overhanging a river or dam. Its eggs are four in number for a sitting of a uniform pale bluish-green colour. Dimensions of a set taken by Mr. K. H. Bennett, at Mossgiel in September, are as follows :— length (A) 1·98 x 1·4 inch ; (B) 1·97 x 1·37 inch ; (C) 2·2 x 1·37 inches ; (D) 2·1 x 1·35 inches.

The breeding season commences in September and continues during the three following months.

Hab. Derby, N.W. Australia, Port Darwin and Port Essington, Rockingham Bay, Port Denison, Wide Bay District, Dawson River, Richmond and Clarence Rivers Districts, New South Wales Interior, Victoria, and South Australia, Tasmania, West and South-West Australia. (*Ramsay.*)

Genus DEMIEGRETTA, *Blyth.*

DEMIEGRETTA SACRA, *Gmelin.*
Blue Reef Heron.
Gould, Handbk. Bds. Aust., Vol. ii., sp, 555, p. 307. XVIII ♂

" This species, says Mr. Macgillivray, inhabits the islands of the north-east coast of Australia and Torres Straits. The nest is usually placed on a tree, but on those islands where there are none, such as Raine's Islet and elsewhere, it breeds among the recesses of the rocks ; where the trees are tall, as on Oomāga or Keats Island the nests are placed near the summit ; on Dugong Island they were placed on the root of a tree, on a low stump, or halfway up a low bushy tree ; they are shallow in form eighteen inches in diameter, and constructed of small sticks, lined with twigs ; the eggs are two in number, and of a pale bluish-white, one inch and seven-eights long by one inch and a quarter broad." (*Gould, Handbk. Bds. Aust.,* Vol. ii., p. 307.)

Specimens of the eggs of this bird in the Dobroyde Collection, taken by Mr. J. A. Boyd, on the 1st of September 1879, measures as follows :—length (A) 1·96 x 1·33 inch ; (B) 1·89 x 1·35 inch.

Mr. E. D. Atkinson found this bird breeding on a small island about an acre in extent, off the north-west coast of Tasmania on the 6th of November, 1888. The nest was composed of sticks and twigs, and sheltered by a rock ; it contained three eggs of a pale bluish-green in a very advanced state of incubation, one of which was unfortunately broken in transit, the two remaining eggs measure as follows :—length (A) 1·96 x 1·4 inch ; (B) 1·95 x 1·41 inch. Mr. Atkinson who flushed the female, informs me that after watching it from an adjacent rock, the male who joined her and flew away in company, was pure white. There were no other Herons on the island.

Hab. Port Darwin and Port Essington, Gulf of Carpentaria, Cape York, Rockingham Bay, Port Denison, Wide Bay District, South Coast New Guinea. (*Ramsay.*)

Genus NYCTICORAX, *Stephens.*

NYCTICORAX CALEDONICUS, *Latham.*

Nankeen Night Heron.

Gould, Handbk. Bds. Aust., Vol. ii., sp. 557, p. 311. XVIII /.

"The nest is a loose structure of a few sticks placed crosswise over forks on the branches of trees overhanging creeks &c.; the specimens under consideration were taken by Mr. Alex. Morton, from some low bushes on Schnapper Island, near Port Stephens, they are two in number of a pale bluish-green, in length (A) 2·1 x 1·55 inches; (B) 2·1 x 1·47 inches in breadth." *(Ramsay, P.L.S., N.S.W.,* Vol. vii., p. 55.)

Two nests of this species taken by Mr. K. H. Bennett in the Lachlan River district in 1879, contained four eggs each, of the same pale bluish-green colour, two average specimens of which measure as follows:—length (A) 2·1 x 1·56 inches; (B) 2·15 x 1·55 inches; one in the Australian Museum Collection, taken on the 30th of November 1880, at Port Stephens, is smaller, measuring 1·96 inch in length by 1·45 inch in breadth.

Hab. Derby, N.W. Australia, Port Darwin and Port Essington, Gulf of Carpentaria, Cape York, Rockingham Bay, Port Denison, Wide Bay District, Richmond and Clarence Rivers Districts, New South Wales, Victoria and South Australia, Tasmania, South Coast New Guinea. *(Ramsay.)*

Genus BOTAURUS, *Stephens.*

BOTAURUS POICILOPTERUS, *Wagler.*

Australian Bittern.

Gould, Handbk. Bds. Aust., Vol. ii., sp. 558, p. 313. XVIII 3

This bird is found over the greater portions of Australia and Tasmania. It resorts to the margins of lakes and rivers for the purposes of breeding, where it constructs an open nest of reeds and other aquatic herbage, which is concealed by the rushes &c, usually found growing in such localities. Eggs three or four in number for a sitting of a uniform pale olive-brown. A specimen in the Australian Museum Collection measures long axis 2·09 inches, short axis 1·45 inch.

Two eggs, taken in September 1860, from a lagoon near Fiery Creek in Victoria, measure as follows :—length (A) 2·12 x 1·48 inches ; (B) 2·11 x 1·48 inches.

This bird breeds during September and the two following months.

Hab. Gulf of Carpentaria, Rockingham Bay, Port Denison, Wide Bay District, Richmond and Clarence Rivers Districts, New South Wales, Victoria and South Australia, Tasmania, West and South-West Australia. (*Ramsay*.)

GENUS BUTOROIDES, *Blyth.*

BUTOROIDES FLAVICOLLIS, *Latham.*

Yellow-necked Mangrove Bittern.

Gould, Handbk. Bds. Aust., Vol. ii., sp. 559, p. 315. XVIII 4

" The nest is a slight structure of sticks placed in a horizontal branch over the water. Eggs three or four, they are white with a very faint tint of green inside. A set taken by Mr. George Barnard of the Dawson River, Queensland, measures as follows : length (A) 1·88 x 1·38 inch ; (B) 1·72 x 1·35 inch ; (C) 1·8 x 1·33 inch.

" I expected to find these eggs something similar to those of *B. macrorhyncha,* taken by my brothers at Dobroyde in 1860-1-2, but Mr. Barnard assures me there can be no mistake about them, and sent me a skin of the bird." (*Ramsay, P.L.S., N.S.W.,* Vol. vii., p. 55.)

Hab. Derby, N.W. Australia, Port Darwin and Port Essington, Gulf of Carpentaria, Cape York, Rockingham Bay, Port Denison,

U

Wide Bay District, Dawson River, Richmond and Clarence Rivers Districts, New South Wales, Victoria and South Australia, West. and South-West Australia, South Coast New Guinea. *(Ramsay.)*

BUTOROIDES MACRORHYNCHA, *Gould.*
Thick-billed Mangrove Bittern.

Gould, Handbk. Bds. Aust , Vol. ii., sp. 560, p. 316. XVIII   6

This species used to breed freely at one time in the neighbourhood of Sydney.   Mr. John Ramsay found several nests in the Mangroves on the Parramatta River as early as September 1860. The nest is composed of a few sticks placed crosswise on the boughs of a Mangrove.   Eggs two or three in number for a sitting of a pale bluish-green ; length (A) 1·65 x 1·2 inch ; (B) 1·75 x 1·3 inch ; (C) 1·7 x 1·25 inch.   There is a fine series of these eggs in the Dobroyde Collection.

Hab. Wide Bay District, Richmond and Clarence Rivers Districts, New South Wales. *(Ramsay.)*

BUTOROIDES JAVANICA, *Horsfield.*
Little Mangrove-Bittern.

Gould, Handbk. Bds. Aust., Vol. ii., sp. 561, p. 317.

"Although generally this is a solitary species, yet at times it congregates in considerable numbers.   Gilbert found a colony breeding on two small islets in Coral Bay, near the entrance of the harbour of Port Essington.   Their nests, about thirty in number, were built both on the mangroves and on the branches of the yellow-blossomed *Hibiscus;* they were very frail structures, consisting of a few small twigs placed across each other on the horizontal branches, and none of them were more than six feet from the ground, each contained either two young birds, or two eggs of a uniform very pale green, one inch and five-eighths long by one inch and a quarter broad." (*Gould, Handbk. Bds. Aust.*, Vol. ii., p. 317.

Hab. Port Darwin and Port Essington, Gulf of Carpentaria, Cape York, Rockingham Bay, Port Denison, Wide Bay District, Richmond and Clarence Rivers Districts, New South Wales, South Coast New Guinea. (*Ramsay.*)

Family RALLIDÆ.

Genus PORPHYRIO, *Brisson.*

PORPHYRIO MELANOTUS, *Temminck.*

Black-backed Porphyrio.

Gould, Handbk. Bds. Aust., Vol. ii., sp. 563, p. 321.

"The nest of this species is a rough structure of rushes and water-weeds &c., placed among the reeds in the lagoons, at a considerable distance from the edge, just above water-mark. The eggs are from three to five in number, of a light brown or yellowish stone-colour, varying considerably in tint and in the shape of the markings, but usually spotted and blotched with umber, dark reddish-brown and slaty-grey. The young are of a uniform blackish-slate colour on leaving the nest." (*Ramsay, P.Z.S.,* 1877, p. 343.)

Measurements of a set taken by Dr. Ramsay, from the Hexham Swamps, in 1861, are as follows :—length (A) 2·07 x 1·4 inches ; (B) 2·09 x 1·43 inches ; (C) 2 x 1·41 inches ; (D) 2·03 x 1·42 inches ; (E) 2·05 x 1·45 inches.

Hab. Port Darwin and Port Essington, Gulf of Carpentaria, Rockingham Bay, Port Denison, Wide Bay District, Richmond and Clarence Rivers Districts, New South Wales, Interior, Victoria and South Australia, Tasmania. (*Ramsay.*)

Genus TRIBONYX, *Du Bus.*

TRIBONYX MORTIERI, *Du Bus.*

(T. gouldii, Sclater.)

Mortier's Tribonyx.

Gould, Handbk. Bds. Aust., Vol. ii., sp. 565, p. 324. XIX ⁴⁄

This bird is found on the southern portions of the continent of
Australia, but Tasmania is its great stronghold. Like *T. ventralis*
it resorts to the margins of rivers and lakes to breed ; and the
nest is formed of reeds and other aquatic herbage. Eggs seven
in number for a sitting, oval in form, of a light stone colour minutely
freckled and sparingly blotched and spotted with rounded markings
of different shades of chestnut-brown, a few of which appear as if
beneath the surface of the shell ; others have irregularly-shaped
markings and hair-like lines, particularly towards the larger end.
Dimensions of two eggs in the Australian Museum Collection,
length (A) 2·18 x 1·48 inches ; (B) 2·23 x 1·5 inches.*

A set in the Dobroyde Collection measure as follows :—length
(A) 2·2 x 1·57 inches ; (B) 2·2 x 1·47 inches ; (C) 2·19 x 1·47 ʻ
inches ; (D) 2·12 x 1·53 inches ; (E) 2·27 x 1·48 inches ; (F) 2·28ₗ
x 1·5 inches ; (G) 2·25 x 1·54 inches.

This bird breeds during September and the four following months.ₗ
Hab. Victoria and South Australia, Tasmania. (*Ramsay.*)   ⁴

TRIBONYX VENTRALIS, *Gould.*

Black-tailed Tribonyx.

Gould, Handbk. Bds. Aust., Vol. ii., sp. 566, p. 325. XVII ⁴⁄.

" Mr. Gould has I think described the eggs of some other water-
fowl, probably those of *G. tenebrosa*, under this name ; they are
certainly not those of the present bird, which are very distinct,
the nest is like that of a *Gallinula* and similarly placed. They
breed in October and the two following months, also in January
and February. When the back country is flooded these birds
litterally overrun it and breed at almost any time of the year, the

eggs are four or five in number of a pale greenish tint with roundish spots of light reddish-brown sprinkled all over the surface, and are of an oblong pointed form. Length (A) 1·85 x 1·28 inch ; (B) 1·85 x 1·3 inch ; (C) 1·76 x 1·27 inch ; (D) 1·7 x 1·15 inch ; (E) 1·88 x 1·3 inch." *(Ramsay, P.L.S., N.S.W.,* Vol. vii., p. 59.)

Hab. Derby,N.W. Australia, Gulf of Carpentaria, Port Denison, New South Wales, Interior, Victoria and South Australia, West and South-West Australia. *(Ramsay.)*

GENUS GALLINULA, *Brisson.*

GALLINULA TENEBROSA, *Gould.*

Sombre Gallinule.

Gould, Handbk. Bds. Aust., Vol. ii., sp. 567, p. 328.

This bird frequents the weedy margins of rivers and creeks, and is particularly plentiful on the Richmond and Clarence Rivers. Mr. J. Macgillivray found a nest of this species in a bush on the edge of the latter river, on the 11th of January 1864 ; it was composed of rushes and other aquatic herbage, and contained four fresh eggs, rather rounded in form, of a pale creamy-white ground colour, freckled and blotched all over with reddish-chestnut and lilac spots, the former colour greatly predominating and becoming larger and more thickly disposed towards the thicker end of the egg. Length (A) 1·55 x 1·15 inch ; (B) 1·53 x 1·18 inch ; (C) 1·37 x 1·2 inch ; (D) 1·55 x 1·2 inch.

Upon comparing these eggs with those of *Gallinula ruficrissa,* (Gould) from Northern Australia, and with those of *Amaurornis moluccana,* (Wallace), from New Britain, (the two latter of which are declared to be identical by some ornithologists) it will be seen that there is little or no variation in either their colour or measurements.

Hab. Port Denison, Wide Bay District, Richmond and Clarence Rivers Districts, New South Wales, Victoria and South Australia, and the South Coast New Guinea. *(Ramsay.)*

GALLINULA RUFICRISSA, *Gould.*

Rufous-vented Gallinule. '

Gould, Suppl. Bds. Aust., pl. 79.

"A single egg of this species in the Dobroyde Collection, is of a dull white ground colour, finely freckled all over with light chestnut-red markings, a few nearly obsolete spots of the same colour appearing as if beneath the surface of the shell more particularly towards the larger end. Long axis 1·6 inch, short axis 1·2 inch." *(North, P.LS., N.S.W.*, Vol. ii., 2nd Series, p. 446.)

Hab. Port Denison, Wide Bay District, South Coast New Guinea. (*Ramsay.*)

Genus FULICA, *Linnæus.*

FULICA AUSTRALIS, *Gould.* . ♦

Australian Coot.

Gould, Handbk. Bds. Aust., Vol. ii., sp. 568, p. 329.

This species is widely distributed over the continent of Australia and is also found in Tasmania. It frequents the lagoons near the coast, and is very abundant in the interior during wet seasons. In the neighbourhood of Melbourne I found it breeding very freely, commencing in August and continuing till the end of February. The nest is built on a platform of rushes just above the surface of the water, and is composed of various aquatic plants. Four nests of this species taken on the 28th of October 1873, each contained seven eggs, which is the usual number for a sitting; the eggs are of a pale stone colour, minutely freckled and spotted all over the surface of the shell with rounded purplish-black markings.

Dimensions of two eggs in the Australian Museum Collection are as follows :—length (A) 1·9 x 1·35 inch; (B) 1·98 x 1·4 inch.

A set of eight in the Dobroyde Collection, taken by Mr. Edward Lord Ramsay at Wattagoona, near Louth, New South Wales, measure as follows :—length (A) 1·92 x 1·33 inch ; (B) 1·95 x 1·35' inch ; (C) 1·94 x 1·33 inch ; (D) 1·83 x 1·36 inch ; (E) 1·97 x 1·36 inch; (F) 1·95 x 1·33 inch ; (G) 1·9 x 1·35 inch. Specimens in my own collection give the same average measurements.

Hab. Derby, N.W. Australia, Gulf of Carpentaria, Cape York, Rockingham Bay, Port Denison, Wide Bay District, Richmond and Clarence River District, New South Wales, Interior, Victoria and South Australia, Tasmania, West and South-West Australia. (*Ramsay.*)

Genus PARRA, *Latham.*

PARRA GALLINACEA, *Temminck.*

Comb-crested Parra.

Gould, Handbk. Bds. Aust., Vol. ii., sp. 569, p. 330. XVI 8

" The eggs of this species are among the most beautiful of any laid by our Australian birds. The curious labyrinthine markings which characterize them, however, are not altogether confined to the eggs of the *Parra;* as those of at least three of our species of *Pomatostomus* are beautifully veined and marbled in the most delicate manner." . . . "They vary in form, being quite oval and pointed equally at both ends, to almost round, or pyriform as in some of the Plovers. When of this last shape, they are usually placed in the nest with their small ends pointing inwards. In length they are from 13·5 lines to 14·5 lines, and in breadth from 10 to 11 lines. The ground colour is a light olive-yellow, becoming with time much darker. The whole surface is crossed and recrossed with irregularly curved and rather broad black lines, turning and twisting in every direction, and, in some examples, with shorter lines, making various ill-shapen letters or figures, while in others these markings take the form of blotches. Appearing beneath the shell are deep yellowish-brown streaks and hair-lines recrossing those on the surface. Some specimens are more numerously streaked than others, and have the broader black lines predominating ; in others the fine hair-lines and those of yellowish-brown are more visible.

The eggs are four in number ; and the nest, which is composed of
sedges, grass, and aquatic plants, is placed close to the water's
edge, or upon any bunches of weeds or grass growing in the water
which may be sufficiently strong enough to bear its weight. This
Parra is tolerably abundant throughout the swampy regions
which abound over the eastern portion of Queensland and north-
eastern parts of New South Wales. I have obtained specimens
as far south as the Clarence River in New South Wales, its most
southern limit, and as far north as the Herbert River, in the
Rockingham Bay district. It is found most plentiful in the
Rockhampton district wherever the swamps and lagoons occur ;
the leaves of the gigantic *Nymphœa* and *Nelumbium* affording a safe
retreat for this species. I know of few more interesting or more
pleasing sights than a troop of this handsome *Parra* wandering
among the bright blue and crimson blooms of the giant waterlilies
which abound in almost every sheet of water of any extent in
North-eastern Queensland." *(Ramsay, Ibis*, 1865, Vol. i., p. 306,
iid. 1867, Vol. iii., New Series, p. 417, pl. viii., fig. 3.)

Count Salvadori states that the *Parra* from New Guinea is the
true *P. gallinacea* of Temminck, upon Dr. Ramsay pointing out
distinction, Count Salvadori described the Australian species
under the name of *P. novœ-hollandiœ.* (See Salvad., Orn. Pap. et
Mollucc. III., 308-9.)

Hab. Rockingham Bay, Port Denison, Wide Bay District,
Richmond and Clarence Rivers Districts, New South Wales.
(Ramsay.)

Genus HYPOTÆNIDIA, *Reichenbach.*

HYPOTÆNIDIA PHILIPPENSIS, *Linnæus.*

Pectoral Rail.

Gould, Handbk. Bds. Aust., Vol. ii., sp. 570, p. 334.

The nest of this species, when built in swampy localities, is
generally composed of the débris left from floods ; it is slightly

hollowed out on the top and measures about nine inches across externally. Those situations that are covered with a growth of low *Melaleuca* are favourite breeding places of this species, also the margins of rivers and creeks and neglected plantations. Eggs four to six in number for a sitting, of a creamy-white, blotched and spotted all over, but more particularly towards the larger end with irregular shaped blood-red markings, a few nearly obsolete spots of faded lilac appearing as if beneath the surface of the shell.

Dimensions of a set in the Australian Museum Collection. Length (A) 1·42 x 1·09 inch ; (B) 1·5 x 1·13 inch ; (C) 1·42 x 1·09 inch ; (D) 1·44 x 1·1 inch.

A set in the Dobroyde Collection measure as follows :—length (A) 1·43 x 1·09 inch ; (B) 1·44 x 1·11 inch ; (C) 1·4 x 1·1 inch ; (D) 1·47 x 1·14 inch.

September and the three following months constitute the breeding season of this species.

Hab. Port Darwin and Port Essington, Gulf of Carpentaria, Cape York, Rockingham Bay, Port Denison, Wide Bay District, Richmond and Clarence Rivers Districts, New South Wales, Interior, Victoria and South Australia, West and South-West Australia. (*Ramsay.*)

Genus RALLUS, *Linnæus.*

RALLUS BRACHIPUS, *Swainson.*

Lewin's Water-Rail.

Gould, Handbk. Bds. Aust., Vol. i., sp. 571, p. 336.

This bird resorts to the margins of rivers and lagoons for the purposes of breeding, the nest is usually placed in a tuft of rushes, and is composed of grass and various aquatic plants.

A set of the eggs of this bird in Dr. James C. Cox's Collection two in number, are of a dull white, finely freckled and blotched with chestnut-brown and superimposed markings of bluish-grey, all of which predominate at the larger end of the egg. Length (A) 1·4 x 1·03 inch; (B) 1·42 x 1·02 inch.

Hab. Wide Bay District, Richmond and Clarence Rivers Districts, New South Wales, Victoria and South Australia, Tasmania, West and South-West Australia. (*Ramsay.*)

RALLINA TRICOLOR, *Gray.*
Red-necked Rail.
Gould, Suppl. Bds. Aust., p. 78.)   XVII 5.

"I found this fine species of Rail by no means rare in the dense scrubs which fringe the rivers and creeks of the coast range near Rockingham Bay; but although tolerably plentiful, they are always very difficult to obtain, on account of the nature of the localities they frequent, and their retiring disposition. They are seldom to be seen without lying in wait for them, and not always then can one obtain a shot, except, perhaps, at such close quarters as would entirely destroy them. They move about more in the evenings and early morn, and at night may be heard calling to one another as they traverse the dense masses of rank vegetation which abound in those districts. I never met with them out of these scrubs, although thick swampy grass-beds close by were frequented by allied species. They seem very local in their habits a pair frequenting the same spot for many months or perhaps the whole year round, and breeding near the same place year after year; the young soon begin to take care of themselves, and leave the parents before they are well able to fly. I found them some four or five months old in pairs. The note resembles a hoarse croak quickly repeated in a somewhat mournful tone, and a quick "cluck cluck" when come upon suddenly. I was not fortunate enough to find the nest and eggs myself, but shortly after I left the Herbert River, I received a fine set of these eggs from Inspector Robert Johnstone, to whom the bird is well known, and who assures me that after finding the nest and eggs he left it until he had twice seen the bird sitting thereon, that he might be perfectly sure there could be no mistake as to their identity. The eggs forwarded by Inspector Johnstone, have a pale cream or whitish ground colour, sprinkled all over, but more thickly at the larger end of some, with irregular shaped spots of light reddish-chestnut, and a few of lilac tinge appearing as if beneath the surface of the shell, having the characteristic form, markings, and colour of all true Rail's eggs. They are four in number, in length 1·55 to 1·65 inch. The nest was composed of a few leaves and grass and hidden among thick débris at the root of a tree in a dense part of the scrub near Mr. Johnstone's camp. The young on leaving the egg are covered with a sooty-black down, having a dark plumbeous tinge on the under surface. The young at about five months old have the upper surface of a dull dark brown tinged with olive and washed with light rufous-brown on the back of the neck ; the under surface is of a duller and more plumbeous brown, with a faint wash of rufous-brown on the chest and under tail-coverts, which latter have two pale rufous bars on each feather ; the under surface of the wings blackish dull brown, a band of white spots near the base, and a similar band about the middle of the quill-feathers ; bill olive-brown ; legs greenish-olive ; iris reddish-brown. Total length 7 inches, wing 3·6, tail 1·5 tarsus 2 inches, bill 0·9 inch." *(Ramsay, P.Z.S.*, 1875, p. 603.)

The above set of eggs measures as follows:—length (A) 1·55 x 1·1 inch; (B) 1·6 x 1·12 inch; (C) 1·62 x 1·1 inch; (D) 1·65 x 1·08 inch.

Hab. Cape York, Rockingham Bay, South Coast New Guinea. (*Ramsay.*)

Genus EULABEORNIS, *Gould.*

EULABEORNIS CASTANEIVENTRIS, *Gould.*
Chestnut-bellied Rail.
Gould, Handbk. Bds. Aust., Vol. ii., sp. 572, p. 338.

" This large and fine species of Rail inhabits the low muddy shores and mangrove swamps of the north coast of Australia. The

eggs are rather lengthened in form, of a pale pinky-white, dotted
all over with reddish-chestnut, the spots being thinly dispersed,
and some of them appearing as if beneath the surface of the shell
giving them a darker tint, two inches and one-eighth long, one
inch and five-eighths broad." (*Gould, Handbk. Bds. Aust.*, Vol.
ii., p. 338.)

Hab. Port Darwin and Port Essington, Cape York. (*Ramsay.*)

Genus PORZANA, *Vieillot.*

PORZANA PALUSTRIS, *Gould.*
Little Water Crake.
Gould, Handbk. Bds. Aust, Vol. ii., sp. 574, p. 340.

" The nest is an irregular loose structure of dry grass and weeds
&c., rather scanty, placed on the ground among the grass and
reeds in the vicinity of water, they are plentiful on the margins
of lagoons in the neighbourhood of Lake George; I also found
them breeding during October to January on the Clarence River
near Grafton; the eggs are three in number of a uniform dark
olive brown, average length 1·07 x 0·81 inch. (*Mus. Dobr.*)"
(*Ramsay, P.L.S., N.S.W.*, Vol. vii., p. 56.)

Hab. Wide Bay District, Richmond and Clarence Rivers
Districts, New South Wales, Victoria and South Australia,
Tasmania, West and South-West Australia. (*Ramsay.*)

Genus ERYTHRA, *Reichenbach.*

ERYTHRA QUADRISTRIGATA, *Horsfield.*
White-eyebrowed Water Crake.
Gould, Handbk. Bds. Aust., Vol. ii., sp. 576, p. 343. XVI 7.

This bird is found in the swamps and thickets of Mangroves
that skirt the mouths of the rivers of Northern Queensland. A
single egg of this species in the Dobroyde Collection, taken by
Mr. E. Spalding from the oviduct of a bird, while engaged on a

collecting tour in the vicinity of Rockingham Bay, for Dr. Ramsay in 1867 is the only specimen that I have seen in any collection. I give Dr. Ramsay's description of it which appeared in the Proc. Zool. Soc. for 1868, p. 388 :—

"*Erythra quadristrigata*, ♀.—An egg taken from the oviduct of this specimen is of a dirty greenish-white, the ground colour almost obscured by dots, spots, and a few blotches of brownish-red and yellowish-brown, many of the larger markings appearing as if beneath the surface ; length 1·08 x 0·86 inch."

Mr. Gould writes as follows in reference to this species in his Handbook to the Birds of Australia, Vol. ii., p. 344 :—"As the nest and eggs of this species have not yet been discovered, they form some of the desiderata to which I would call the attention of the rising ornithologists of Australia ; and I can assure them that the study of the eggs will greatly assist them in assigning the birds to which they belong to their proper genus."

It will be observed that Mr. Gould was right in adopting a different generic term for this bird, as the egg materially differs from those of the typical *Porzanæ*.

Hab. Port Darwin and Port Essington, Cape York, Rockingham Bay, Port Denison, Wide Bay District. (*Ramsay.*)

Order NATATORES.
Family ANATIDÆ.
Genus CYGNUS, *Linnæus*.
CYGNUS ATRATUS, *Latham*.
Black Swan.

Gould, Handbk. Bds. Aust., Vol. ii., sp. 577, p. 346.

This bird is found over the eastern and southern portions of the Australian continent, and likewise the greater part of Tasmania. It is particularly plentiful on the southern coast of Victoria where

it resorts to breed in the numerous inlets and bays, also on the
islands adjacent to the mainland. Capital sport is to be had
during the moulting season, by chasing down and capturing these
birds by means of a fast sailing yacht. It breeds from September
to January, and constructs a large nest of reeds and other aquatic
herbage, laying from five to nine eggs of a pale green, the shell
of which is rather rough, and stained with brown.

Dimensions of four eggs taken from a swamp near the Lachlan
River, taken by Mr. K. H. Bennett are as follows :—length (A)
4·1 x 2·6 inches ; (B) 4·03 x 2·65 inches ; (C) 3·9 x 2·55 inches ;
(D) 3·9 x 2·6 inches.

Hab. Derby, North-West Australia, Rockingham Bay, Port
Denison, Wide Bay District, Richmond and Clarence Rivers
Districts, New South Wales, Victoria and South Australia,
West and South-West Australia. (*Ramsay.*)

Genus CEREOPSIS, *Latham*.

CEREOPSIS NOVÆ-HOLLANDIÆ, *Latham.*

Cereopsis Goose.

Gould, Handbk. Bds. Aust., Vol. ii., sp. 578, p. 350.

This bird is found on the southern coast of Australia, and the
shores of Tasmania, but the islands of Bass's Straits is its great
stronghold and breeding ground.   It breeds readily in
confinement, and I am indebted to Dr. Sinclair of Gladesville, for
a description of the nest and eggs of this species, who kindly
allowed me to examine those in his possession.   One, of several
pairs of these birds had chosen for the site of their nest a clump
of bamboo canes which was growing in a small inclosure in one of
the paddocks. The nest was made on the ground and was composed
of the dried leaves and strips of stiff paper-like débris of the
bamboo, intermingled with down plucked off the breasts of the
birds.   It measured sixteen inches across externally, and

contained three eggs of a dirty-white colour ; the shell of which was rather rough. Dimensions of the eggs are as follows :—length (A) 2·93 x 2·12 inches ; (B) 2·94 x 2·12 inches ; (C) 3·15 x 2·11 inches. Both birds vigorously defended their nest, and showed every sign of resentment at the intrusion on their domain. Dr. Ramsay who had them breeding at Dobroyde for several years informs me that five is the full number of eggs for a sitting.

A specimen received from Mr. E. D. Atkinson taken on Flinders Island in Bass's Straits, is much finer in the texture of the shell, and measures 3·02 x 2·2 inches.

A specimen in the Macleayan Museum Collection, taken on one of the islands of Bass's Straits, is of a pale creamy-white and measures, long diameter 3·63 inches, short diameter 2·48 inches.

Hab. Victoria and South Australia, Tasmania. *(Ramsay.)*

Genus ANSERANAS, *Lesson.*

ANSERANAS MELANOLEUCA, *Latham.*

Semipalmated Goose.

Gould, Handbk. Bds. Aust., Vol ii., sp. 579, p. 352.

This bird is found plentifully dispersed over the northern, eastern, and southern portions of the Australian continent, and is also found inhabiting the large swamps and reed beds found in the back-waters of many of the rivers of the interior during the wet seasons, where it resorts to breed in great numbers in September and the three following months. The nest is composed of sedges and other aquatic herbage, and is built in the rushes that fringe the edge of rivers and watercourses ; at other times being placed in the marshes and lagoons. Eggs usually five in number for a sitting, but sometimes as many as eleven have been taken, probably the result of two birds laying in the same nest; they are of a creamy or dull white colour, the surface of the shell being smooth and slightly glossy.

Dimensions of a set taken by Mr. J. A. Boyd and now in the Australian Museum Collection are as follows :—length (A) 2·86 x 2·18 inches ; (B) 2·89 x 2·06 inches; (C) 3·05 x 2·16 inches ; (D) 2·8 x 1·98 inch ; (E) 2·83 x 2·1 inches.

Hab. Derby, N.W. Australia, Port Darwin and Port Essington, Gulf of Carpentaria, Rockingham Bay, Port Denison, Wide Bay District, Richmond and Clarence Rivers Districts, New South Wales, Interior, Victoria and South Australia. (*Ramsay.*)

Genus CHLAMYDOCHEN, *Bonaparte.*

CHLAMYDOCHEN JUBATA, *Latham.*

Maned Goose.

Gould, Handbk. Bds. Aust., Vol. ii., sp. 580, p. 354.

This handsome bird is dispersed over the greater portion of Australia, but more particularly over the eastern and southern parts, it is also found in Tasmania. It breeds in the hollow limbs of large trees, laying from seven to eleven creamy white eggs, the texture of shell being very fine, and slightly glossy. Two average specimens in the Australian Museum Collection measure, length 2·25 x 1·55 inches ; (B) 2·27 x 1·57 inches.

An average specimen in the Dobroyde Collection taken by Mr. James Ramsay, at Cardington on the Bell River, on the 5th of September 1861, measures, long axis 2·47 inches, short axis 1·63 inch. Specimens taken by Mr. George Barnard of Coomooboolaroo, Queensland, give the following measurements :—length (A) 2·31 x 1·55 inches ; (B) 2·3 x 1·57 inches.

This bird commences to breed in August and continues till the end of January.

Hab. Derby, N.W. Australia, Rockingham Bay, Port Denison, Wide Bay District, Dawson River, Richmond and Clarence Rivers Districts, New South Wales, Interior, Victoria and South Australia, West and South-west Australia. (*Ramsay.*)

Genus NETTAPUS, *Brandt.*

NETTAPUS PULCHELLUS, *Gould.*

Pygmy Goose.

Gould, Handbk. Bds. Aust., Vol. ii., sp. 581, p. 357.

"Gilbert found a nest of this species at Port Essington near the margin of a lake, it was built up in the long grass about a foot above the water, the bottom of the nest resting on its surface, it was composed of long dried grasses, slightly hollowed for the reception of the eggs, the nest in this instance was destitute of any kind of lining, but one afterwards brought him by the natives was interiorly constructed with feathers and contained six eggs, which are white, one inch and seven-eighths long by one inch and three-eighths broad." (*Gould, Handbk. Bds. Aust.,* Vol. ii., p. 357.)

Hab. Derby, N.W. Australia, Port Darwin and Port Essington, Gulf of Carpentaria, Cape York, Rockingham Bay, South Coast New Guinea. (*Ramsay.*)

NETTAPUS ALBIPENNIS, *Gould.*

White-quilled Pygmy Goose.

Gould, Handbk. Birds Aust., Vol. ii., sp. 582, p. 359.

This species is found on the eastern portions of the Australian continent, and is rather abundant on the Richmond and Clarence Rivers; an egg taken from the oviduct of a bird of this species shot by Mr. J. Macgillivray at South Grafton, during October 1864 is of a faint creamy-white colour, the texture of the shell being very fine and smooth to the touch, but without any gloss. Long axis 1·79 inch, short axis 1·4 inch.

Hab. Rockingham Bay, Port Denison, Wide Bay District, Richmond and Clarence Rivers Districts, New South Wales. (*Ramsay.*)

v

Genus TADORNA, *Leach.*

TADORNA RADJAH, *Garnot.*

Radjah Shieldrake.

Gould, Handbk. Bds. Aust., Vol. ii., sp. 583, p. 360.

This fine species ranges all over the northern portions of the continent of Australia, and has lately been received from Derby, on the north-western coast; it is likewise found on the south coast of New Guinea. "A set of the eggs of this bird taken from the hollow branch of a tree during 1875, are five in number, of a rich creamy-white, the texture of the shell being fine and the surface smooth. Length (A) 2·2 x 1·63 inches; (B) 2·2 x 1·58 inches; (C) 2·2 x 1·59 inches; (D) 2·13 x 1·61 inches; (E) 2·17 x 1·58 inches." *(North, P.L.S., N.S.W.,* Vol. ii., 2nd Series, p. 446.)

Dr. Ramsay found this species breeding on the Burnett River, Queensland, during the months of November, December and January 1873-4.

Hab. Derby, N.W. Australia, Port Darwin and Port Essington, Gulf of Carpentaria, Rockingham Bay, Port Denison, Wide Bay District, South Coast Guinea. *(Ramsay.)*

Genus CASARCA, *Bonaparte.*

CASARCA TADORNOIDES, *Jardine.*

Chestnut-coloured Shieldrake.

Gould, Handbk. Bds. Aust., Vol. ii., sp. 584, p. 361.

Dr. Ramsay remarks, Mr. Whittell informs me that he found the nest of this species placed on the ground behind a mass of *Polygonum* bushes, it was made of grass and débris with a few sticks; the eggs were eight in number and covered over with the grass lining of the nest. The colour of the eggs is of a light cream, dull white, or whity-brown, rough to the touch, oval, in length 2·7 inches by 1·92 inch in short diameter. I have never taken the eggs of this bird myself, but Mr. P. Faithful informed me of a nest similarly placed on the banks of a creek on his estate near Goulburn. *(Ramsay.)*

An average sized specimen in the Macleayan Museum Collection measures 2·67 x 1·9 inch.

Specimens taken from a hollow stump near the margin of a lagoon in the Western District of Victoria, give the same average measurements, but the texture of the shell is much finer and the surface smooth.

August and the four following months constitute the breeding season of this species.

Hab. New South Wales, Victoria and South Australia. *(Ramsay.)*

GENUS DENDROCYGNA, *Swainson.*

DENDROCYGNA VAGANS, *Eyton.*

(D. gouldi, Bonaparte)

Whistling Tree-Duck.

Gould, Handbk. Bds. Aust., Vol. ii., sp. 591, p. 374.

This species is distributed over the greater portion of the Australian continent, according to Gilbert, "some eggs brought to the settlement by the natives, and said to belong to this bird, were taken early in March, from nests built in long grass on the small islands adjacent to the harbour at Port Essington ; they are of a-creamy-white, one inch and seven-eighths long by one inch and a-half in breadth." *(Gould, Handbk. Bds. Aust.,* Vol. ii., p. 374.)

Hab. Derby, N.W. Australia, Port Darwin and Port Essington, Gulf of Carpentaria, Cape York, Rockingham Bay, Port Denison, Wide Bay District, Richmond and Clarence Rivers Districts, New South Wales, Interior, South Coast New Guinea. *(Ramsay.)*

DENDROCYGNA EYTONI, *Gould.*

Eyton's Tree-Duck.

Gould, Handbk. Bds. Aust., Vol. ii., sp. 592, p. 375.

"An egg of this species taken from the oviduct by Mr. J. Rainbird of Port Denison, measures 1·95 inch in length by 1·5 inch in breadth ; it is of a light creamy-white, and slightly ovate, and appears to be comparatively a very small egg for a bird of this size to lay." (*Ramsay, P.Z.S.*, 1877, p. 347.)

Hab. Gulf of Carpentaria, Cape York, Rockingham Bay, Port Denison, Wide Bay District, Richmond and Clarence Rivers Districts, Victoria and South Australia, West and South-west Australia. (*Ramsay.*)

GENUS STICTONETTA, *Reichenbach.*

STICTONETTA NÆVOSA, *Gould.*

Freckled Duck.

Gould, Handbk. Bds. Aust., Vol. ii., sp. 587, p. 367.

This remarkable species is found in the Interior and in the southern and western portions of Australia, likewise in Tasmania. It is tolerably abundant during certain seasons on the Gippsland Lakes and on Lake Colac in Victoria, and is frequently exposed for sale in the poulterers shops of Melbourne. A nest of this species taken in October 1868 near the margin of a swamp in the Western District of Victoria, contained seven eggs of a pale creamy-white, with a greenish tinge inside ; and like most of the eggs of the family (*Anatidæ*), the shell is of a fine texture, and the surface smooth and slightly glossy. Two average specimens measure as follows :—length (A) 2·26 x 1·65 inches ; (B) 2·3 x 1·7 inches. A specimen in the Macleayan Museum measures 2·67 inches by 1·9 inch.

Hab. Richmond and Clarence Rivers Districts, New South Wales, Interior, Victoria and South Australia, West and South-west Australia, (*Ramsay.*)

Genus ANAS, *Linnæus.*

ANAS SUPERCILIOSA, *Gmelin.*

Australian Black Duck.

Gould, Handbk. Bds. Aust., Vol. ii., sp. 585, p. 363.

" The eggs of this species vary in number from six to ten for a sitting. The nest is often placed at some distance from the water among herbage on the ground, which hides the bird from view when sitting. Often a small 'run' through the long grass and herbs leads to the nest itself. A great variety of situations is chosen for the nest, and the eggs are always covered over with down and feathers of the parent bird when she leaves the nest. The colour is a pale cream tint, sometimes with a greenish shade. One egg I have seen has a round green spot, but this must be looked upon as quite accidental. Average length 2·2 x 1·9 inches in breadth." (*Ramsay, P.L.S., N.S.W.,* 2nd Series, Vol. i., p. 1152.)

Hab. Derby, N.W. Australia, Port Darwin and Port Essington, Gulf of Carpentaria, Cape York, Rockingham Bay, Port Denison, Wide Bay, Dawson River, Richmond and Clarence Rivers Districts, New South Wales, Interior, Victoria and South Australia, Tasmania, West and South-west Australia, South Coast New Guinea. (*Ramsay.*)

ANAS CASTANEA, *Eyton.*

(A. punctata, Cuvier.)

Australian Teal.

Gould, Handbk. Bds. Aust., Vol. ii., sp. 586, p. 365.

This well known bird is widely dispersed over the Continent of Australia, and the greater portion of Tasmania. It usually resorts to the hollow limbs of trees to breed, enveloping its eggs in a mass of down plucked from the breast of the parent birds; but occasionally its nest is found in the long grass or rushes bordering the margins of rivers and lagoons. Eggs eight or nine in number for a sitting, creamy-white, the texture of the shell being fine and the surface smooth.

A set in the Dobroyde Collection, taken by Mr. James Ramsay at Yanda, in the interior of New South Wales, on the 15th of November 1879, measure as follows :—length (A) 1·9 x 1·42 inch; (B) 1·9 x 1·43 inch; (C) 1·97 x 1·42 inch; (D) 1·95 x 1·44 inch; (E) 1·91 x 1·43 inch; (F) 1·94 x 1·4 inch; (G) 1·9 x 1·43 inch; (H) 1·95 x 1·43 inch.

This species commences to breed in August and continues the four following months.

Hab. Derby, N.W. Australia, Rockingham Bay, Port Denison, Wide Bay District, Richmond and Clarence Rivers Districts, New South Wales, Victoria and South Australia, Tasmania, West and South-west Australia. (*Ramsay.*)

ANAS GIBBERIFRONS, *Müller.*

Müller, Verhandelingen Land en Volkenkunde, 1839-41, p. 159.

" There has been much discussion about this species which had always been looked upon in Australia until the last few years, as the female of *A. castanea (A. punctata* of Gould's Bds. Aust.) I have not been able to find any good characteristics between the females of these species up to the present time. But the males may at once be known, as in *A. gibberifrons,* the sexes are alike in plumage ; in *A. castanea* the male has a rich chestnut-red breast and a glossy green head when adult, and even on the young male the chest is tinged with rufous. The eggs are usually six to ten in number, and are laid in the hollow branches of trees, &c. Creamy-white. Length 2·15 x 1·45 inches." (*Ramsay, P.L.S., N.S.W.,* Vol. i., 2nd Series, p. 1151.)

A fine set of these eggs taken by Mr. K. H. Bennett at Ivanhoe, on the 9th of September 1885, from the hollow limb of a tree, and which were enveloped in a mass of down, measure as

follows :—Length (A) 1·88 x 1·45 inch ; (B) 1·92 x 1·47 inch ;
(C) 1·94 x 1·45 inch ; (D) 1·92 x 1·47 inch; (E) 1·91 x 1·47 inch;
(F) 1·95 x 1·47 inch.

This bird usually commences to breed in August and continues
till theend of of November, and very often deposits its eggs under
the shelter of a cotton-bush on the plains far away from any water,
but like all the members of this family the breeding season is
greatly·influenced by the rains.

Hab. Gulf of Carpentaria, Wide Bay District, Dawson River,
Richmond and Clarence Rivers Districts, New South Wales,
Interior, Victoria and South Australia. *(Ramsay.)*

It has also been recorded from New Zealand under the name
of *A. gracilis.*

Genus SPATULA, *Boie.*

SPATULA RHYNCHOTIS, *Latham.*

Australian Shoveller.

Gould, Handbk. Bds. Aust., Vol. ii., sp. 588, p. 368.

This species ranges over the southern half of the continent of
Australia, and is likewise found in Tasmania and has been recorded
from New Zealand. It is rather difficult to procure specimens of
this bird, as it is extremely shy, and takes to flight upon the
least approach of danger. The nest is generally built beneath the
shelter of a bush or tuft of rank grass at no great distance from
water. Eggs seven or nine in number for a sitting, creamy-white
with a faint greenish shade pervading through the shell. Dimensions
of a set of three eggs taken at Wilcannia, in September 1883 :—
Length (A) 2·3 x 1·54 inches ; (B) 2·27 x 1·55 inches ; (C) 2·24
x 1·53 inches.

Three specimens in the Macleayan Museum Collection measure
as follows :—length(A) 2·12 x 1·53 inches; (B) 2·08 x 1·52 inches;
(C) 2·2 x 1·6 inches.

Hab. Wide Bay District, Richmond and Clarence Rivers Districts, New South Wales, Victoria and South Australia, Tasmania, West and South-West Australia. *(Ramsay.)*

Genus MALACORHYNCHUS, *Swainson.*

MALACORHYNCHUS MEMBRANACEUS, *Latham.*
Pink-eyed Duck.

Gould, Handbk. Bds. Aust., Vol. ii., sp. 590, p. 372.

" For a member of the *Anatidæ,* this bird certainly selects the most unique spots imaginable in which to make its nests. The first instance was brought under my notice by Mr. K. H. Bennett, of Yandembah, a most enterprising naturalist, to whom the Australian Museum is indebted for several rare specimens. Mr. Bennett informs me that having occasion to visit a nest of the White-fronted Heron, *Ardea novæ-hollandiæ,* he was much surprised to find it much altered in appearance, and from the mass of down which covered the whole of the upper part of the Heron's nest the duck flew off, leaving two eggs, which with the nest have been transmitted to the Museum ; the eggs unfortunately were broken in transit, this deficiency however, is supplied by specimens taken by Mr. Whittell from a similar mass of dark slaty-grey down, which was placed on a flattened portion of a thick horizontal bough about ten feet from the ground, overhanging the water, on the bank of the Darling River near Wilcannia ; in this instance the eggs were six in number of a rich light cream colour, rather pointed ovals. Length (A) 1·85 x 1·3 inch ; (B) 1·82 x 1·3 inch. The beautiful structure above mentioned, sent by Mr. Bennett, consisted of the platform of sticks, which formed the nest of the Heron, being thickly covered with down, which formed a rim four inches in height, a large quantity of down was worked in among the sticks and covered the greater part of the sides, it closed over the eggs above in an elastic mass, quite hiding them." *(Ramsay, P.L.S., N.S.W.,* Vol. vii., p. 58.)

A set of eggs taken by Mr. K. H. Bennett on the 12th October 1885 at Ivanhoe, from the hollow limb of a tree, and which were enveloped in a similar mass of down, measure as follows:—length (A) 1·9 x 1·33 inch ; (B) 1·86 x 1·33 inch ; (C) 1·93 x 1·33 inch ; (D) 1·88 x 1·36 inch ; (E) 1·95 x 1·31 inch.

This bird commences to breed about the end of September, and continues the two following months. *(From Mr. K. H. Bennett's Collection.)*

Hab. Derby, N.W. Australia, Port Denison, Wide Bay District, Richmond and Clarence Rivers Districts, New South Wales, Interior, Victoria and South Australia, Tasmania, West and Southwest Australia. *(Ramsay.)*

Genus NYROCA, *Fleming.*

NYROCA AUSTRALIS, *Gould.*

White-eyed Duck.

Gould, Handbk. Bds. Aust., Vol. ii., sp. 593, p. 377.

This bird is widely distributed over the Australian continent, and is also found in Tasmania, frequenting secluded bays, and likewise the lakes of the interior. The nest is concealed beneath a bush or tuft of long grass, in the vicinity of water, or among the rushes close to the water's edge, it is composed of aquatic herbage and dried grasses, lined inside with feathers and down. Eggs, seven to nine in number for a sitting, creamy white, varying in form from swollen to elongated ovals, the texture of the shell being fine and the surface smooth and slightly glossy.

Specimens taken in the Western District of Victoria, in October 1878, measure as follows:—length (A) 2·55 x 1·88 inches ; (B) 2·45 x 1·9 inches.

Hab. Rockingham Bay, Port Denison, Wide Bay District Richmond and Clarence River Districts, New South Wales, Interior, Victoria and South Australia, Tasmania, West and Southwest Australia *(Ramsay.)*

Genus ERISMATURA, *Bonaparte.*

ERISMATURA AUSTRALIS, *Gould.*

Blue-billed Duck

Gould, Handbk. Bds. Aust., Vol. ii., sp. 594, p. 379.

This bird is dispersed over the southern and western portions of the continent of Australia, and is tolerably abundant on the Gippsland Lakes in Victoria; it has also recently been found on Lake Buddah, near Trangie in New South Wales.

According to Mr. Gould " it breeds in September and October, constructing a nest very like that of the *Biziura lobata,* and laying from two to nine or ten eggs, which are of a large size, and of a uniform bluish-white, with a very rough surface; two inches and five-eighths long by two inches broad." *(Gould, Handbk. Bds. Aust.,* Vol. ii., p. 379.)

Hab. New South Wales, Victoria and South Australia, West and South-west Australia. *(Ramsay.)*

Genus BIZIURA, *Leach.*

BIZIURA LOBATA, *Shaw.*

Musk Duck.

Gould, Handbk. Bds. Aust., Vol. ii., sp. 595, p. 381.

" The nest of this species is placed among the rushes, reeds, and weeds on the banks of the small islands in the lakes and the lagoons. It is composed of aquatic plants, leaves of the reeds, flags, and the like, and lined with a few feathers. The eggs are usually two in number, of a pale olive colour, 3·2 inches in length by 2·1 inches in breadth. The shell is minutely granulated, rough, and very strong. The breeding season begins in August and continues to the end of November." *(Ramsay, Ibis,* 1867, Vol. iii., New Series, p. 413, pl. viii., fig. 1.)

Hab. Wide Bay District, Richmond and Clarence Rivers Districts, New South Wales, Interior, Victoria and South Australia. *(Ramsay.)*

Family PODICIPITIDÆ.

Genus PODICEPS, *Latham*.

PODICEPS CRISTATUS, *Linnæus*.

(P. australis, Gould.)

Australian Tippet-Grebe.

Gould, Handbk. Bds. Aust., Vol. ii., sp. 665, p. 511.

This bird is found breeding during wet seasons on the back waters of many of the inland rivers of New South Wales and Victoria. The nest like all others of the genus is composed of sedges and other aquatic herbage, and is attached to reeds growing in the water, the top of the nest being nearly level with the surface. Eggs usually five in number for a sitting, in form elongated ovals, slightly pointed at both ends, of a pale bluish-white colour, which is usually entirely hidden with a thin coating of lime, except in some places where a few deep scratches reveals the true colour of the shell underneath. A set taken on the Lachlan during November 1886, measures as follows :— length (A) 2·2 x 1·35 inches ; (B) 2·23 x 1·35 inches ; (C) 2·11 x 1·32 inches.

Hab. Wide Bay District, Richmond and Clarence Rivers Districts, New South Wales, Interior, Victoria and South Australia, Tasmania. *(Ramsay.)*

PODICEPS NESTOR, *Gould.*

Hoary-headed Grebe.

Gould, Handbk. Bds. Aust., Vol. ii., sp. 666, p. 512.

This bird is dispersed over the eastern and south-eastern portions of the continent of Australia, and the whole of Tasmania. It forms a floating nest of aquatic herbage and sedges, which is attached to two or three reeds in the water, and is only a few inches above the surface. Eggs five in number for a sitting are of a pale bluish-white colour when freshly laid, covered with a thin coating of lime, which in some places gives the shell a roughened appearance, the eggs however become quickly discoloured with the decaying vegetable matter of which the nest is composed, and before they are hatched often become of a reddish-brown tint. Dimensions of a set taken in October 1883, are as follows :—length (A) 1·6 x 1·09 inch ; (B) 1·62 x 1·07 inch; (C) 1·65 x 1·08 inch ; (D) 1·55 x 1·05 inch.

This species commences to breed in October and continues during the three following months.

Hab. Wide Bay District, Richmond and Clarence Rivers Districts, New South Wales, Interior, Victoria and South Australia, Tasmania, West and South-West Australia. (*Ramsay.*)

PODICEPS NOVÆ-HOLLANDIÆ, *Stephen.*

(P. gularis, Gould.)

Black-throated Grebe.

Gould, Handbk. Bds. Aust., Vol. ii., sp. 667, p. 513.

Unlike the preceding species, this Grebe is widely dispersed over the whole continent of Australia, and is particularly abundant on the swamps and lagoons of Victoria, and the inland waters of New South Wales.   The nest is similar to that of the preceding species, being composed of sedges and other aquatic herbage, and attached to a few reeds in the water ; they are often placed within a few feet of each other.   While sitting, the female covers herself over with the outer portions of the nest, her head and neck alone being visible ; when leaving the nest, she covers her eggs over, and dives at once, reappearing about ten or fifteen yards away.   During 1873 many nests of this species were procured in a single afternoon from the Albert Park lake near Melbourne.   The eggs are bluish-white when first laid, thinly coated with lime, but quickly becoming soiled with the wet and decaying weeds of which the nest is formed.   Eggs usually five in number for a sitting, although six are occasionally found, they are oval in form, slightly pointed at both ends. Dimensions of a set in the Australian Museum Collection :—length (A) 1·37 x 0·98 inch; (B) 1·4 x 0·98 inch; (C) 1·38 x 0·95 inch; (D) 1·34 x 0·95 inch.

Specimens in the Dobroyde Collection taken by Mr. James Ramsay, on the Merule, on the 30th of September 1867, measures as follows :—length (A) 1·47 x 1 inch; (B) 1·46 x 0·98 inch; (C) 1·35 x 0·96 inch ; (D) 1·5 x 0·97 inch.

A set taken at Albert Park, on the 28th of October 1874, gave the following measurements:—length (A) 1·5 x 1 inch; (B) 1·48 x 0·98 inch ; (C) 1·47 x 0·98 inch ; (D) 1·42 x 0·95 inch ; (E) 1·4 x 0·97 inch.

Eggs of this species have been taken in the early part of September and also in the latter end of February, November and December however are the usual months for breeding.

Hab. Derby, N.-W. Australia, Gulf of Carpentaria, Rockingham Bay, Port Denison, Wide Bay District, Dawson River, Richmond and Clarence Rivers Districts, New South Wales, Interior, Victoria and South Australia, Tasmania, South Coast New Guinea. (*Ramsay.*)

Family SPHENISCIDÆ.

Genus EUDYPTULA, *Bonaparte.*

EUDYPTULA MINOR, *Forst.*

Little Penguin.

Gould, Handbk. Bds. Aust, Vol. ii., sp., 669, p. 518.

This bird is plentifully dispersed over the south coast of Australia, the islands of Bass's Straits, and Tasmania. In company with the Rev. J. D. Nicholson of Cowes, Phillip Island, Western Port, I procured a great number of the eggs of this species on the 25th of October 1883, near the "Nobbys," about ten miles distant from Cowes. At that time some of the burrows contained fresh eggs, and others young birds, resembling balls of slaty-black down, while a few were nearly ready to take to the water, it was worthy of note that the burrows close to the edge of the cliffs contained either young birds or eggs nearly incubated, while those that were about one hundred and fifty yards from the water all contained fresh eggs. These tunnels were about three feet in length, and the entrance in most cases was hidden by a bush, the parent bird in every instance ably defending her eggs or young ones, and inflicting a smart blow with the bill, on the hand of the intruder. Should these birds be taken out of their burrows they return to them again at once seldom trying to make their escape. Eggs sometimes three in number for a sitting, but usually only two; in form swollen ovals, rather sharply pointed at the smaller apex, when fresh they are of a pale bluish-white, and the shell semi-transparent, but when partly incubated the egg becomes soiled and is of a dull lead-white colour.

Dimensions of two sets are as follows:—length (A) 2·3 x 1·7 inches; (B) 2·32 x 1·71 inches; (C) 2·27 x 1·72 inches; (D) 2·12 x 1·66 inches; (E) 2·15 x 1·67 inches; (F) 2·17 x 1·68 inches.

They breed during September and the three following months.

Hab. Richmond and Clarence Rivers Districts, New South Wales, Victoria and South Australia, Tasmania, West and South-West Australia. (*Ramsay*)

EUDYPTULA UNDINA, *Gould.*

Fairy Penguin.

Gould, Handbk. Bds. Aust., Vol. ii., sp. 670, p. 512.

This bird is found on the Tasmanian coast and on the islands adjacent thereto. It closely resembles in every respect the preceding species, *E. minor*, but is slightly smaller; the number and colour of the eggs are similar. A set taken at Anderson's Bay on the north-west coast of Tasmania, measures as follows:— length (A) 2·08 x 1·63 inches; (B) 2·05 x 1·63 inches; (C) 2·06 x 1·65 inches. Taken during the month of October 1882.

Hab. Tasmania and adjacent islands,

Family LARIDÆ.

Genus LARUS, *Linnæus*.

LARUS PACIFICUS, *Latham*.

Pacfic Gull.

Gould, Handbk. Bds. Aust. Vol. ii., sp. 596, p. 385. ‾X‾X‾ /.

This bird ranges over the eastern and southern shores of Australia, but is found more plentifully on the islands of Bass's Straits and Tasmania, where it breeds in colonies on the low islands adjacent to the mainland, depositing its eggs three in number, upon the ledges of the rocks or on shingly beaches. Two eggs taken by the late Mr. S. White of the Reed Beds near Adelaide, are of a light olive, spotted and blotched all over with umber- and blackish-brown, a few nearly obsolete markings appearing as if beneath the surface of the shell. Length (A) 2·82 x 1·89 inches ; (B) 2·85´x 2 inches.

A set taken on the Tasmanian coast in November 1882, are of a darker olive, and have the markings larger and more thickly disposed. Length (A) 2·87 x 1·98 inches ; (B) 2·84 x 1·95 inches ; (C).2·9 x 1·93 inches.

The months of October, November, and December constitute the breeding season of this species.

Hab. Rockingham Bay, Port Denison, Wide Bay District, Richmond and Clarence Rivers Districts, the Coasts of New South Wales, Victoria and South Australia, Tasmania, West and South-west Australia. (*Ramsay.*)

Genus XEMA, *Leach*.

XEMA NOVÆ-HOLLANDIÆ, *Stephens*.

Jameson's Gull.

Gould, Handbk. Bds. Aust., Vol. ii., sp. 597, p. 387. ‾X‾X‾ ⁴⁄

This Gull is found all over the eastern and southern portions of Australia, also on the islands of Bass's Straits and Tasmania. It breeds in colonies, making a slight nest of grasses or sea-weeds, placed upon the ledges of rocks, at other times upon the bare headlands or low sandy beaches. Phillip and French Islands in Western Port Bay are favourite breeding localities of this species, also King Island in Bass's Straits. The eggs, usually three in number for a sitting vary considerably in the colour and disposition of their markings. A set taken on the 27th November 1882, at Phillip Island, have a pale green ground colour heavily blotched with dark umber-brown markings, some of a lighter shade appearing as if beneath the surface of the shell. Length (A) 2·15 x 1·49 inches ; (B) 2·15 x 1·5 inches ; (C) 2·2 x 1·47 inches.

A set taken by the late Mr. S. W. White, of the Reed Beds, near Adelaide, in 1878, are of a light olive-brown streaked and blotched with different shades of umber-brown, in three specimens the markings are scattered all over the surface of the shell, in the others they are confined to the larger end where they form ill-defined zones. Length (A) 2·18 x 1·57 inches : (B) 2·15 x 1·5 inches ; (C) 2·1 x 1·55 inches ; (D) 2·09 x 1·55 inches.

The breeding season is during the months of October, November and December.

Hab. Port Denison, Wide Bay District, Richmond and Clarence Rivers Districts, New South Wales, occasionally on the rivers and swamps of the Interior, Victoria and South Australia, and Tasmania. *(Ramsay.)*

Family STERNIDÆ.

Genus SYLOCHELIDON, *Brehm*.

SYLOCHELIDON CASPIA, *Pall.*

Caspian Tern.

Gould, Handbk. Bds. Aust., Vol. ii., sp. 600, p. 392. ~~XX~~ 3

This bird ranges over the greater portion of the Australian Coast, the islands of Bass's Straits and Tasmania. It deposits from two to three eggs on the bare ground, usually on the headlands and low promontories of the coast. A set of three taken by the late Mr. S. White of the Reed-beds, South Australia, are of a light stone colour, spotted and blotched all over with irregular shaped markings of umber and blackish-brown, some of which appear as if beneath the surface of the shell, particularly in one specimen (A), where they become confluent towards the larger end. Length (A) 2·61 x 1·68 inches ; (B) 2·62 x 1·75 inches ; (C) 2·41 x 1·73 inches.

Two eggs taken on King Island, in Bass's Straits, in November 1878, are of a stone colour thickly covered with longitudinal markings of blackish-brown, others with bluish-black markings appearing as if beneath the surface of the shell. Length (A) 2·47 x 1·68 inches ; (B) 2·35 x 1·65 inches.

Two eggs in the Macleayan Museum Collection, taken on Bountiful Island, in the Gulf of Carpentaria, give the following measurements :—length (A) 2·56 x 1·7 inches ; (B) 2·63 x 1·75 inches.

The breeding season commences in September and continues throughout December.

Hab. Gulf of Carpentaria, Cape York, Rockingham Bay, Port Denison, Wide Bay District, Richmond and Clarence Rivers Districts, New South Wales, Victoria and South Australia, Tasmania. (*Ramsay.*)

Genus HYDROCHELIDON, *Boie.*

HYDROCHELIDON HYBRIDA, *Pallas.*

(*H. fluviatilis,* Gould.)

Marsh-Tern.

Gould, Handbk. Bds. Aust., Vol. ii., sp. 610, p. 406.

This Tern is widely dispersed over the Australian Continent, and is found frequenting the inland rivers and lagoons, it constructs a nest of sedges and other aquatic herbage, and attaches it to weeds

W

or rushes growing in the water. Two eggs taken from a nest in a lagoon in the Loddon District of Victoria, in November 1883, are of a light greenish-grey, blotched, spotted and dotted all over with dark umber and wood-brown, but chiefly towards the larger end where the markings become more thickly disposed, and in a few places confluent. Length (A) 1·52 x 1·07 inch ; (B) 1·5 x 1·07 inch.

Hab. Gulf of Carpentaria, Rockingham Bay, Wide Bay District, Richmond and Clarence Rivers Districts, New South Wales, Interior, Victoria and South Australia, West and South-West Australia. *(Ramsay)*

Genus STERNA, *Linnæus.*

STERNA BERGII, *Lichtenstein.* •

(S. pelecanoïdes, King; *T. poliocercus,* Gould.)

Torres Straits Tern.   XIX 2

Gould, Handbk. Bds. Aust., Vol. ii., sps. 601, 602, pp. 394, 396.

This Tern is found all over the coast of Australia, the islands of Torres Straits and the South coast of New Guinea. The eggs are extremely variable in their markings, in a number of specimens now before me, no two are similar, they are oval in form and somewhat sharply pointed at the thinner end. I give a description of four eggs taken by the late Mr. S. White in Torres Straits, to attempt to describe all the different varieties would be an endless task.

Var. A. Ground colour stone-grey with a few large blotches of umber scattered over the surface of the shell, clouded markings of bluish-grey appearing beneath the shell's surface ; length 2·23 x 1·56 inches.

Var. B. Ground colour stone-grey with long black lines, twisted and curved in every direction over the surface of the shell ; length 2·32 x 1·55 inches.

W—2

Var. C. Ground colour light brown, with a large coalesced patch of brownish-black on the thicker end; length 2·3 x 1·58 inches.

Var. D. Ground colour light reddish-buff heavily blotched all over the surface of the shell with dark blackish-red markings; length 2·29 x 1·57 inches.

These eggs were taken in the month of July.

Hab. Port Darwin and Port Essington, Gulf of Carpentaria, Cape York, Rockingham Bay, Port Denison, Wide Bay District, Richmond and Clarence Rivers Districts, New South Wales, Victoria and South Australia, Tasmania, West and South-West Australia, South Coast of New Guinea. (*Ramsay.*)

STERNA ANGLICA, *Mont.*

(Gelochelidon macrotarsa, Gould.)

Long-legged Tern.

Gould, Handbk. Bds. Aust., Vol. ii., sp. 608, p. 403. XVII 2

This species is found on the eastern and south-eastern portions of Australia, it likewise ranges very far inland, where it resorts to breed. It formerly bred in large colonies near the shores of Lake Bolac in Victoria, from whence numerous specimens were obtained. Mr. K. H. Bennett also found them breeding in the interior of New South Wales; from his MSS. I extract the following:

" On two occasions (1870 and 1872) I have known *S. anglica,* to breed in the Ivanhoe district. In both instances the sites chosen were similar, viz., a sandy bank rising some two or three feet above the surrounding plain, and thickly covered with dwarf saltbush ; these breeding places were about forty miles apart, in one case close to a wide sheet of water, and in the other quite two miles away from the nearest water. At both places hundreds of the birds were breeding, and the eggs, two in number for a sitting, were deposited on the bare ground, and so closely together that care was required when walking so as not to step upon them."

Eggs of a buffy-white or whitey-brown sparingly marked with light umber, the remainder of the surface being boldly splashed and spotted with purplish-red and purplish-grey, some with obsolete patches of a lighter tint appearing as if beneath the surface of the shell. Length 2 inches by 1·42 inch. Taken by Mr. Bennett in November 1872. Other specimens examined are similar in their markings but vary very much in size. Length (A) 2 x 1·5 inches; (B) 1·85 x 1·4 inch.

Hab. Rockingham Bay, Wide Bay District, Richmond and Clarence Rivers Districts, New South Wales, Interior. (*Ramsay.*)

STERNA MELANAUCHEN, *Temminck.*
Black-naped Tern.
Gould, Handbk. Bds. Aust., Vol. ii., sp. 606, p. 400.

This bird inhabits the northern coast of Australia, and the islands of Torres Straits, it has also been obtained on the New South Wales coast, and Islands of the South Pacific. It deposits its eggs two in number in a slight depression in the sand, they are of a pale creamy-white, blotched and spotted, chiefly towards the larger end with dark umber and bluish-grey markings, the latter colour predominating and appearing as if beneath the surface of the shell. Length (A) 1·53 x 1·13 inch; (B) 1·4 x 1·07 inch.

Specimens taken by Mr. J. A. Boyd on the 18th of June 1882, are similar in colour and measure as follows:—length (A) 1·6 x 1·15 inch; (B) 1·58 x 1·15 inch.

Hab. Port Darwin and Port Essington, Cape York, New South Wales, South Coast New Guinea. (*Ramsay.*)

STERNA ANÆSTHETA, *Scop.*
(*Onychoprion panayensis*, Gmelin.)
Panayan Tern.
Gould, Handbk. Bds. Aust., Vol. ii., sp. 612, p. 411.

This species ranges over the greater portion of the Australian coast and the islands of Torres Straits, depositing its single egg in the cleft of a rock, without any attempt at forming a nest for its reception. Specimens in the Dobroyde Collection, taken by Mr. Macgillivray on Bramble Cay in 1862, are of a creamy-white ground colour, spotted and blotched all over with irregularly shaped markings of dark reddish-brown and bluish-grey, the latter colour being nearly obsolete, and appearing as if beneath the surface of the shell; in one specimen, the markings are confluent on the larger end and form a coalesced patch. Dimensions of three average specimens are as follows :—length (A) 1·67 x 1·27 inch; (B) 1·87 x 1·21 inch; (C) 1·73 x 1·25 inch.

Hab. Port Darwin and Port Essington, Cape York, Rockingham Bay, Port Denison, Wide Bay District, New South Wales, Victoria and South Australia, West and South-West Australia, South Coast New Guinea. (*Ramsay.*)

STERNA FULIGINOSA, *Gmelin.*
Sooty Tern.

Gould, Handbk. Bds. Aust., Vol. ii., sp. 611, p. 408. XXI /.

This bird is found all over the coast line of Australia, but particularly on the northern and western shores. Its single egg is deposited in a slight depression in the sand; specimens in the Dobroyde Collection, taken by Mr. Macgillivray in 1862 on Bramble Cay, whither this bird resorts to breed in great numbers, vary considerably in their markings, the most usual variety found being white or of a creamy-white ground colour, spotted and blotched with irregular shaped markings of dark chestnut, reddish-brown and bluish-grey, the latter colour being very faint and hardly visible. Four average sized specimens measure as follows :—length (A) 2·1 x 1·4 inches; (B) 2·09 x 1·39 inches; (C) 2 x 1·47 inches; (D) 2 x 1·45 inches.

The breeding season of this species in Torres Straits is during the months of May, June, and July, and at Lord Howe Island, from September till the end of January.

Hab. Port Darwin and Port Essington, Cape York, Rockingham Bay, Wide Bay District, New South Wales, Victoria and South Australia, West and South-West Australia, South Coast New Guinea. (*Ramsay.*)

Genus STERNULA, *Boic.*

STERNULA NEREIS, *Gould.*
Fairy Tern.
Gould, Handbk. Bds. Aust., Vol. ii., sp. 607, p. 402.

This little Tern is found all over the south coast of Australia, the islands of Bass's Straits and Tasmania. The late Mr. S. White obtained a number of the eggs of this species in South Australia, they were deposited in a slight depression in the sand, and were two in number for a sitting, of a pale bluish-stone colour, marked all over but particularly towards the thicker end with dark umber-brown and nearly obsolete spots of bluish-grey, the latter colour in some instances predominating and appearing as if beneath the surface of the shell. Length (A) 1·43 x 0·98 inch; (B) 1·44 x 1 inch.

This species breeds during the months of November and December. (*Dobr. Mus. Coll.*)

Hab. Wide Bay District, New South Wales, Victoria and South Australia, Tasmania, West and South-West Australia. (*Ramsay.*)

Genus ANOUS, *Leach.*
ANOUS STOLIDUS, *Linnaeus.*
The Noddy.
Gould, Handbk. Bds. Aust., Vol. ii., sp. 613, p. 413.   *XXI* 2.

The Noddy ranges over the whole of the Australian coast line, and is particularly plentiful on the islands in Torres Straits, where Mr. Macgillivray found it breeding in great numbers during 1862. The nest is composed of small twigs and seaweed and is placed on the top of a bush, at other times upon the sand, it never contains more than a single egg. Like all the Terns' eggs they are extremely variable in their markings, the most usual variety found is of a creamy-white ground colour, with clouded spots and blotches of chestnut-red and faint bluish-grey, the latter colour appearing as if beneath the shell's surface ; these markings are more thickly disposed towards the larger end of the egg, and in some specimens form an irregular zone. The average length of a number of specimens is 1·98 x 1·39 inch.

A specimen in the Macleayan Museum Collection taken on Bramble Cay on the 13th of August 1875, measures 2·1 inches by 1·47 inch.

Hab. Port Darwin and Port Essington, Gulf of Carpentaria, Cape York, Rockingham Bay, Port Denison, Wide Bay District, Richmond and Clarence Rivers Districts, New South Wales, Victoria and South Australia, West and South-West Australia, South Coast New Guinea. (*Ramsay.*)

ANOUS TENUIROSTRIS, *Temminck.*

(A. melanops, Gould.)

The Lesser Noddy.

Gould, Handbk. Bds. Aust., Vol. ii., sp. 614, p. 417.

Gilbert found this species breeding in great numbers on South Island, Houtmann's Abrolhos, forming a nest of seaweed on the branches of the mangroves and placed at a height of from four to ten feet from the ground. " Like its near ally, *A. stolidus*, it commences the task of incubation in December, and lays but a single egg. The egg is of a pale stone or cream colour, marked all over with large irregular-shaped blotches of dull chestnut-red and dark brown, the latter colour appearing as if beneath the surface of the shell ; the blotches are thinly dispersed except at the larger end, where they are largest and most numerous ; it is one inch and three quarters long, by one inch and five sixteenths broad." *(Gould, Handbk. Bds. Aust.,* Vol. ii., p. 417.)

Hab. Cape York, Rockingham Bay, New South Wales, Victoria and South Australia, West and South-West Australia, South Coast New Guinea. *(Ramsay.)*

Family PROCELLARIIDÆ.

GENUS NECTRIS, *Bonaparte.*

NECTRIS BREVICAUDUS, *Brandt.*

Short-tailed Petrel.

Gould, Handbk. Bds. Aust., Vol. ii., sp. 636, p. 459.

This Petrel is found on the eastern and southern coasts of Australia, and is particularly plentiful on the islands of Bass's Straits and Tasmania. On Phillip Island, in Western Port Bay, Victoria, it arrives in thousands, usually on the evening of the 24th of November, to deposit its single egg in a burrow in the earth, which is often from three to four feet in length. On the morning of the 25th and for several days after, the place is like a fair, the fishermen from the Bay and usually a number of visitors from Melbourne and elsewhere, being busily engaged in extracting the eggs from the burrows, which to a novice is no easy task. During a recent visit to the breeding grounds of this species two experienced fishermen took no less than sixty dozen eggs in one day. Notwithstanding the great amount of eggs and birds annually taken, there seems to be no diminution in their numbers, and they resort to the same places to breed regularly year after year.

The eggs are pure white and vary in form from true ovals to elongated and pointed ovals. Six selected from over a hundred specimens measure as follows :—length (A) 2·81 x 1·98 inches ; (B) 2·95 x 1·9 inches ; (C) 2·87 x 1·9 inches ; (D) 2·97 x 1·75 inches; (E) 2·7 x 1·85 inches; (F) 2·75 x 1·85 inches.

Hab. Wide Bay District, New South Wales, Victoria and South Australia, Tasmania, West and South-West Australia. (*Ramsay.*)

NECTRIS CARNEIPES, *Gould.*

Pink-footed Petrel.

Gould, Handbk. Bds. Aust., Vol. ii., sp. 637, p. 465.

" This species of Petrel resorts to the small islands off Cape Leeuwin in Western Australia, for the purpose of breeding. Its single white egg is about two inches and seven-eighths long by nearly two inches wide." (*Gould, Handbk. Bds. Aust.*, Vol. ii., p. 465.

Hab. West and South-West Australia.

Genus FREGETTA, *Bonaparte.*

FREGETTA MELANOGASTER, *Gould.*

Black-bellied Storm-Petrel.

Gould, Handbk. Bds. Aust., Vol. ii., sp. 647, p. 479.

This bird is found on the eastern and southern coasts of Australia, also Tasmania. Eggs taken in November 1884, on Cliffy Island, Bass's Straits, are oval in form, pure white, and measure as follows :—length (A) 1·48 x 1·25 inch ; (B) 1·52 x 1·27 inch ; (C) 1·49 x 1·27 inch.

Hab. Wide Bay District, New South Wales, Victoria, and South Australia, Tasmania, West and South-West Australia. (*Ramsay.*)

Genus PELAGODROMA, *Reichenbach.*

PELAGODROMA FREGATA, *Linnaëus.*

White-faced Storm-Petrel.

Gould, Handbk. Bds. Aust., Vol ii., sp. 649, p. 482.

An average specimen from a number of eggs of this species taken on Swan Island, near Queenscliff, Victoria, in November 1878, measures 1·45 x 1·08 inch ; it is oval in form and pure white.

A specimen taken by Mr. E. D. Atkinson on a small island off the north-west coast of Tasmania, measures 1·35 x 1·07 inch.

Hab. Victoria and South Australia, West and South-West Australia. (*Ramsay.*)

Family PELECANIDÆ.

Genus PHAËTON, *Linnæus.*

PHAËTON RUBRICAUDA, *Bodd.*

(*P. phænicurus,* Gmelin.)

Red-tailed Tropic Bird.

Gould, Handbk. Bds. Aust., Vol. ii., sp. 660, p. 501. ☒☒ *l.*

This bird ranges over the northern and north-eastern coasts of Australia, the islands of Torres Straits and the south coast of New Guinea. It usually deposits its eggs, two in number, under the shelter of a projecting ledge of rock. Specimens taken on Raine's Islet, lying off the North-eastern coast of Australia, are of a reddish-buff ground colour thickly freckled and spotted all over with reddish-brown markings, but more particularly towards the thicker end, where the markings become confluent and in some instances forming a coalesced patch on the larger apex. Length (A) 2·65 x 1·85 inches ; (B) 2·6 x 2·85 inches.

Hab. Port Darwin and Port Essington, Cape York, Rockingham Bay, New South Wales, South Coast New Guinea. (*Ramsay.*)

Genus PLOTUS, *Linnæus.*

PLOTUS NOVÆ-HOLLANDIÆ, *Gould.*

New Holland Darter.

Gould, Handbk. Bds. Aust., Vol. ii., sp. 657, p. 496.

This bird is widely dispersed over the Australian continent, it frequents rivers and lagoons where it obtains its food, consisting chiefly of fish. The nest is built of sticks and twigs often placed on a branch of a tree overhanging the water, and usually contains three or four eggs of a pale bluish-white, often thickly covered with lime like to those of *Graculus*, and *Podiceps*. An average specimen measures 2·35 x 1·55 inch.

Hab. Port Darwin and Port Essington, Gulf of Carpentaria, Cape York, Rockingham Bay, Port Denison, Wide Bay District, Dawson River, Richmond and Clarence Rivers Districts, New South Wales, Interior, Victoria and South Australia, and the South Coast New Guinea. (*Ramsay.*)

Genus SULA, *Brisson.*

SULA SERRATOR, *Banks.*

(*S. australis*, Gould.)

The Australian Gannet.

Gould, Handbk. Bds. Aust., Vol. ii., sp. 661, p. 504.

This species ranges over the seas of the eastern and southern portions of the continent of Australia, the islands of Bass's Straits and the coast of Tasmania. The nest is built upon the ground and is composed of twigs and dried marine débris, it usually contains two eggs, elongated in form, of a pale bluish-white colour with a rough coating of lime. An average specimen taken on one of the islands off the Tasmanian coast, measures as follows :— length 3·05 inches, breadth 1·85 inch.

Hab. Port Denison, Wide Bay District, Richmond and Clarence Rivers Districts, New South Wales, Victoria and South Australia, Tasmania, West and South-West Australia. (*Ramsay.*)

SULA LEUCOGASTRA, *Bodd.*

(S. fiber, Linnæus.)

Brown Gannet.

Gould, Handbk. Bds. Aust., Vol. ii., sp. 663, p. 507.

The members of the "Chevert Expedition" (fitted out by the Hon. Wm. Macleay), found this bird breeding in great numbers on Bramble Cay, during 1875.

The nests were built upon the ground and consisted of a few dried sticks and grasses, and were placed so close together, that it was difficult to walk without treading upon them, most of them contained two eggs of a bluish-white colour thickly coated with lime, which in some parts is scratched off evidently by the feet of the bird, revealing the true colour of the shell ; they are oval in form and vary considerably in size. Two eggs in the Australian Museum Collection measure as follows:—length (A) 2·33 x 1·42 inches ; (B) 2·45 x 1·65 inches.

Specimens in the Macleayan Museum Collection give the following measurements :—length (A) 2·35 x 1·48 inches ; (B) 2·4 x 1·45 inches.

Hab. Port Darwin and Port Essington, Gulf of Carpentaria, Cape York, South Coast New Guinea. (*Ramsay.*)

Genus TACHYPETES, *Vieillot.*

TACHYPETES MINOR, *Gmelin.*

Small Frigate-Bird.

Gould, Handbk. Bds. Aust., Vol. ii., sp. 659, p. 499.

This bird is found breeding in colonies on the islands of Torres Straits ; the nest is composed of sticks and twigs, and placed on the ground or in low bushes, the eggs are one or two in number for a sitting, of a chalky-white colour. An average specimen in the Australian Museum Collection, taken on Raine's Islet, by Mr. Macgillivray, measures 2·5 inches in length by 1·7 inch in breadth.

Hab. Port Darwin and Port Essington, Cape York, Rockingham Bay, Wide Bay District, South Coast New Guinea. (*Ramsay*.)

Genus GRACULUS, *Linnæus*.

GRACULUS NOVÆ-HOLLANDIÆ, *Stephens*.

Australian Cormorant.

Gould, Handbk. Bds. Aust., Vol. ii., sp. 652, p. 488.

This bird is very common in nearly every part of Australia and the whole of Tasmania. It builds a nest of sticks and aquatic herbage, often placed on a *Polygonum* bush or *Casuarina* over-hanging the water. Eggs two in number for a sitting, although frequently one only is found, in form they are elongated ovals, of a pale bluish-white, the surface of the shell being almost entirely hidden by a rough coating of lime. Length (A) 2·45 x 1·6 inches; (B) 2·47 x 1·65 inches.

Specimens in the Macleayan Museum Collection give the following measurements:—length (A) 2·38 x 1·45 inches; (B) 2·52 x 1·43 inches.

Hab. Cape York, Rockingham Bay, Port Denison, Wide Bay District, Richmond and Clarence Rivers Districts, New South Wales, Victoria and South Australia, Tasmania, West and South-West Australia. (*Ramsay*.)

GRACULUS VARIUS, *Gmelin*.

Pied Cormorant.

Gould. Handbk. Bds. Aust., Vol. ii., sp. 653, p. 490.

The late Mr. S. White, of the Reed Beds, South Australia, found this species breeding in great numbers during the month of November, on some islands off the South Australian coast, the nidification is similar to that of the foregoing species, the eggs

being of a pale bluish-white, with the surface of the shell partially
obscured by a coating of lime ; they vary both in form and size ;
five specimens in the Dobroyde Collection, taken by the late Mr. S.
White, give the following measurements :—length (A) 2·11 x 1·43
inches ; (B) 2·37 x 1·5 inches ; (C) 2·35 x 1·4 inches ; (D) 2·39 x
1·41 inches ; (E) 2·37 x 1·35 inches

Hab. Wide Bay District, New South Wales, Victoria and
South Australia, West and South-West Australia. (*Ramsay.*)

GRACULUS LEUCOGASTER, *Gould.*
White-breasted Cormorant.

Gould, Handbk. Bds. Aust., Vol. ii.; sp. 654, p. 492.

The White-breasted Cormorant ranges over the southern portions
of the continent of Australia, the islands of Bass's Straits, and
Tasmania.   It builds a nest of sticks and sea-weed on the rocky
islets off the coast of Tasmania and South Australia, and lays two
eggs of a pale bluish-white, thickly coated with lime.   An average
specimen from South Australia measures 2·1 inches in length by
1·35 inch in breadth.

Hab. Port Denison, Wide Bay District, Richmond and Clarence
Rivers Districts, New South Wales, Victoria and South Australia,
Tasmania, West and South-West Australia. (*Ramsay.*)

GRACULUS MELANOLEUCUS, *Vieillot.*
Little Black and White Cormorant.

Gould, Handbk. Bds. Aust., Vol. ii., sp. 655, p. 493.

This species is distributed over the greater portions of Australia
and Tasmania, and is particularly plentiful on the inland rivers
and lagoons.   In new South Wales it breeds during the months
of October and November, constructing a nest of sticks on some

suitable bush standing in or near the water. Eggs two to four in number for a sitting, oval in form, of a pale bluish-white, and usually coated more or less with lime. Measurements of a set in the Dobroyde Museum Collection are as follows:—length (A) 1·9 x 1·25 inch; (B) 1·93 x 1·25 inch.

Specimens in the Macleayan Museum Collection, and others taken by Mr. K. H. Bennett give the same average measurements.

Hab. Rockingham Bay, Port Denison, Wide Bay District, Dawson River, Richmond and Clarence Rivers Districts, New South Wales, Victoria and South Australia, Tasmania, West and South-West Australia, South Coast New Guinea, and Lord Howe Island. (*Ramsay.*)

GRACULUS STICTOCEPHALUS, *Bonaparte.*
Little Black Cormorant.

Gould, Handbk. Bds. Aust, Vol. ii., sp. 656, p. 495.

This species is widely distributed over the Australian Continent and is found inhabiting most of the lakes and rivers of the interior of New South Wales. Eggs in form elongated ovals, of a pale bluish-white, thickly coated with lime. Length (A) 2·12 x 1·47 inches; (B) 2·09 x 1·48 inches.

Three specimens in the Macleayan Museum Collection, taken on the Darling River in 1882, give the following dimensions:— length (A) 2·1 x 1·42 inches; (B) 2·05 x 1·42 inches; (C) 2·18 x 1·45 inches.

October and November comprise the breeding season of this species.

Hab. Cape York, Rockingham Bay, Port Denison, Wide Bay District, Dawson River, Richmond and Clarence Rivers Districts, New South Wales, Interior, Victoria and South Australia, West and South-West Australia. (*Ramsay.*)

Genus PELECANUS, *Linnæus*.

PELECANUS CONSPICILLATUS, *Temminck*.

Australian Pelican.

Gould, Handbk. Bds. Aust., Vol. ii., sp. 651, p. 486.

This species is universally dispersed over the whole of Australia. They construct large nests of sticks and aquatic herbage, often placed close together on the top of inaccessible rocky islands off the coast, or on some of the islets in our large inland lakes. The eggs two in number for a sitting, are of a dirty yellowish-white, thickly coated with lime, and varying considerably in their size and form. Specimens in the Dobroyde Collection taken by the late Mr. S. White of the Reed Beds, South Australia, vary from swollen to elongated ovals. I give the measurements of three of them :— length (A) 3·34 x 2·31 inches ; (B) 3·59 x 2·2 inches ; (C) 3·99 x 2·25 inches.

An average specimen in my own collection taken at the Gippsland Lakes, Victoria, measures 3·38 inches in length by 2·2 inches in breadth.

Hab. Port Darwin and Port Essington, Gulf of Carpentaria, Cape York, Rockingham Bay, Port Denison, Wide Bay District, Dawson River, Richmond and Clarence Rivers Districts, New South Wales, Interior, Victoria and South Australia, Tasmania, West and South-West Australia, South Coast New Guinea. (*Ramsay.*)

APPENDIX.

No authentic information has yet been recorded of the breeding of the following species in Australia, many of which are only occasional visitors to our shores, although it may be only question of time, for when the vast breeding-grounds in the back waters and reed beds of some of our inland rivers and lakes, are properly explored, the nests and eggs of several recorded here will probably be found. The young of *Ibis falcinellus* have been found in a very early stage of plumage in the Interior, and in December 1888, I saw a young bird in the Sydney market obtained near Brisbane a few days before. *Herodias alba* and several others may also be found breeding.

In this list I have given the locality or country where authentic eggs of each species have been taken, and a reference showing where an accurate description of each may be found. Australian Oologists are greatly indebted to the researches of Messrs. Seebohm and Harvie-Brown, and especially to Mr. Allan Hume, who in his work on the " Nests and Eggs of Indian Birds," gives a most interesting account of the nidification of many birds which are visitors to Australia. Coming nearer home the labours of Sir Walter Buller, Professor Hutton, and the late Mr. T. H. Potts, have contributed largely to a knowledge of the breeding and habits of the birds of New Zealand, a country closely connected with Australia as regards its marine avi-fauna ; many of which breed there and farther south, and which are also residents or visitors to Australia. Dr. W. M. Crowfoot has also contributed an interesting paper on the breeding of certain sea birds frequenting Norfolk Island and the adjacent islets, many of which are found inhabiting our coasts.

In conclusion I have to acknowledge my indebtedness to the authorities above quoted, from which the localities are taken, and to whose works I would refer anyone requiring a full account of the nidification, eggs, and habits of these species.

x

Esacus magnirostris, *Geoffroy*, Small islands off the Indian Coast (*Hume, Nests and Eggs of Indian Birds*.)

Charadrius fulvus, *Gmelin*, *(C. pluvialis orientalis*, Temm. ; *C. longipes*, Temm.), Eastern Siberia, *(Seebohm)*.

Charadrius helveticus, *Linnæus*, Petchora River, Northern Russia. *(Seebohm and Harvie-Brown, Ibis,* 1876).

Ægialitis geoffroyi, *Wagler*, Island of Formosa, *(Swinhoe)*.

Ægialitis mongolus, *Pallas, (Hiaticula inornata*, Gould,) Young birds found at Thibet, *(Legge)*.

Ægialitis bicinctus, *Jard. & Selb.*, New Zealand, *(Potts)*; Young birds were obtained in Tasmania by the Museum Collectors.

Ægialitis hiaticula, *Linnæus*, British Isles, *(Dresser.)*

Totanus stagnatalis, *Temminck,* Lapland and Hungary,(*Dresser*).

Totanus glottis, *Linnæus*, Scotland, Lapland, (*Seebohm, Hist. Brit. Bds.*)

Actitis hypoleucus, *Linnæus*, Europe, (*Seebohm*), Cashmere, (*Hume*).

Actitis bartramius, *Wilson*, British North America, *(Coues)* *(Seebohm)*.

Cinclus interpres, *Linnæus*, Within the Arctic Circle, (*Seebohm*).

Herodias garzetta, *Linnæus*, *(H. immaculata*, Gould). Valley of the Danube, Europe, (*Seebohm*) ; Ceylon, (*Legyc*) ; India, (*Hume*).

Herodias asha, Ceylon, Dahlak Archipelago, Red Sea, (*Legge*).

Herodias alba, *Linnæus*, India, (*Hume*) ; Ceylon, (*Legge*).

Herodias intermedia, *Hasselq.*, *(H. plumiferus*, Gould), India, (*Hume*).

Ardea minuta, *Linnæus*, Cashmere, (*Hume, Nests and Eggs of Indian Birds*).

Ardea cinerea, *Linnæus*. Europe, Asia, Africa, (*Hume*) *(Seebohm)*.

Terekia cinerea, *Gmelin*, Northern Siberia, (*Dresser, Birds of Europe*) ; and Russia, (*Harting, P.Z.S.*, 1874).

x—2

SPATULA CLYPEATA, *Linnæus.* This bird has never been recorded from Australia, except on the occasion referred to by Mr. Gould (Handbook, Vol. ii., p. 370) if the specimen was really identical with the *S. clypeata* of Europe, it could only have been a straggler, and should not be included in the Australian avi-fauna ; its nest and eggs are fully described by various writers on British Birds.

GLAREOLA ORIENTALIS, *Leach,* Pegu, Brit. India, (*Hume, Nests and Eggs of Indian Birds*).

IBIS FALCINELLUS, *Linnæus,* Europe, (*Seebohm*) ; India, (*Doig*).

CATARRACTES CHRYSOCOME, *Latham,* New Zealand, and Chatham Island (*Hutton*).

STERCORARIUS ANTARCTICUS, *Lesson,* Kerguelen's Lands (*Hutton*).

STERNA DOUGALLI, *Mont., (S. gracilis,* Gould), North America (*Brewer*)· This bird was reported to Gilbert as having been found breeding on Houtmann's Abrolhos.

STERNA MEDIA, *Horsfield, (Thalasseus bengalensis,* Lesson ; *T. torresii,* Gould), Islands of the Persian Gulf, (*Legge*); America, (*Brewer*).

STERNULA SINENSIS, *Gmelin, (S. placens,* Gould), Ceylon, (*Legge*); Formosa,(*Swinhoe*). Young birds were obtained by Dr. Ramsay near Ballina at the mouth of the Richmond River in 1867.

ANOUS LEUCOCAPILLUS, *Gould.* Eggs taken by Mr. J. A. Boyd, and Mr. Ralph Hargrave in the New Hebrides.

DIOMEDEA EXULANS, *Linnæus,* Auckland Islands, (*McCormick, Voy. H.M.S. Erebus*).

MAJAQUEUS PARKINSONI, *Gray,* Little Barrier Island off the coast of New Zealand, (*Buller*).

PRION VITTATA, *Forst.,* Island of St. Paul, Indian Ocean,(*Gould*).

PRION TURTUR, *Smith,* Small islands off the coast of New Zealand (*Potts*).

ÆSTRELATA COOKII, *G. R. Gray,* Fiji, (*Ramsay*), (*Boyd*).

TACHYPETES AQUILUS, *Linnæus,* Islands off the coast of British Honduras, (*Salvin*).

NESTS AND EGGS OF BIRDS,

FOUND BREEDING ON

LORD HOWE AND NORFOLK ISLANDS.

OUR knowledge of the nesting and eggs of the Lord Howe Island birds is still very limited, and until an excursion fitted out by the Trustees of the Australian Museum in September 1887, very little had been done towards recording authentic information relative to the breeding season, or the eggs of the birds found there. Of. the sixty species recorded in Dr. Ramsay's "List of Birds found on Lord Howe Island," only eleven are strictly peculiar, and of only one of these, is the nest and eggs known; much remains therefore to be done, and it is to be hoped that any one favorably situated for acquiring further information will not fail to make notes on this interesting subject.*

HALCYON VAGANS, *Lesson.*

(H. norfolkiensis, Tristram.)

New Zealand Kingfisher.

Buller, Bds. New Zealand, p. 69.

Mr. E. H. Saunders, who has lately returned from Lord Howe Island, states that he found this bird breeding freely during the month of November 1887, in the hollow limbs of trees. The eggs, five in number for a sitting, are rounded in form, and of a beautiful pearly-white tint. The dimensions of a set are as follows :—length (A) 1·14 x 0·92 inch (this specimen is somewhat sharply pointed at one end); (B) 1·08 x 0·91 inch; (C) 1·12 x 0·91 inch; (D) 1·1 x 0·9 inch; (E) 1·08 x 0·92 inch.

Hab. Lord Howe, and Norfolk Islands, New Zealand.

---

* Lord Howe Island Report, p. 45, (Oology; *North.*)

APLONIS FUSCUS, *Gould.*

Fuscous Aplonis.

According to Mr. Saunders, the nidification of this bird is entirely different from that of the allied genus *Calornis* of the Australian continent, the birds resorting to the hollow branches of trees to construct their nests, several were found with young birds, but only one contained eggs. In every instance these nests were built of dried grasses, and placed in a hollow at the end of a branch. The eggs are usually four in number for a sitting, and vary somewhat in form, even in the same set ; two eggs of one set are swollen ovals being thickest at the centre, and slightly pointed towards each end, the other two are long ovals, slightly tapering at one end only ; in colour they are of a pale bluish-green, freckled with markings of a reddish- and wood-brown tint, equally disposed over the surface of the shell, some of the markings are very indistinct. Length (A) 1·07 x 0·78 inch ; (B) 1·06 x 0·76 inch, (thick ovals); (C) 1·14 x 0·76 inch ; (D) 1·12 x 0·75 inch.

Hab. Lord Howe, and Norfolk Islands.

CHALCOPHAPS CHRYSOCHLORA, *Wagler.*

Little Green Pigeon.

Gould, Handbk. Bds. Aust., Vol. ii., sp. 459, p. 118.

The nest of this bird is composed of a few thin twigs placed crosswise on the horizontal branch of a tree, not far from the ground. The eggs are two in number, oval in form, of a light creamy-white colour, and give an average measurement of 1·47 x 0·8 inch. This bird previously plentiful upon the island, has already become very scarce, and will probably with other species, soon be exterminated by the islanders.

Hab. Lord Howe Island and Australia.

GYGIS CANDIDA, *Gmelin.*

White Tern.

Gould, Handbk. Bds. Aust., Vol. ii., sp. 609, p. 405. ~~XXI~~ 4.

The single egg laid by this bird for a sitting is deposited in a slight cavity or roughened surface on the bare branch of a tree, usually at a considerable distance from the shore in some valley or sheltered situation. Unlike *Anous melanogenys*, (G. R. Gray) this bird does not breed in colonies, but returns season after season to the same place and tree to deposit its egg, although they may be repeatedly taken. The trees usually selected are *Lagunaria patersoni, Noteloea longifolia,* and *Baloghia lucida.** Two eggs taken by Dr. Metcalfe are oval in form and nearly equal in size at both ends ; one is of a stony-buff ground colour, thickly freckled, spotted, and splashed all over with different shades of brown and greyish-black, the latter colour appearing as if beneath the surface of the shell, in some places these markings are confluent forming large irregular-shaped patches on the shell ; length 1·67 x 1·24 inch. The other has a light greenish-grey ground colour, which is almost obscured by thick irregularly-shaped linear markings and smears of umber-brown, and nearly obsolete dashes of deep bluish-grey. Length 1·71 x 1·22 inch.

Hab. Norfolk Island.

STERNA FULIGINOSA, *Gmelin.*

Sooty Tern.

Gould, Handbk. Bds. Aust., Vol. ii., sp. 611, p. 408. ~~XXI~~ 1.

This bird was found breeding on the rocky ledges and flat parts of the cliffs, but more often on the bare sand ; little or no attempt was made at forming a nest, except in a few instances where a small portion of débris was found scraped around the single egg laid by this bird for a sitting. Mr. Saunders who visited the island during the breeding season, collected a large number of the eggs during

---

* Crowfoot, Ibis, 1885, p. 267.

November. In a series of over one hundred eggs examined, there is a great variation in the size, colour, and disposition of the markings. The predominant form is oval, tapering slightly towards the thin end; the colour a dull white, some being nearly devoid of markings, others uniformly freckled and spotted over the whole surface of the shell with reddish-brown, others have large irregularly-shaped confluent blotches of purplish-red and slaty-grey, the latter appearing as if beneath the shell, these markings predominate in some towards the larger end of the egg, out many have rounded spots of rich-red evenly distributed over the surface of the shell. in a few instances the markings assume the form of a zone. Length (A) 2·13 x 1·42 inches; (B) 2 x 1·45 inches; (C) 1·85 x 1·4 inch; (D) 2·15 x 1·45 inches; (E) 2·11 x 1·4 inches; (F) 2·04 x 1·34 inches; (G) 2·02 x 1·47 inches; (H) 2·14 x 1·48 inches.

An egg of this species recently received from Dr. Metcalfe, and taken on Norfolk Island, is white, and without markings of any kind.

Hab. Lord Howe, Norfolk, and the Islands of the Pacific generally.

ANOUS STOLIDUS, *Linnæus.*
Noddy.

Gould, Handbk. Bds. Aust., Vol. ii., sp. 613, p. 413. XXI 2.

This species was found breeding during October and November, its single egg is deposited on a flat nest of sticks, twigs, and sea-weed, placed upon the sandy beaches. The eggs usually oval in form, are slightly pointed at one end, and vary in colour from white to creamy-white, some being minutely spotted all over with brownish-black, others being closely blotched, more particularly towards the larger end, with blood-red markings and nearly obsolete spots of the same colour. Two average specimens measure as follows:—length (A) 2·03 x 1·45 inches; (B) 2·18 x 1·47 inches.

Hab. Lord Howe and Norfolk Islands, and the Islands of the Pacific Ocean.

ANOUS MELANOGENYS, *G. R. Gray.*
Black-cheeked Noddy.

Gray, P.Z.S., 1876, p. 670.   ~~XXI~~   *5.*

The eggs of this species have been kindly sent by Dr. P. H. Metcalfe, who collected them on Norfolk and Philip Islands, during the months of October and November. This bird, he states, builds a rather neat cup-shaped nest of fresh sea-weed which is usually placed on the branches of the *Lagunaria patersoni*, in some secluded spot, many nests are often built on the same tree. Occasionally the nests are built upon trees growing upon the coast. Only one egg is laid for a sitting, they are oval in form, slightly pointed at one end and of a faint creamy-white, spotted and blotched with dull reddish-brown, many of the markings being nearly obsolete. In some instances the markings are equally dispersed over the surface of the shell; others again have them confined to the larger end. Three specimens taken during November 1885 on Philip Island, measure as follows :—length (A) 1·78 x 1·25 inch; (B) 1·68 x 1·22 inch ; (C) 1·83 x 1·25 inch.

Hab. Norfolk and Philip Islands.

ANOUS CINEREUS, *Gould.*
Grey Noddy.

Gould, Handbk. Bds. Aust., Vol. ii., sp. 616, p. 420.   ~~XXI~~ *6.*

This, a somewhat rare species, was found breeding in the early part of September, also during the month of November. The eggs were rather difficult to obtain, as for the purposes of breeding the birds usually resort to almost inacessible ledges of rocks, but some-times they deposit a single egg on the bare sand. In form the eggs are nearly true ovals, being but slightly tapered at one end, of a dull creamy-white ground colour, sparingly freckled and spotted with faint reddish-brown and slaty-grey markings, the latter colour predominating in some instances, and appearing as if beneath the surface of the shell ; others have short thick wavy markings, resembling ill-shapen letters and figures, equally distributed over the surface of the shell, which although not thickly disposed yet are in some places confluent, and more indistinct than is usually found on the eggs of other allied species. There is very little variation in their size and shape, two average specimens measure as follows :—length (A) 1·63 x 1·16 inch; (B) 1·67 x 1·2 inch.

Hab. Lord Howe, and Norfolk Islands.

PUFFINUS SPHENURUS, *Gould.*
Wedge-tailed Petrel.
Gould, Handbk. Bds. Aust., Vol. ii., sp. 638, p. 466.

During the months of November and December, this bird was found breeding in great numbers, and like most of the *Procellariidœ*, they dig a long tunnel or burrow in the sand or soft earth, many of these burrows are several feet in length, and a single egg is deposited at the extremity, which when fresh is snow-white, but soon becomes stained and soiled. There is great variation in the shape and size, true ovals, lengthened and swollen ovals predominating, some terminating abruptly at one end, others being sharply pointed. Length (A) 2·35 x 1·67 inches; (B) 2·45 x 1·6 inches; (C) 2·45 x 1·68 inches; (D) 2·57 x 1·64 inches.

Dr. Metcalfe remarks that on Norfolk Island he has frequently found the egg of this bird under the shelter of an overhanging rock.

Hab. Lord Howe, and Norfolk Islands.

PUFFINUS NUGAX, *Soland.*
Allied Petrel.
Gould, Handbk. Bds. Aust., Vol. ii., sp. 635, p. 458. $\overline{XXI}$ 3

Like other members of the family this bird deposits its single egg upon the bare sand, but either in a hole or under the shelter of some projecting rock. Two eggs taken by Dr. Metcalfe from some rocks off the coast of Norfolk Island during August 1887, are oval in form slightly pointed at one end, in colour pure white. Length (A) 2·01 x 1·39 inch ; (B) 2·11 x 1·46 inch.

Hab. Norfolk Island.

NECTRIS BREVICAUDUS, *Brandt.*
Short-tailed Petrel.
Gould, Handbk. Bds. Aust., Vol. ii., sp. 636, p. 459.

This bird was likewise found breeding in great numbers during the months of November and December. The mode of nidification is so precisely similar to that of *P. sphenurus,* Gould, that a separate description is not necessary. Like all Petrel's eggs they are white and have a peculiar musky odour, which they always retain, even when empted of their contents and kept for many years. Only one egg is laid for a sitting ; six specimens measure as follows :—length (A) 2·63 x 1·78 inches ; (B) 2·8 x 1·73 inches; (C) 2·78 x 1·8 inches ; (D) 2·65 x 1·81 inches ; (E) 2·82 x 1·72 inches ; (F) 2·87 x 1·81 inches.

Hab. Lord Howe Island, and South Pacific.

PHÆTON RUBRICAUDA, *Bodd.*
Red-tailed Tropic-bird.
Gould, Handbk. Birds Aust., Vol. ii., sp. 660, p. 501.   XIX /.

This bird is found breeding during November and December, its single egg is laid under the shelter of projecting ledges, of almost inaccessible rocks, on the face of cliffs, and are consequently very difficult to procure. The eggs are oval in form, being thickest at the centre and tapering slightly at one end, of a dull reddish-brown colour, which is nearly obscured by minute freckles and spots of purplish-brown and grey, in some instances they are blotched and smeared, not unfrequently on the smaller end. Two specimens obtained are nearly white, and entirely devoid of markings. Length (A) 2·6 x 1·85 inches; (B) 2·65 x 1·9 inch ; (C) 2·78 x 1·95 inch.

Hab. Lord Howe, Norfolk and Solomon Islands, and South Pacific.

SULA CYANOPS, *Sundevall.*
Masked Ganet.

Gould, Handbk. Bds. Aust., Vol. ii., sp. 662, p. 506.

The Masked Gannet was found breeding from September to December ; little or no attempt is made at forming a nest, the eggs, two in number, usually being deposited on the bare ground, when newly laid they are of a pale greenish-white colour, which in most instances is covered with a thick coating of lime ; after being sat upon for a few days the eggs become soiled and assume a dirty brown hue. In form they vary from short to long ovals. Length (A) 2·47 x 1·84 inches ; (B) 2·62 x 1·48 inches ; (C) 2·47 x 1·9 inches ; (D) 2·65 x 1·81 inches ; (E) 2·6 x 1·87 inches ; (F) 2·57 x 1·9 inches.

Hab. Lord Howe, and Norfolk Islands, and the South Pacific generally.

INDEX.

ERRATA.

Page 16, line 18, *for* " sexeral" *read* " several."

" 48, " 22, " " PARDALOTINÆ " *read* " PARDALOTIDÆ."

" 159, " 3, " " Horsefield's" *read* " Horsfield's."

" 191, " 26, " " Fower" *read* " Flower."

" 192, " 14, " " basilis " *read* "basalis."

" 222, " 15, " " SANGUINOLENTA " *read* " SANGUINEOLENTA."

" 227, " 23, " " Cacomantus" *read* " Cacomantis."

" 243, " 17, " " CULCULIDÆ" *read* " CUCULIDÆ."

" 246, " 31, " " Sericornis " *read* " Smicrornis."

APPENDIX II.

THE delay caused in printing off the Plates has been instrumental in furnishing additional information on the nidification of some of the foregoing species, and descriptions of nests and eggs of others not previously given ; as heretofore the names of correspondents or collectors from whom the specimens were obtained will be found prefixed to each description. The early and exceptionally wet period in New South Wales during last year (1889) rendered it one of the best seasons for collecting since 1870, an advantage which Mr. K. H. Bennett, of Yandembah, availed himself of, and procured the nests and eggs of several rare species of water-fowl, as well as making some interesting notes relative to their breeding habits. The nests and eggs of two species were also obtained last year which had not been previously recorded from Australia, viz., *Ibis falcinellus*, and *Sternula sinensis*.

In some parts of Australia the breeding time is greatly influenced, either by periods of excessive drought, when many species do not visit their usual breeding grounds at all, or by the rainy season being early or late, the birds, more especially waterfowl commencing to breed immediately after the autumnal rains ; the breeding season however is more constant in the southern portions of Australia. In every instance where the breeding time is given in the preceding pages unless otherwise stated it applies only to the normal season.

ASTUR RADIATUS, *Latham.*
Radiated Goshawk.

Gould, Handbk. Bds. Aust., Vol. i., sp. 16, p. 40.

In 1883 Mr. George Barnard of Coomooboolaroo, Dawson River, Queensland, obtained a nest of this species in a lofty *Eucalyptus*, locally known as the "Moreton Bay Ash," the nest was about thirty feet from the ground, and was similar to that of *A. approximans*, a large structure composed of sticks, lined inside with finer ones and *Eucalyptus* leaves, and contained two eggs, one of which was forwarded to Dr. Ramsay, and described by him in the Proceedings of the Linnean Society of N.S.W.* Mr. Barnard has since forwarded to me the other egg for description, it is rounded in form, and is of a uniform dull bluish-white, but unlike the one previously sent, it is entirely free from smears and blotches of blackish-brown ; length 2·18 inch; breadth 1·83 inch.

Mr. Barnard informs me that his sons obtained another nest on the 27th of August, 1889, "built on a flat fork of a projecting limb of a lofty *Eucalyptus citriodora*, at least fifty feet from the ground, the eggs of which, two in number have a decided bluish tinge." In a subsequent letter Mr. Barnard writes as follows :— "A rather singular occurrence took place about the Radiated Goshawk's nest, when my sons found it there were two eggs in it, and one of them shot the male ; about a month after being up that way again, one of them climbed the tree and found another egg in the nest, laid after the first eggs were taken and the male bird shot."

Hab. Wide Bay District, Dawson River, Richmond and Clarence River Districts, New South Wales, Interior. (*Ramsay.*)

MILVUS AFFINIS, *Gould.*
Allied Kite.

Gould, Handbk. Bds. Aust., Vol. i., sp. 21, p. 49.

Mr. K. H. Bennett, of Yandembah, New South Wales, has kindly communicated the following notes relative to the nidification of this species :—

---

* P.L.S., N.S.W., 2nd Ser., Vol. i., p. 1141.

" Up to the present season (1889) the instances of this bird breeding here have been very few, in fact only in two cases has it come under my notice during a residence in this locality of over twenty-five years, but the present season has been an exception, for within a radius of a few miles I have in the past three or four months found no less than four nests ; the birds also have been much more numerous than they have been for many years previous, indeed for several preceding the present one they have been entirely absent.   The first nest I found on the 8th of October, the last one on the 20th of December.

" The nest is a rough structure, very similar to that of *Circus assimilis*, Jard. & Selby, composed outwardly of sticks, and in the four I have examined, lined with small pieces of sheep's skin with the wool on, picked up from carcasses of dead sheep scattered over the plains.   The nests are placed as a rule in the tops of Pine trees *Callitris* sp., where the topmost branches divide forming a three or more pronged fork or division, which securely holds the rough flat structure in position.   In two instances this year the disused nests of *Hieracidea orientalis*, were taken possession of, (from one of which in October last I took the *Hieracidea's* eggs) the only additions being the pieces of sheep's skin lining.   The number of eggs for a sitting never so far as my experience goes exceeds two. As a rule the prey of this bird consists of insects, small reptiles, &c., to which offal is added whenever obtainable, but this year the prey judging from the quantity of remains in the nests as well as on the ground beneath consists chiefly of rabbits of all sizes, which considering the comparative weakness of this bird's talons is somewhat singular."

Two eggs in Mr. Bennett's collection, taken on the 8th of Oct. are of a dull white on the outer surface, green inside when held up to the light, one specimen (A) is ovoid in form, and has hair lines, freckles and dots of rusty-brown scratched over the surface of the shell, but particularly towards the larger end where a few irregular shaped blotches appear; the other specimen (B) is rounded in form, and is more sparingly and finely marked with the same colour, but has no blotches, and the markings are confined principally to the smaller end. Length (A) 2·07 x 1·57 inch ; (B) 2·05 x 1·55 inch.

Another set taken by Mr. Bennett on the 28th of November, the female of which was shot from the nest, are rounded in form, of a dull white, one specimen (A) being blotched towards the smaller end with dark umber-brown, freckles and dots of the same colour being evenly dispersed over the remainder of the shell ; the other specimen (B) is thickly mottled and marked all over, but particularly towards the larger end with indistinct fleecy markings of dull reddish-brown, where also a few obsolete markings of dull purple appear. Length (A) 1·87 x 1·58 inch : (B) 1·83 x 1·56 inch. In the eggs of this species there is great variation in their size, shape, and the colour and disposition of their markings. They closely resemble some of the varieties of *Aquila morphnoides*, and *Haliastur sphenurus*, but are much smaller.

Hab. Derby, N.W. Australia, Cape York, Rockingham Bay, Port Denison, Wide Bay District, Dawson River, Richmond and Clarence River Districts, New South Wales, Interior, Victoria and South Australia, West and South-west Australia. (*Ramsay.*)

NINOX MACULATA, *Vigors & Horsfield.*
Spotted Owl.

Gould, Handbk. Bds. Aust., Vol. i., sp. 37, p. 76.

For an opportunity of examining an egg of this species I am indebted to Mr. E. D. Atkinson of Table Cape, Tasmania. Like its near ally *N. boobook*, it deposits its eggs in the hollow limb of a tree. The egg is rounded in form, dull white, the texture of the shell being fine and slightly glossy ; in the specimen examined a few small limy excrescences appear on either end of the shell. Length 1·58 x 1·37 inch. Mr. Atkinson states the above specimen is one of a set of four taken by Mr. Massey at Bridgewater, Tasmania, on the 7th of October, 1886.

Hab. Wide Bay District, New South Wales, Victoria and South Australia, Tasmania. (*Ramsay.*)

CHERAMŒCA LEUCOSTERNUM, *Gould*.

White-breasted Swallow.

Gould, Handbk. Bds. Aust., Vol. i., sp. 57, p. 115. ___XIII___ /5.

Mr. Edward Lord Ramsay informs me that during several years residence on Wattagoona Station, near Louth, in the interior of New South Wales, he found many nests of this species. In favourable situations they breed in small communities, boring a tunnel from eight inches to two feet in length in the loose loamy soil of the bank of a dry creek or dam, at the extremity of which a chamber is hollowed out, and on the bottom a small saucer-shaped nest is formed of a thick layer of dead "Mulga" leaves (*Acacia aneura*). In a number of nests examined five eggs was the usual number laid for a sitting, in one instance only did he find a nest containing six. The eggs are pure white and invariably true ovoids in form. The set of six referred to, taken on the 28th September, 1887, measure as follows :—length (A) 0·68 x 0·5 inch ; (B) 0·7 x 0·5 inch ; (C) 0·68 x 0·5 inch ; (D) 0·69 x 0·5 inch ; (E) 0·7 x 0·49 inch ; (F) 0·68 x 0·51 inch.

Another set of four taken by Mr. Ramsay on the same date, measure (A) 0·67 x 0·5 inch ; (B) 0·67 x 0·48 inch ; (C) 0·67 x 0·48 inch ; (D) 0·67 x 0·49 inch.

Hab. New South Wales, Interior, South Australia, West and South-west Australia. (*Ramsay.*)

ALCYONE DIEMENENSIS, *Gould*.

Tasmanian Kingfisher.

Gould Handbk. Bds. Aust., Vol. i., sp. 70, p. 141.

Mr. Atkinson of Tasmania, has kindly forwarded the eggs of this species for description with the accompanying data :—

"The eggs of *Alcyone diemenensis*, were taken on the 11th of January, 1890, by Mr. M. Ford, who lives near me, from a hole in the bank of Seabrook Creek in this neighbourhood ; I have measured the hole, finding it including the chamber at the end

twenty inches in length ; there was nothing in the way of a nest,
only the bare ground and a few small fish bones and scales, it
contained two eggs perfectly fresh.　This is late in the season."

The above eggs are nearly round in form, pearly-white, the texture of the shell fine, and the surface glossy.　Length (A) 0·92 x
0·77 inch ; (B) 0·93 x 0·8 inch.

Dr. L. Holden of Circular Head, has also forwarded a description as follows :—" Once only have I found the nest of *Alcyone
diemenensis*, and that on the 29th of October, 1889, near the
mouth of Detention River, on the North-west coast of Tasmania ;
the nidification in all respects is like that of the English Kingfisher,
a hole in a bank over a river and a scanty collection of tiny fish
bones at the end of it.　The hole would not admit my hand without enlargement and sloped a little upwards, the nest chamber
was the length of my forearm from the orifice, and might hold
two small fists.　The nest contained six hard set nearly round,
glossy-white eggs."

Hab. Tasmania.

PARDALOTUS ORNATUS, *Temminck.*

(P. striatus, Vigors & Horsfield.)

Striated Pardalote.

Gould, Handbk. Bds. Aust., Vol. i., sp. 84, p. 161.

This bird was found breeding freely both by Mr. James Ramsay,
and Mr. Edward Lord Ramsay, on Wattagoona Station near Louth,
a variety of situations being chosen as a site for its nest, about
the buildings they often took possession of the deserted nest of
Lagenoplastes ariel, and on one occasion constructed their dome-
shaped nest between the ceiling and roof of the house ; another
pair worked very assiduously for some time at the mortar in a
crevice of the stonework, but finally had to abandon it, in fact any
situation is utilized by this bird where it is possible to construct
its nest under cover.　In the paddocks Mr. E. L. Ramsay, obtained

their nests from the sites usually chosen by this bird, the hollow limbs of trees, and on several occasions found them breeding in company with *Cheramœca leucosternum*, in a hole in the side of a bank of a creek, they prefer however to tunnel a hole where the earth is harder than the site usually chosen by the White-breasted Swallow for its nest. When resorting to the bank of a creek Mr. Ramsay informs me the nest is cup-shaped, with a short spout and is composed entirely of wiry rootlets and grasses, neither bark or feathers being used as when placed in the hollow limb of a tree, and that the burrows of the Pardalote can easily be detected from those of the White-breasted Swallow by being smaller and rounder.

The eggs are ovoid in form, and pure white ; a set of four taken by Mr. E. L. Ramsay on the 6th November, 1889, from the end of a tunnel two feet six inches in length in the bank of a creek in Wattagoona horse paddock, measure as follows :—length (A) 0·72 x 0·56 inch; (B) 0·7 x 0·55 inch ; (C) 0·71 x 0·57 inch ; (D) 0·72 x 0·55 inch.

Hab. Port Denison, Dawson River, New South Wales, Interior, Victoria and South Australia, West and South-west Australia. (*Ramsay.*)

COLLYRIOCINCLA PARVISSIMA, *Gould.*
Smaller Rufous-breasted Thrush.

Gould, Ann. & Mag., Nat· Hist., Vol. x., p. 114.   *VIII*   /

Mr. J. A. Boyd of the Herbert River, Queensland, has kindly sent the following notes relative to the nidification of this species together with the bird and two sets of the eggs for description :— "The nest of this species internally is cup-shaped and is a sub-stantially built structure, composed principally of fibrous bark with a few leaves woven in, it is usually built in a fork about a couple of feet from the ground, I saw one however about eight feet up in a *Mango ;* the last nest I found was in a stunted *Dracœna.* This season (1889-90) *C. parvissima*, commenced to breed in September, the first nest I found being on the 14th of that month, which contained three eggs, and the last with two fresh eggs in it on the 2nd of January."

Eggs three in number for a sitting, ovoid in form, pearly-white one set being evenly spotted over the surface of the shell with umber-brown and slaty-grey markings. Length (A) 0·98 x 0 74 inch ; (B) 1 x 0·73 inch; (C) 0·97 x 0·73 inch. In the other set the markings are mostly confined to the larger end, being heavily blotched with umber-brown and superimposed markings of dark slaty-grey.

Hab. Gulf of Carpentaria, Cape York, Rockingham Bay, South Coast of New Guinea. (*Ramsay.*)

HETEROMYIAS CINEREIFRONS, *Ramsay.*
Ashy-fronted Flycatcher.
" *Win-dan.*" Aborigines of Cairns District. •

Ramsay, Proc. Zool. Soc., 1875, p. 588.

During September and October of 1889, several nests of this species were obtained by Messrs. Cairn and Grant, in the scrubs of the Herberton tableland, in every instance they were found in the " lawyer vines " (a species of *Calamus*), about four or five feet from the ground, several of these nests now before me are built between the forked stems, or where several vines cross each other, in other instances they are placed at the base of leaves on the thin horizontal stems, to which the nest is attached. The outside of the nest is formed of thin twigs, wiry rootlets, skeletons of leaves, and the fibre of the " lawyer vine," the inside which is cup-shaped, being neatly lined with finer materials, while the exterior portion of the nest is ornamented with mosses and lichens, which gives" it a pleasing appearance. Exterior diameter 4·5 inches, depth 4 inches, internal diameter 2·75 inches, depth 1·1 inch. The eggs are two in number for a sitting, and closely resemble in shape and colour large specimens of *Artamus superciliosus*, being of a dull buffy-white ground colour, thickly covered, especially towards the larger end with clouded markings of umber-brown, in some instances they are more clearly defined and boldly blotched, and have markings of deep bluish-grey appearing as if beneath the surface of the shell. A set taken on the 18th September measures as follows :—Length (A) 1·05 x 0·75 inch ; (B) 1·07 x 0·77 inch.*

Hab. Rockingham Bay. (*Ramsay.*)

ACANTHORNIS MAGNA, *Gould.*

Gould, Handbk. Bds. Aust., Vol. i., sp. 228, p. 373.

Mr. E. D. Atkinson has sent the following notes relative to the nidification of *A. magna*, together with the eggs :—" A nest of this species found by Mr. G. H. Hinsby on the 29th of October 1886, at Kangaroo Valley about five miles from Hobart, was nearly round in form with an entrance in the side, and similar to that of *Sericornis humilis;* it was outwardly composed of strips of bark, dried grasses, and leaves, being neatly lined inside with feathers, and hair, and was placed in a low " Native Currant" bush (*Coprosma microphylla.*)"

Eggs three in number for a sitting ; an egg taken from the above nest is rather swollen in form, being thickest in the centre and tapering gradually towards each end, which are nearly equal in size ; it is white with fine freckles of dull red particularly towards one end where they form an irregularly shaped zone. Length 0·71 x 0·56 inch in width. Another egg taken by Mr. Hinsby's brother is ovoid in form, white with light red and reddish-brown markings which are mostly confined towards the larger end of the egg. Length 0·75 x 0·54 inch. In the position and construction of its nest this bird approaches that of the genus *Sericornis*, but the egg is like that of the typical *Acanthizæ.*

Hab. Tasmania.

---

* North, Records Aust. Mus., Vol. i., pt. i., p. 37.

NEOCHMIA PHAETON, *Hombron & Jacquinot.*
Crimson Finch.
Gould, Handbk. Bds. Aust., Vol. i., sp. 256, p. 415.

Mr. Boyd has kindly forwarded a set of the eggs of *N. phaeton* for description, together with the following note :—

"After several attempts resulting either in young birds or empty nests I obtained last Monday, December 9th, a nest of *N. phaeton*, containing eight eggs, all more or less incubated, seven of which I emptied successfully. These Finches seem to build exclusively among the leaves of the *Pandanus* trees this season ; the nest is of a dome-shaped form and is composed of dry blades of grasses lined with downy tops of grass seeds and a few feathers."

The eggs are white, varying in form from ovals to lengthened ovals, slightly pointed at one end.   Length (A) 0·65 x 0·45 inch ; (B) 0·65 x 0·45 inch ; (C) 0·65 x 0·46 inch ; (D) 0·7 x 0·43 inch; (E) 0·68 x 0·45 inch ; (F) 0·65 x 0·47 inch ; (G) 0·64 x 0·45 inch.

Hab. Derby, N.W. Australia, Port Darwin and Port Essington, Gulf of Carpentaria, Cape York, Rockingham Bay, Port Denison, South Coast New Guinea. (*Ramsay.*)

GLYCIPHILA FASCIATA, *Gould.*
Fasciated Honey-eater.
Gould, Handbk., Bds. Aust., Vol. i., sp. 303, p. 499.   *XIII. 9.*

From Mr. George Barnard of Coomooboolaroo, Dawson River, Queensland, I have received the following note relative to the nidification of this species, together with the eggs :—"The nest of *G. fasciata*, is a large dome-shaped structure, with a hole in the side, and is composed entirely of the papery-like bark of the *Melaleuca*, coarse strips outside and finer inside, and is fastened to the thin twigs of the same species of tree overhanging and within three or four feet of the water, always as far as we have found over a waterhole.   The breeding season commences late in the month of November."

It will be observed that both *G. fasciata*, and *G. modesta*, whose habitat is confined to Northern Australia, build dome-shaped nests, while *G. fulvifrons*, and *G. albifrons*, which are found in the southern portions of the continent, build open cup-shaped structures, the former deep cup-shape, the latter very shallow.

The eggs of *G. fasciata*, are two in number for a sitting, in form elongated ovoids, white with innumerable freckles and dots of light reddish-brown, which in some instances predominate towards the thicker end where they become larger and confluent, and form an ill-defined zone. Length (A) 0·8 x 0·55 inch ; (B) 0·78 x 0·53 inch. In another set the markings are paler, larger, and more bran like, and are sparingly distributed over the shell, but predominate as usual towards the thicker end.

Hab. Port Darwin and Port Essington, Gulf of Carpentaria, Cape York, Rockingham Bay, Port Denison, Dawson River. (*Ramsay.*)

GLYCIPHILA MODESTA, *Gray.*
(G. subfasciata, Ramsay.)
Plain-coloured Honey-eater.
Gray, *Proc. Zool. Soc.*, 1858, pp. 174, 190. $\overline{XIII}$ *10.*

Mr. J. A. Boyd has forwarded me several sets of the eggs of this Honey-eater, together with the following note :—" The nests of this species on the Herbert River are always built in *Melaleuca* swamps ; they are hanging dome-shaped structures with a small verandah or hood over the opening in the side, and are composed entirely of the paper-like bark of the *Melaleuca*. This bird breeds from September to the end of February." Eggs two in number for a sitting varying in form from ovoid to lengthened ovoids, white, with very minute but distinct purplish black dots scattered over the surface of the shell predominating as usual towards the larger end. Three sets measure as follows :—length (A) 0·8 x 0·53 inch, (B) 0·78 x 0·53 inch ; (C) 0·83 x 0·5 inch ; (D) 0·8 x 0·51 inch ; (E) 0·78 x 0·52 inch ; (F) 0·8 x 0·55 inch.

Hab. Cape York, Rockingham Bay, South Coast New Guinea. (*Ramsay.*)

PTILOTIS LEUCOTIS, *Latham.*

White-eared Honey-eater.

Gould, Handbk. Bds. Aust., Vol. i., sp. 311, p. 510.

A nest of this species, in the Australian Museum Collection, taken by Dr. Hurst at Cabramatta, New South Wales, on the 1st of September, 1888, is a deep cup-shaped structure composed of strips of bark, bark fibre, and spider's cocoons matted up together, and lined inside at the bottom with cow-hair; it measures exteriorly three inches and a half in width by three inches and a half in depth. The eggs were two in number, fleshy-white, with small reddish-chestnut dots and spots sparingly scattered over the larger end of the egg. This nest is similar to others I have seen, but it was placed unusually high, being attached to the topmost leafy twigs of a *Melaleuca,* about eighteen feet from the ground.

At Dobroyde I have also observed that the Yellow-breasted Robins, *Eopsaltria australis,* towards the latter end of last season, probably after having been repeatedly robbed, had taken to build their nests on the horizontal boughs of the *Eucalyptus,* and *Syncarpia* at a height from twenty to thirty feet. This bird usually places its nest within a few feet of the ground.

Hab. Derby, North-West Australia, Port Darwin and Port Essington, Gulf of Carpentaria, Wide Bay District, Richmond and Clarence Rivers Districts, New South Wales, Interior, Victoria and South Australia, West and South-West Australia. (*Ramsay.*)

PTILOTIS FLAVA, *Gould.*

Yellow Honey-eater.

Gould, Handbk. Bds. Aust., Vol. i., sp. 317, p. 518.

Mr. J. A. Boyd has kindly forwarded the nest and eggs of this species, which he found breeding on his plantation on the Herbert River, Northern Queensland, on January 16th, 1890. The nest is of the usual cup-shaped form built by the members of this genus, and is mostly composed of the hair-like fibre of the Cocoanut palm, with a few narrow strips of bark and spider's webs; it measures exteriorly three inches in diameter by two inches and three quarters deep, internally two inches and a half, by two inches deep. The nest was suspended by the rim to the thin leafy twigs of a "Cumquat" orange tree, one of the leaves being worked into the side, and was within hand's reach of the ground. Eggs two in number for a sitting, ovoid in form, of a pale reddish-white ground colour; one specimen being thickly blotched towards the larger end with reddish chestnut markings, the remainder of the surface being sparingly but evenly marked with spots and dots of the same colour; the other has an obsolete band of confluent purplish-grey blotches on the larger end with a few markings of deep reddish-chestnut on the exterior portion of the shell, several penumbral blotches and minute dots of the same colour appearing towards the smaller end. Length (A) 0·87 x 0·62 inch; (B) 0·9 x 0·65 inch.

Mr. Boyd states, "All the nests taken by me were mostly composed of Cocoanut fibre. I cannot say what material this bird used for building its nest before Cocoanut trees were planted here, but it could easily obtain supplies from decaying Palms and wild Bananas. Two nests were built in a species of *Ficus*, and were eighteen feet from the ground; another was built in a *Mango* tree about eight feet from the ground."

Mr. Boyd obtained a nest of this species on the 10th December, containing one young one and one perfectly fresh egg.

Hab. Gulf of Carpentaria, Cape York, Rockingham Bay, Port Denison. (*Ramsay.*)

MELITHREPTUS MELANOCEPHALUS, *Gould.*
Black-headed Honey-eater.
Gould, Handb. Bds., Austr., Vol. i., sp. 352, p. 573. XII /8.

Dr. Holden, of Circular Head, Tasmania, writes as follows :—
"I have found this Honey-eater's nest in December. One was

commenced on the 7th of that month, and had three fresh eggs in
it on the 25th ; another nest found on the 8th of December had
three half fledged ones in it, and one found on the 31st December,
1888, had been lately left by a brood of young.   Another
nest at Circular head was taken on the 27th November, containing
three newly hatched young.

"All the nests were built among the leaves of small gum
trees, and were fastened to the leaves and their stems at the rim
and sides at a height from twenty-five to fifty feet from the ground
which renders them very difficult to procure, being at the end of thin
branches.   One nest was made of green moss, mixed with wool,
firmly felted together and lined with a little hair, on the exterior
a little lichen and cobwebs ; the nest was a deep cup-shaped
structure, and measured externally three inches and a half in
length by three inches in width, depth inside two inches and a
quarter.

"Another nest had no wool, and was chiefly composed of green
moss and cobwebs, with a lining of flower seeds.   The colour of
the nests renders them very difficult to be found among the leaves
to which they are attached.

"The eggs are flesh coloured, with a ring of a darker tinge at
the thicker end, and a good many dull red spots, all more or less
in a ring together with a few faint purplish spots.   The eggs are
in colour more like those of *Lichmera* than of *Meliornis*, which
are the two common Tasmanian Honey-eaters, but the reverse as
to shape.   They are not easily distinguished from those of either
of these birds."

Eggs three in number for a sitting, an average specimen received
from Mr. E. D. Atkinson of Table Cape, Tasmania, agrees very
well with the description given by Dr. Holden, but in this specimen the markings are more evenly distributed towards the larger
end, and do not assume the form of a zone.   Long diameter 0·72   *13,3*
inch, short diameter 0·53 inch.

Hab. Tasmania.

CINNYRIS FRENATA, *Müller.*

Australian Sun-bird.

Gould, Handb. Bds. Austr., Vol. i., sp. 359, p. 584.

Regarding the nidification of this species, Mr. Boyd under date 31st December, 1889, writes as follows :—" We have on the estate three houses with verandahs, and in each verandah a pair of *Cinnyris*, have built ; it is strange why this little bird should seek man's society, one pair has bred for years in a verandah nearly always occupied by three children and four kangaroo dogs. One pair that for the last two seasons have built by the side of the house, came round to the front door on the 23rd November and selected a piece of rope that pulled up the bamboo verandah blind, and began building. I at once nailed the rope so that it could not be moved, and have since kept them under observation. Their first proceeding was to cover the cord for about eighteen inches with a layer of bark, cobweb, moss, &c., until it was about two inches in thickness ; on the 28th the bottom of the nest and the little verandah were begun, and with the sides were almost completed on the following day. On the 5th December I saw the female in the nest, on the 17th I looked in the nest and saw two eggs, on the 21st there were young ones.''

Mr. Boyd informs me in a subsequent letter that the young birds left the nest on the 4th January, which was forty-three days from the date of commencing the nest.

This bird usually selects the twigs of a low shrub as a site to attach its domed and hood-covered nest.

Hab. Cape York, Rockingham Bay, Port Denison, South Coast New Guinea. (*Ramsay.*)

ORTHONYX SPALDINGI, *Ramsay.*

Spalding's Orthonyx.

" *Chowchilla.*" Aborigines of Cairns District.

Ramsay, Proc, Zool. Soc., 1868, p. 386.

This species has recently been met with rather freely dispersed through the dense brushes of the coastal range, chiefly in the neighbourhood of the Mulgrave and Russell Rivers, in North-Eastern Queensland. Mr. Cairn, who found several nests of this species, states they are usually built in the tangled roots of "lawyer vines," but not unfrequently on the top of the elk's-horn fern, as high as twelve feet from the ground. The nest is a large bulky dome-shaped structure with an entrance on one side, it is composed of twigs, roots and mosses, chiefly a species of *Hypnum*, so loosely put together that it will not bear removal. Unlike its southern ally *O. spinicaudus*, it appears that only one egg is laid for a sitting. A nest found near "Boar Pocket," on the 20th June last, contained but one egg in an advanced state of incubation, others were found as late as the middle of August. The breeding season this year (1889) would appear to be from May till the end of September, young birds being procured in June, but as in other parts of Australia the breeding season of birds is greatly influenced by the rains. The eggs, which are pure white, vary from elongated to swollen ovals, some being equal in size at each end. Two average sized specimens measure :—(A) 1·45 x 1 inch ; (B) 1·38 x 1·1 inch.*

Hab. Rockingham Bay. (*Ramsay.*)

TRICHOGLOSSUS CHLOROLEPIDOTUS, *Kuhl*
Scaly-breasted Lorikeet.

Gould, Handbk. Bds. Aust., Vol. ii., sp. 446, p. 96.

Mr. George Barnard of Coomooboolaroo, writes as follows relative to the nidification of this species :—" Last season (1889) my sons found two nests of *Trichoglossus chlorolepidotus*, each with two eggs in them, these are the only occasions that I have known the nests to contain more than one egg for a sitting, all other nests (seven in number) found previously of this species had only one egg in, sometimes fresh, at other times heavily incubated, it may have been accidental and struck me as being strange at the time, but I think two eggs must be the proper number for a sitting."

---

* North, Rec. Aus. Mus., Vol. i., part i., p. 38.

Hab. Rockingham Bay, Port Denison, Wide Bay District, Dawson River, Richmond and Clarence Rivers District, New South Wales, Interior, Victoria and South Australia. (*Ramsay.*)

EUDROMIAS AUSTRALIS, *Gould.*
Australian Dotterel.

Gould, Handbk. Bds. Aust., Vol. ii., sp. 505, p. 227.   *XVI* 3.

Mr. K. H. Bennett has sent the following notes upon the nidification of this species :—" 26th April; 1889.—Found to-day a nest of *Eudromias australis,* containing three eggs ; this is unusually early, for hitherto I have never known this bird to breed before September or October. The eggs were placed on a small natural mound of earth some four or five inches in diameter, and about the same height above the surrounding ground, and were completely covered with small dried sticks some two or three inches in length. I disturbed the bird from the nest on which she was sitting, and noticing only the sticks I at first thought that in consequence of the ground all round being covered in water to the depth of two or three inches—the result of recent heavy rains—that the bird in this particular instance had departed from the usual custom, and had constructed a kind of nest, and that she had not yet deposited her eggs, but on closer examination I found the eggs on the bare ground, and that the sticks had been placed carefully over them as a safeguard from the keen-eyed Crow, as whenever the old bird should leave her nest without this covering, situated as they were, they would have been very conspicuous, as the little mound in which they were placed was the only dry spot for fifty or sixty yards around."

In his notes Mr. Bennett also records finding another nest of this species with two eggs in it on the 29th April, covered in a similar manner with small sticks, and another on the 3rd May with two eggs ; in the latter instance they were not covered, but were simply deposited on the loose earth on high dry ground.

Hab. New South Wales, Interior, South Australia. (*Ramsay.*

IBIS FALCINELLUS, *Linnæus.*
(*F. igneus*, G. R. Gray.)
Glossy Ibis.

Gould, Handbk., Bds. Aust., Vol. ii., sp. 540, p. 286.

Mr. K. H. Bennett has forwarded the following notes respecting the nidification of this bird :—" On the 22nd of October, 1889, whilst swimming about in a large depression on the plains, filled with water by the late heavy rains, and thickly overgrown with " Box " trees, (a species of *Eucalyptus*) in quest of the eggs of *Platalea flavipes*, I noticed a Glossy Ibis, (*Ibis falcinellus*) fly off a nest, but as I had never known or heard of this bird breeding here I did not take much notice of the occurrence, thinking that the Ibis had been merely perched on the nest, although I thought at the time that it appeared very different from those of the Herons and Spoonbills. After swimming about for some time and obtaining several Spoonbill's eggs, I returned to land and in doing so passed the tree in which I had noticed the Ibis, and again saw it fly off the nest, and at once concluded it was the nest of the Ibis after all. On ascending the tree (the branch on which the nest was placed being not more than eight or nine feet from the water) I found that such was the case, and that it contained one freshly laid egg, which I unfortunately broke whilst swimming to land. One the 2nd November, I again visited this swamp or depression in the hope of obtaining more Ibises' eggs, and was so fortunate as to obtain six, three of which were from the nest from which I took one on the 22nd ultimo ; to my surprise and gratification on nearing the tree I observed the bird fly off the nest, and on examination I found it contained three eggs. A further search revealed another nest which also contained three eggs, but which are considerably larger than those previously obtained, so much so that had I not seen the bird fly off the nest I should have been in doubt as to their identity, but on this point there was no possibility of mistake, 'for the eggs being in a somewhat advanced stage of incubation, the old bird evinced a great reluctance to quit the nest, and allowed me to approach almost within arm's length before she did so. The two nests were placed in

three or more upright pronged forks of the branches of small
" Box" trees, and were both composed of bunches of "Box"
leaves piled up in the forks to the height of about a foot, the top
being slightly hollowed out, but without any other lining. On
the 26th of November I again visited this swamp and found two
more Ibises' nest, both of which contained young lately hatched,
(one three, the other four) covered with black down. One of the
nests from which I had taken the eggs on the 2nd instant, had in
the meantime been appropriated by the little Pink-eared Duck
(*Malacorhynchus membranaceus,*) and now contained five Duck's
eggs enveloped in the usual manner in a mass of down."

In the letter accompanying the above description Mr. Bennett
writes as follows :—" You will see at the conclusion of the
description, that had I continued my search at the time I found
the eggs, the probability is that instead of getting six eggs I
should have procured thirteen, but I was so benumbed with cold
swimming about for hours with my clothes on that it was with
great difficulty I reached the land, and had I been half an hour
longer in the water, the chances are that the first recording of
this bird's eggs in Australia would have fallen to some other
person."

A set of the above eggs are lengthened ovals in form, and are
of a deep greenish-blue colour, the shell being slightly rough in
texture and lustreless ; they measure as follows, length (A) 1·94
x 1·33 inch ; (B) 1·95 x 1·35 inch ; (C) 1·97 x 1·31 inch. A set
in the Australian Museum taken on the same date, vary from
pyriform to a lengthened oval, one specimen being somewhat
sharply pointed at one end. Length (A) 2·16 x 1·48 inch ; (B)
2·21 x 1·4 inch ; (C) 2·2 x 1·47 inch. The eggs of the Glossy
Ibis can readily be distinguished from those of any other Australian bird, by the intensity and depth of their colouring.

Hab. Port Darwin and Port Essington, Gulf of Carpentaria,
Cape York, Rockingham Bay, Port Denison, Wide Bay District,
Richmond and Clarence Rivers Districts, New South Wales,
Interior, Victoria and South Australia. (*Ramsay.*)

HERODIAS ALBA, *Linnæus.*
Australian Egret.
Gould, Handbk. Bds. Aust., Vol. ii., sp. 549, p. 301.

Mr. E. D. Atkinson of Tasmania, has forwarded a set of the eggs of this bird for description together with the following note: " Mr. John Wright found a small colony of *Herodias alba,* breeding in a species of *Eucalyptus,* overhanging a river on the East coast of Tasmania, during 1883. The eggs I send you were from a nest containing four, one of which was unfortunately broken in transit." The eggs vary considerably in shape, one specimen (A) is an elongated oval tapering slightly towards the smaller end, (B) is nearly a true oval in form, and (C) a swollen oval; they are of a delicate sea-green in colour, one specimen (B) having a slight limy covering on one side, giving the egg a blanched appearance. The surface of the shell is smooth and lustreless, but all have more or less minute indistinct shallow pittings. Length (A) 2·13 x 1·43 inch ; (B) 2·02 x 1·43 inch ; (C) 1·95 x 1·48 inch.

Hab. Derby, N.W. Australia, Port Darwin and Port Essington, Gulf of Carpentaria, Cape York, Rockingham Bay, Port Denison, Wide Bay District, Richmond and Clarence Rivers Districts, New South Wales, Victoria and South Australia, Tasmania, West and South-West Australia. (*Ramsay.*)

DENDROCYGNA VAGANS, *Eyton.*
(D. gouldi, Bonaparte.)
Whistling Tree-Duck.
Gould, Handbk. Bds. Aust., Vol. ii., sp. 591, p. 374.

Mr. George Barnard, of the Dawson River, Queensland, has kindly supplied the following information regarding the nidification of this species :—

"Coming home with cattle on the 25th May, 1890, my sons flushed a Duck of some sort off a nest in the grass too hurriedly to see what it was, they left it till next day when one of them rode out to identify the species, it proved to be a " Whistler,' *D. vagans,* Eyton. The nest was made in the grass, and without any lining of feathers or down, and contained fifteen eggs in an early stage of incubation, several of which he took. This Duck is very common in the neighbourhood, and is found frequenting the large swamps, but this is the first time we have obtained the nest."

The eggs in form, are an ellipse tapering sharply to each end, which are pointed and of equal size. They are of a pale creamy-white, and in the specimens forwarded have light reddish-purple markings on one end appearing as if beneath the surface of the shell, these markings are I think abnormal, one specimen having only a few spots on the side. Length (A) 2·09 x 1·43 inch ; 2·13 x 1·42 inch.

In a subsequent letter, Mr. Barnard writes as follows:—"Nearly all the Whistling-Duck's eggs taken had markings on one end, but most of those left in the nest were without them, I do not think the markings are typical, but only the effect of the season, as I have noticed the markings on the Butterflies and Moths were darker and richer this past season than in ordinary ones."

Hab. Derby, N. W. Australia, Port Darwin and Port Essington, Gulf of Carpentaria, Cape York, Rockingham Bay, Port Denison, Wide Bay District, Richmond and Clarence Rivers District, New South Wales, Interior, South Coast New Guinea. (*Ramsay.*)

SPATULA RHYNCHOTIS, *Latham.*
Australian Shoveller.
Gould, Handbk. Bds. Aust., Vol. ii., sp. 588, p. 368.

Mr. Bennett has found several nests of this species last season, so I take this opportunity of giving the measurements of a full set of the eggs, together with the following extract from one of his letters :—"September 22nd, found nest of *Spatula rhynchotis,* containing eleven eggs, all quite fresh. I shot the female as she flew off. The nest was composed of a few stems of grasses &c., and was placed in a slight hollow in a bunch of herbage on the plain."

In several sets examined they are elongated ovals in form. A set of eleven taken by Mr. Bennett at Yandembah, on the 22nd September, 1889, measures as follows :—Length (A)

1·96 x 1·48 inch ; (B) 1·97 x 1·47 inch ; (C) 1·95 x 1·47 inch ;
(D) 2·1 x 1·5 inch ; (E) 2 x 1·46 inch ; (F) 2 x 1·5 inch ; (F) 2 x 1·5
inch ; (G) 2·03 x 1·52 inch ; (H) 2·07 x 1·52 inch ; (I) 1·95 x 1·47
inch ; (J) 2 x 1·49 inch ; (K) 1·97 x 1·5 inch.

Hab. Wide Bay District, Richmond and Clarence Rivers
Districts, New South Wales, Victoria, South Australia, Tasmania,
West and South-west Australia. (*Ramsay.*)

MALACORHYNCHUS MEMBRANACEUS, *Gould.*

Pink-eared Duck.

Gould, Handbk. Bds. Aust., Vol. ii., sp. 590, p. 372.

The Trustees of the Australian Museum have lately received
from Mr. K. H. Bennett several nests of the Pink-eared Duck,
taken at Yandembah during 1889, one of them is.placed on the
deserted nest of *Geronticus spinicollis*, which was built on the top
of a *Polygonum* bush about eighteen inches above the water. The
nest of *G. spinicollis* is a flat structure composed of thorny sticks
and twigs interlaced through one another, and measures eighteen
inches in width, by five in height. That of *M. membranaceus* is
elliptical in form, and is composed entirely of down plucked from
the breast of the parent bird, and measures twelve inches in width
by five inches in height. To another nest Mr. Bennett has
attached the following note :—"Taken at Yandembah, 26th
August, 1889. The nest is placed on an old disused nest of
Tribonyx ventralis, built on the lower dead horizontal stems of a
Polygonum bush about a foot above the water, and was screened
from view in a great measure by the overhanging green top of
the bush. The eggs were placed as now in the nest and were
completely covered by the down."

The eggs from the above nest are six in number, of a rich creamy-
white, and measure as follows :—Length (A) 2 x 1·43 inch ; (B)
1·97 x 1·47 inch ; (C) 2·02 x 1·45 inch ; (D) 1·98 x 1·47 inch ;
(E) 1·99 x 1·46 inch ; (F) 1·96 x 1·42 inch.

Mr. Bennett also records finding a nest of this species on 30th July in a slight hollow of a stump standing in the water, containing six eggs enveloped as usual in a mass of down, and another on 12th September, in the disused nest of *Corone australis*, on the branches of a tree about twelve or fourteen feet above the water.

Hab. Derby, N.W. Australia, Port Denison, Wide Bay District, Richmond and Clarence Rivers Districts, New South Wales, Interior, Victoria, South Australia, Tasmania, West and South-west Australia. (*Ramsay.*)

ERISMATURA AUSTRALIS, *Gould.*

Blue-billed Duck.

Gould, Handbk. Bds. Aust., Vol. ii., sp. 594, p. 379.

Several instances of this bird having bred in New South Wales have recently come under my notice; during 1888 I saw four young birds that were obtained near Trangie, and this year (1889) Mr. Bennett in his notes records the finding of a nest on the 2nd of November at Yandembah, New South Wales, while engaged in looking for the nests of the Glossy Ibis. Mr. Bennett writes as follows :—" On a further examination of this large sheet of water I discovered a nest of the Blue-billed Duck, *Erismatura australis*, containing fragments of egg shells showing that some had already been hatched, and one egg with a live. young one just upon the point of hatching, so it is evident that this rare duck occasionally breeds here. The nest was a rather neatly made cup-shaped structure, composed of rushes, and was placed in a low *Polygonum* bush a few inches above the water. The old female I disturbed from the nest, whilst the male was a few feet away in company with several newly hatched young ones."

Hab. New South Wales, Victoria and South Australia, West and South-west Australia. (*Ramsay.*)

HYDROCHELIDON HYBRIDA, *Pallas.*
Marsh-Tern.
Gould, Handbk. Bds. Aust., Vol. ii., sp. 610, p. 406.

Mr. K. H. Bennett of Yandembah, has recently obtained a fine and varied series of the eggs of this inland Tern, which he has forwarded with the accompanying interesting notes :—" On the 31st of October I discovered the breeding place of the little Marsh-Tern, *Hydrochelidon hybrida*, in a swamp overgrown with dwarf *Polygonum* bushes. About a week previously when riding around this swamp I was led to the conclusion that these birds intended breeding there, as numbers were flying about above the water whilst many others were perched on the slender tops of the dwarf *Polygonum* bushes which projected a few inches above the water, and I also noticed that several of the birds flying about were carrying rushes in their bills. I made a careful search at the time, but beyond finding a few green rushes placed in a loose careless manner on the top of one of the *Polygonum* bushes, I saw nothing else to indicate that it was a contemplated breeding site. On visiting the place to day I observed numbers of the birds on the tops of the bushes, but not more than one on each bush, whilst numbers were also flying about in an excited manner and as I neared the edge of the swamp, kept up a continuous croaking. On wading in for a closer examination, I found that each bird was sitting on a nest (if nest such a structure could be called) each of which contained from one to three eggs, the latter number apparently being the full set. These nests were simply a few green rushes, in most cases quite flat, and the whole structure rising and falling with the motion of the water, caused by a slight breeze, and it was a mystery to me how the birds managed to leave or return to the nests without knocking the eggs off. Although this swamp is of considerable extent and similar throughout, the breeding place was confined to a space of not more than twenty yards square, showing that like *Sterna anglica*, they breed in companies."

On the 11th of December, Mr. Bennett writes as follows :—
"To day I passed the swamp in which I obtained the Marsh-

Tern's eggs, at the end of October and beginning of November, and noticed they were in far greater numbers than on the previous occasion, and that they were breeding all over the swamp, and had not only constructed fresh nests, but had utilized the ones from which I had taken the eggs, and also the disused ones of *Tribonyx ventralis*, and other birds. I examined a great number of nests all of which contained eggs."

The eggs are two or three in number for a sitting, usually the latter, and vary in shape from oval to pyriform, the ground colour varies from bright green to pale olive brown, but the most usual variety found is of a dull greenish-grey, some specimens being boldly blotched and spotted with penumbral markings of blackish-brown and umber-brown, particularly towards the larger end, others have freckles and dots of the same colour over the entire surface of the shell, in some instances a few large under-lying blotches of sepia appear, others are uniformly dotted and spotted with rounded markings of the same colour appearing as if beneath the shell, the latter variety closely resembling small eggs of *Sarciophorus pectoralis*. Two average sized sets taken on the 8th of November, measure as follows :—length (A) 1·57 x 1·1 inch ; (B) 1·55 x 1·07 inch ; (C) 1·53 x 1·12 inch . (D) 1·51 x 1·11 inch ; (E) 1·53 x 1·05 inch ; (F) 1·48 x 1·07 inch.

Hab. Gulf of Carpentaria, Rockingham Bay, Wide Bay District, Richmond and Clarence Rivers Districts, New South Wales, Interior, Victoria and South Australia, West and South west Australia. (*Ramsay.*)

STERNA FRONTALIS, *Gray.*

(*S. melanorhyncha*, Gould.)

White-fronted Tern.

Gould, Handbk. Bds. Aust., Vol. ii., sp. 604, p. 398.

Mr. Atkinson has forwarded the eggs together with the following note regarding the breeding of this bird :—" The eggs of *Sterna melanorhyncha*, were taken on Actæon Island in D'Entrecasteaux

Channel, South-east Tasmania, by Mr. J. Graves. They were laid just above high water mark, and like our other Terns in a slight hollow in the bare ground. The eggs are two for a sitting, and large numbers of these birds were nesting close together in the same locality." The eggs are lengthened ovals slightly pointed at the smaller end, and are of a stone-grey ground colour, one specimen (A) being thickly covered all over with rounded dots and spots of different shades of olive-brown and dark umber, obsolete markings of the same colour appearing as if beneath the surface of the shell; the other specimen (B) is irregularly blotched and streaked with short wavy markings of the same colour, becoming confluent in some places towards the larger end, all the markings being larger than on the previous specimen, but not so thickly dispersed over the surface of the shell. Length (A) 1·83 x 1·3 inch; (B) 2 x 1·34 inch. Another specimen is smaller, and is of a light coffee-brown, with irregular shaped spots of rich umber-brown scattered over the surface of the shell; a few large blotches and fine streaks, together with obsolete markings of the same colour appear towards the larger end. Length 1·75 x 1·23 inch.

Hab. Derby, N. W. Australia, Port Darwin, Port Essington, Cape York, Rockingham Bay, Port Denison, Wide Bay District, Richmond and Clarence Rivers Districts, New South Wales, Victoria, South Australia, Tasmania. (*Ramsay.*)

STERNULA NEREIS, *Gould.*

Fairy Tern.

Gould, Handbk. Bds. Aust., Vol. ii., sp. 607, p. 402.

An egg of this species taken by Mr. E. D. Atkinson on 14th November, 1889, from a slight hollow in a loose shelly sand bank near the shore of Mosquito Sound, Walker's Island, in Bass's Straits, is similar to that of the following species, *S. sinensis*, but is slightly larger. It is a swollen oval in form, somewhat sharply pointed at the smaller end, of a pale stone-grey ground colour, with minute freckles, dots, spots, and a few penumbral blotches of different shades of umber-brown uniformly dispersed over the surface of the shell, obsolete markings of the same colour, and deep bluish-grey also appear. Length 1·33 x 1·01 inch. Mr. Atkinson obtained another nest in the same locality together with the bird, each of which contained but a single egg, but two is the usual number laid for a sitting,

Hab. Wide Bay District, New South Wales, Victoria, South Australia, Tasmania, West and South west Australia. (*Ramsay.*)

STERNULA SINENSIS, *Gmelin.*

(*S. placens*, Gould.)

Chinese Tern, White-shafted Ternlet, &c.

Gould, Ann. & Mag. Nat. Hist. (4), Vol. viii., p. 192.

This bird was found breeding by Messrs. Grime & Yardley, during a visit to the Tweed River Heads on the 7th October, 1889. The eggs two in number for a sitting were laid in a slight depression in the sand, all the eggs taken at that time being in an advanced stage of incubation; in form they vary from true ovals to swollen ovals, terminating somewhat abruptly at one end, some of them being of a stone-grey ground colour, others a light coffee-brown, with rounded spots and irregular shaped penumbral blotches of umber-brown and dark slaty-grey, the latter colour in some instances predominating and appearing as if beneath the surface of the shell. Four specimens measure as follows:—(A) 1·28 x 0·95 inch; (B) 1·27 x 0·94 inch; (C) 1·18 x 0·95 inch; (D) 1·25 x 0·9 inch. Skins of the parent birds were obtained and forwarded with the eggs for identification.*

Hab. Gulf of Carpentaria, Cape York, Rockingham Bay, Port Denison, Wide Bay District, Richmond and Clarence Rivers Districts, New South Wales, South Coast of New Guinea. (*Ramsay.*)

---

* North, Records Australian Museum, (1890) Vol. I., p. 39.

PRION TURTUR, *Smith.*

Dove-like Prion.

Gould, Handbk. Bds. Aust., Vol. ii , sp. 641, p. 472

Mr. Atkinson has forwarded two eggs of this bird for description together with the following note :—" The eggs of *Prion turtur,* which I send you were taken by Mr. G. H. Hinsby of Hobart, from off the Friar's Rocks at the extreme south of South Bruny Island on the South-east coast of Tasmania, on the 10th and 11th of December 1887. There was a colony of these birds at an elevation of about one hundred feet up the side of the steep little island. The burrows were from one foot to two feet long in the soft earth, and just large enough for one to pass the hand into, and had a chamber at the end about eight inches in diameter, where the single egg laid by this bird for a sitting was deposited. Several of the birds were brought alive to Hobart, identified and then set at liberty."

One egg (A) is in form a slightly lengthened oval ; the other (B) is an ellipse in form, compressed slightly in the centre. The eggs are white, the texture of the shell being fine and lustreless, and have that musky odour peculiar to the eggs of the *Procellaridæ.* Length (A) 1·81 x 1·28 inch ; (B) 1·8 x 1·26 inch.

Hab. Wide Bay District, New South Wales, Victoria and South Australia, Tasmania, West and South-west-Australia. (*Ramsay.*)

NESTS AND EGGS OF BIRDS

FOUND BREEDING ON

LORD HOWE AND NORFOLK ISLANDS.

THESE remote insular dependencies of New South Wales, situated in the Pacific Ocean, possess a great interest to students of Australian Ornithology, as within their limited areas several genera of birds are found that are represented in the Australian and New Zealand regions. Both islands, however, in regard to their avifauna decidedly belong to the Australian region, as only three genera have been met with typical of New Zealand that are not found in Australia, viz. :—*Nestor*, *Notornis*, and *Ocydromus*, the former two however are now extinct, and the genus *Ocydromus* is confined to Lord Howe Island. The genera found in these Islands and represented in Australia, but not in New Zealand, are *Haliætus*, *Haliastur*, *Strepera*, *Pachycephala*, *Myiagra*, *Cacomantis*, *Chalcophaps*, *Onychoprion*, *Anous*, *Gygis*, and *Phæton*, to these may be added *Eurystomus* and *Sula*, which only occur as stragglers in New Zealand. There are no genera peculiar to either Island, but several distinct and well-defined species, a list of which will be found on reference to the Appendix of Dr. Ramsay's "Tabular List of Australian Birds," p. 37.

For information on the habits of the birds found breeding on Norfolk Island, I am indebted to Dr. P. Herbert Metcalfe, the Resident Medical Officer, who has obtained with one or two exceptions the nests and eggs of all birds found breeding there, the eggs of which he has kindly forwarded on loan for description, together with the notes on their nidification and breeding seasons.

The descriptions of the nests and eggs of two birds peculiar to Lord Howe Island are also given, (*Merula vinitincta*, and *Ocydromus sylvestris*,) taken from specimens lately acquired by Mr. T. R. Icely, the Visiting Magistrate, on behalf of the Trustees of the Australian Museum.

HALCYON VAGANS, *Lesson.*
(Var. *norfolkiensis,* Tristram.)
Norfolk Island Kingfisher.

Tristram, Ibis, 1885, p. 49.

Dr. Metcalfe states that this bird breeds during September and the two following months, excavating with its powerful bill a tunnel from eight to twelve inches in length in the side of a bank or in the trunk of a tree fern, at the extremity of which a small chamber is formed, and the eggs laid either on the bare soil or soft pulverised débris. The eggs are four or five in number for a sitting, rounded in form, and pearly-white, the texture of the shell being fine and glossy. An egg taken on the 17th December, 1889, measures 1·01 x 0·85 inch; another taken the day after measures 1·05 x 0·9 inch; while a third specimen taken on 27th December, 1887, is not quite so rounded, and measures 1·1 x 0·88 inch.

Hab. Norfolk Islands.

SYMMORPHUS LEUCOPYGIUS, *Gould.*

Gould, Proc. Zool. Soc., 1837, p. 145.

This species was found breeding by Dr. Metcalfe during the month of September. In its mode of nidification and the number and colour of its eggs, it closely approaches that of the genus *Lalage,* the nest being a round shallow structure, outwardly composed of mosses, lichens, and fibrous roots, lined inside with similar material of a finer description. The eggs are two in number for a sitting, oval in form, compressed slightly towards the thinner end, of a pale green ground colour slightly tinged with grey, with thick irregular shaped longitudinal markings of different shades of olive-brown, and a few minute freckles scattered over the surface of the shell, on the larger end several nearly obsolete markings of dull bluish-grey appear. Length 0·88 x 0·67 inch.

Hab. Norfolk Island,

PACHYCEPHALA XANTHROPROCTA, *Gould.*

Gould, Proc. Zool. Soc., 1837, p. 149.

The nest of the Norfolk Island Thickhead closely resembles that of the well known *P. gutturalis* of Australia. Dr. Metcalfe informs me it is an open cup-shaped structure composed entirely of twigs, and lined inside with dried grasses, and is usually placed in lemon, and other low trees, and that the eggs are three or four in number for a sitting. Of two eggs forwarded by Dr. Metcalfe, one is an elongated oval, of a rich creamy-white ground colour, with spots, freckles and dots of dark umber-brown, and nearly obselete markings, of sepia on the thicker end, becoming confluent and forming a broken zone. Length 1·05 x 0·72 inch. The other specimen is oval and nearly equal in size at both ends, the ground colour being of a darker shade than in the previous specimen, with rich umber-brown spots, freckles and dots, and spots of bluish-grey on the larger end appearing as if beneath the surface of the shell, forming an irregular shaped cap or patch. Length 1 x 0·75 inch. These eggs could not be distinguished from some of the varieties of *P. gutturalis.* September and the two following months constitute the breeding season of this species.

RHIPIDURA PELZELNI, *G. R. Gray.*

Pelzeln's Fantail.

G. R. Gray, Ibis, 1862, p. 226.

The nidification of Pelzeln's Flycatcher is similar to that of other members of the genus *Rhipidura*, found in Australia and elsewhere. Dr. Metcalfe informs me the nest is a small, round, deeply cup-shaped structure, outwardly composed of mosses, fibrous roots and cobwebs, lined inside with finer grasses and a few feathers, and is usually placed on the branch of a low tree. Eggs three or four in number for a sitting, oval in form, white, with dull coloured light brown markings, and several blotches and spots of pale bluish-grey on the thicker end, appearing as if beneath the surface of the shell. Length (A) 0·63 x 0·46 inch; (B) 0·63 x 0·47 inch.

Hab. Norfolk Island.

GERYGONE MODESTA, *Pelzeln.*
The Ashy-fronted Gerygone.
Gray, Ibis, 1862, p. 221.

The nidification of this species is somewhat similar to other members of the same genus found in Australia, New Zealand, and the Austro-Malayan region. The nest is a pointed oval in form, with an entrance in the side near the top, usually protected with a hood; it is composed exteriorly of mosses, grasses, fibrous roots, spiders' webs and nests, and warmly lined inside with feathers, and is suspended by the top to the end of a slender hanging branch. Eggs two to four in number for a sitting, closely resembling those of the Australian species, *G. fusca*, they are oval in form, one specimen sent being a rosy-white, thickly covered over the entire surface of the shell with minute red and pink-red markings, length 0·64 x 0·48 inch; another specimen is of a delicate white, with short scratchy markings more sparingly distributed, but in some places confluent, length 0·65 x 0·48 inch.

Hab. Norfolk Island.

PETRŒCA MULTICOLOR, *Gmelin.*
Norfolk Island Robin.
Gould, Handbk. Bds. Aust., Vol. ii., App., sp. 2, p. 526.

From Dr. Metcalfe's description of the nidification of the Norfolk Island Robin, it will be seen that the nest of this species is precisely similar to that of *P. leggii* and *P. phœnicea*, of Australia and Tasmania. The nest is a round deeply cup-shaped structure, composed exteriorly of mosses, fibre, and dried grasses, beautifully woven together and lined inside with hair, and is placed between and built around a three pronged upright branch, about four feet from the ground. Eggs three or four in number for a sitting, a specimen forwarded is oval in form, of a dull greyish-white ground colour, with irregular shaped markings, freckles, and dots of different shades of greyish and wood-brown, with obsolete markings of dull bluish-grey, the markings as usual predominating on the thicker end. The entire appearance of the egg is dull and lustreless. Length 0·76 x 0·58 inch. September and the two following months constitute the usual breeding season of this species.

Hab. Norfolk Island.

MERULA POLIOCEPHALA, *Latham.*
Grey-headed Ouzel.
Gould, Handbk. Bds. Aust., App., sp. 4, p. 528.

The members of this genus enjoy a wide geographical range, being found in both the northern and southern hemispheres; although different species are distributed throughout the Pacific region, no member of the genus is found in Australia or New Zealand. The Norfolk Island Ouzel breeds from August until December, constructing a nest similar to other members of the genus *Merula*, an open cup-shaped structure, and placed in banks, the tops of stumps or branches of trees, at a height varying from twelve to sixteen feet from the ground. The eggs vary from two to four in number for a sitting. Dr. Metcalfe has forwarded three specimens taken from different nests, one of them is oval in form, having a pale greenish-grey ground colour, with reddish chestnut markings, freckles and dots of light purple appearing as if beneath the surface of the shell. Length 1·18 x 0·83 inch; another specimen (B) inclines to an elongate oval in form, and has a light isabelline ground colour, with irregular shaped chestnut markings, and nearly obsolete freckles and dots of light purple, length 1·21 x 0·82 inch; (C) is an elongate oval of a pale green ground colour, almost obscured by numerous markings of chestnut brown, thickly and uniformly dispersed over the surface of the shell, length 1·19 x 0·8 inch.

Hab. Norfolk Island.

MERULA VINITINCTA, *Gould.*
Vinous-tinted Ouzel.
" Doctor Bird." Inhabitants of Lord Howe Island.

Gould, Handbk. Bds. Aust., Vol. ii.,_App., sp. 5, p. 529.

A nest of this species, taken during the month of October, 1889, is elliptical in form, with a small cup-shaped depression in the top, and is composed throughout of strips of palm leaves and fibre of one of the species of *Kentia* peculiar to the Island, together with skeletons of leaves, but without any special lining, it measures exteriorly seven inches in diameter by five inches in depth ; internally, three inches in depth by two inches and a half across. Mr. Icely states that the nest was built in the branches of a shrub not far from the ground. The eggs are two in number for a sitting, inclining to elongated ovals in form, slightly pointed at the thinner end, of a pale greenish-grey ground colour, with freckles, dots and longitudinal markings of reddish-brown dispersed over the entire surface of the shell, in some places a few nearly obsolete blotches of purplish grey appear. Length (A) 1·15 x 0·77 inch ; (B) 1·12 x 0·77 inch.*

Hab. Lord Howe Island.

ZOSTEROPS TENUIROSTRIS, *Gould.*
Long-billed Zosterops.
Gould, Handbk. Bds. Aust., Vol. ii., App., sp. 9, p. 536.

This very distinct and well marked species Dr. Metcalfe found breeding during the month of September, the nidification is almost similar to that of other members of this widely distributed genus found in Australia and elsewhere. The nests are round

---

*North, Records Australian Museum, (1890) Vol. i., p. 36.

and cup-shaped structures, but shallower and not so compactly formed as those of certain species of *Zosterops* found on the Australian continent; they are composed exteriorly of mosses, fibrous roots and grasses, lined inside with hair, and slightly suspended by the rim from the thin twigs of branches of low trees and shrubs not far from the ground. The eggs are four or five in number for a sitting, in form elongated ovals of a uniform delicate pale blue, the texture of the shell being very fine and slightly glossy. Length 0·88 x 0·61 inch.

Hab. Norfolk Island.

ZOSTEROPS ALBIGULARIS, *Gould.*

White-breasted Zosterops.

Gould, Handbk. Bds. Aust., Vol. ii., App., sp. 8, p. 535.

The nidification of the White-breasted Zosterops is similar to that of the preceding species, *Z. tenuirostris*, a description of which is therefore unnecessary. Eggs four or five in number for a sitting, of a uniform pale blue, two average sized specimens are oval in form, one of which tapers gently to the smaller end. Length (A) 0·81 x 0·59 inch ; (B) 0·81 x 0·58 inch.

Hab. Norfolk Island.

LAMPROCOCCYX PLAGOSUS, *Latham.*

(*C. lucidus*, Gmelin.)

Bronze Cuckoo.

Gould, Handbk. Bds. Aust., Vol. i., sp. 383, p. 623.

Like its Australian congeners the Bronze Cuckoo on Norfolk Island deposits its egg in the nest of one of the smaller birds, upon whom necessarily devolves the task of incubation, and finally the care of the young Cuckoo when hatched. An egg of this species taken by Dr. Metcalfe from a nest of *Gerygone modesta* on the 25th of November, 1885, is an oblong oval nearly equal in size at each end, and is of a uniform pale olivaceous-brown. When moisture is applied to the shell, the colouring rubs off easily disclosing a light sky-blue tint underneath. Length 0·72 x 0·52 inch.

Hab. Norfolk Island, Australia, Tasmania, New Zealand.

OCYDROMUS SYLVESTRIS, *Sclater.*
The Rufous-winged Wood-hen.
" *Wood-hen.*" Inhabitants of Lord Howe Island.

Sclater, Proc. Zool. Soc., 1869, p. 472.

This bird is found breeding in the most rugged and inaccessible portions of the Island, such as the Erskine Valley, between Mount Ledgbird and Mount Gower. Here the rough character of the country, consisting of huge boulders of granite almost hidden in a dense and luxuriant mass of sub-tropical vegetation, affords it a secure retreat. The nest in question was found last October at the head of the Erskine Valley, and consisted merely of a depression in a thick débris of fallen leaves, under the shelter of a low bush. The eggs four in number vary in shape from ovals to lengthened ovals, being slightly pointed at one end, and are of a dull white, with minute dots and large irregular shaped markings of light chestnut-red more or less scattered over the surface of the shell, obsolete markings of the same colour predominating towards the larger end, they are not unlike very large specimens of *Hypotœnidia philippensis* (*Rallus pectoralis*), but the markings are paler and not so well defined. Length (A) 1·9 x 1·32 inch ; (B) 1·88 x 1·36 inch ; (C) 1·95 x 1·3 inch ; (D) 2 inches x 1·32 inch. Mr. Icely remarks that this is the first time that any of the present inhabitants of the Island have seen the eggs of the Wood-hen, or had heard of them being taken.*

Hab. Lord Howe Island.

---

* North, Records Australian Museum, (1890) Vol. 1., p. 37.

ORTYGOMETRA TABUENSIS, *Gmelin.*
Tabuan Water Crake.
Gould, Handbk. Bds. Aust., Vol. ii., sp. 575, p. 341.

This species is dispersed all over the Australian, Austro-Malayan and Pacific regions. Although having such an extensive geographical range it is a very difficult species to procure, as its favourite haunts, the sedgy and reed covered margins of lagoons and rivers, always afford it a tolerably secure retreat on the first approach of danger. On Norfolk Island Dr. Metcalfe informs me he found an old nest of this species with an egg in it, built in rushes, and that it was composed of dead flags and raised above the water similar to that of *Porphyrio melanotus.* The egg is oval in form, rounded at each end which are equal in size, the texture being fine and slightly glossy, of a very pale creamy-brown ground colour, with numerous indistinct fine fleecy markings of light chestnut-brown thickly and uniformly dispersed over the entire surface of the shell. Length 1·15 x 0·91 inch.

Hab. Australian, Austro-Malayan and Pacific regions.

PORPHYRIO MELANOTUS, *Temminck.*
Black-backed Porphyrio.
Gould, Handbk. Bds. Aust., Vol. ii., sp. 563, p. 321.

This species is dispersed over the greater part of the Australian Continent, Tasmania and New Zealand, breeding in swampy places, and constructing a nest of dried flags and weeds slightly cupped at the top, and placed in rushes in the centre of streams several feet above the water. Dr. Metcalfe informs me that on Norfolk Island the number of eggs laid by this bird for a sitting is " twelve or more," this is greatly in excess of the number laid by the same species in Australia, where five, the usual complement laid, is rarely exceeded. An average specimen taken by Dr. Metcalfe on the 2nd of November, 1889, is ovoid in form, of a yellowish-brown ground colour, with large irregular shaped markings of reddish-chestnut and reddish-brown, principally on the apex of the egg, intermingled with others of a purplish and deep bluish-grey appearing as if beneath the surface of the shell. Length 1·95 x 1·44 inch.

Hab. Norfolk and Lord Howe Islands, Australia, New Zealand.

ŒSTRELATA PHILLIPII, *G. R. Gray.*
Phillip's Petrel.

G. R. Gray, Ibis, 1862, p. 246.

This species figured in Governor Phillip's Voyage to Botany Bay, as the Norfolk Island Petrel, and subsequently described by G. R. Gray as *Procellaria phillipii,* Dr. Metcalfe informs me is very difficult to procure on account of its nocturnal habits, and is only to be obtained about January when it resorts to the west side of the Island to breed, depositing a single egg at the end of a burrow in the sandy soil.   During a period of ten years he has only obtained two birds and three eggs, one of the latter of which he has kindly forwarded ; it is ovoid in form, of a dull white, the surface of the shell having numerous shallow pittings, although smooth to the touch and presenting a glossy appearance. Length *5⅜,⅘.* 2·14 x 1·62 inches.

Hab. Norfolk Island.

www.ingramcontent.com/pod-product-compliance
Lightning Source LLC
Chambersburg PA
CBHW021346210326
41599CB00011B/769